EQUATIONS OF
MATHEMATICAL
DIFFRACTION THEORY

Differential and Integral Equations and Their Applications

A series edited by:
A.D. Polyanin
Institute for Problems in Mechanics, Moscow, Russia

Volume 1
Handbook of First Order Partial Differential Equations
A.D. Polyanin, V.F. Zaitsev and A. Moussiaux

Volume 2
Group-Theoretic Methods in Mechanics and Applied Mathematics
D.M. Klimov and V. Ph. Zhuravlev

Volume 3
Quantization Methods in the Theory of Differential Equations
V.E. Nazaikinskii, B.-W. Schulze and B. Yu. Sternin

Volume 4
Hypersingular Integral Equations and Their Applications
I.K. Lifanov, L.N. Poltavskii and G.M. Vainikko

Volume 5
Equations of Mathematical Diffraction Theory
Mezhlum A. Sumbatyan and Antonio Scalia

EQUATIONS OF MATHEMATICAL DIFFRACTION THEORY

Mezhlum A. Sumbatyan
Antonio Scalia

CRC Press
Taylor & Francis Group
Boca Raton London New York

CRC Press is an imprint of the
Taylor & Francis Group, an **informa** business

A CHAPMAN & HALL BOOK

CRC Press
Taylor & Francis Group
6000 Broken Sound Parkway NW, Suite 300
Boca Raton, FL 33487-2742

First issued in paperback 2019

ISBN-13: 978-0-415-30849-6 (hbk)
ISBN-13: 978-0-367-39380-9 (pbk)
Library of Congress Card Number 2004051957

Library of Congress Cataloging-in-Publication Data

Sumbatyan, Mezhlum A.
    Equations of mathematical diffraction theory / Mezhlum A. Sumbatyan, Antonio Scalia.
        p. cm.
    Includes bibliographical references and index.
    ISBN 0-415-30849-6 (alk. paper)
        1. Diffraction—Mathematics. I. Scalia, A. II. Title.

QC415.S95 2004
535'.42'0151—dc22
                                                              2004051957

**Visit the Taylor & Francis Web site at
http://www.taylorandfrancis.com**

**and the CRC Press Web site at
http://www.crcpress.com**

PREFACE

The connection between heuristic and strictly formal methods is seemingly one of the most interesting and debatable questions in modern mathematics. Each of these two different approaches, whose foundations were laid by Socrates and Aristotle, respectively, and in the new history are reflected in discussions and written papers by Descartes, Leibnitz and Bacon, has its own intrinsic merits and restrictions. Moreover, a large number of discoveries in science were made owing to a combination of strict and heuristic methods of investigation.

Apparently, one of the brightest examples in modern mathematical physics is diffraction theory, where the combination of the two approaches would lead so efficiently to such impressive results. Many important and interesting solutions and even some classical theories appeared from heuristic ideas, and the most impressive example was given by Kirchhoff's physical diffraction theory, which is based upon a clear "light and shadow" concept for diffracted wave fields. Subsequently, many of these heuristic results were rigorously substantiated and proved as theorems. On the other hand, unsuccessful attempts to prove some other heuristic ideas caused significant progress in the development of formal methods that yielded correct solutions, different sometimes from those prompted by someone's intuition.

The above specific features have affected the style of presentation of the book. Each section deals with a discussion of heuristic ideas, which as a rule are substantiated (or disproved) with the use of rigorous mathematical methods. Due to limited volume of the book, at some places we give only a brief sketch of the substantiation, referring the reader to the original literature for more details.

One more specific feature of the presented material is connected with the rapid progress in computer technology over the last 20 years, which has significantly changed our viewpoint on what could be accepted as efficient methods of investigation. Only recently, expansion of unknown functions into series in terms of special functions, when a problem reduced to infinite system of linear algebraic equations with respect to coefficients of the expansion, was regarded as a standard method. Such a "semi-analytical" approach was efficient 15–20 years ago, when the evaluation of regularity of the obtained infinite systems seemed to be very important, since this could guarantee accuracy of a solution by retaining only few first equations, which was acceptable for first-generations computers. Nowadays, when there is not much difference between 10×10 and 500×500 systems even for home personal computers, such a viewpoint looks archaic, since the time required to convert the system to a form appropriate for "fast computations" is much greater than that for "slow computations" based on modern direct numerical methods like boundary element method and finite element method. Apparently, it should be agreed that in the cases where direct numerical techniques provide reliable results in an acceptable computational time, they should be regarded as most efficient for the problem in question. It is very important to recognize the cases where one has *a priori* to reject direct numerical methods. These are listed below.

$1°$. Problems where an exact analytical solution or a good approximation to it can be obtained. Diffraction theory shows many examples of this kind.

$2°$. Studying dynamic processes with high frequencies. Here, one has to take at least

10 nodes per wavelength to obtain reliable results by any direct numerical method. As the wavelength decreases (i.e., the frequency increases) within a given frequency range, the total number of nodes increases very rapidly, which results in too large algebraic systems. An impressive example is given by room acoustics. Suppose a sound wave of frequency $f = 2\,\text{kHz}$, whose wavelength in air is $17\,\text{cm}$, propagates in a 17-m long room. For reasonable numerical accuracy, one should hence take at least 1000 nodes along the room length. If the room has a width of $8\,\text{m}$ and a height of $5.1\,\text{m}$, one has to consider $1000 \times 500 \times 300 \approx 10^8$ finite-element nodes and perform complex-valued arithmetic. This cannot be implemented even on the most powerful super computers. Here, a reasonable criterion for acceptability of a numerical approach is its implementability on a PC or similar computer. So, obtaining solutions to such high-frequency problems in exact formulation by direct numerical techniques does not seem to be feasible in the visible future.

$3°$. Studying phenomena of complex qualitative nature. Since direct numerical methods provide only numbers, which are usually tabulated and plotted, it is often very difficult to extract such complex qualitative effects from numerous tables and graphs. Instead, it is preferable to construct an approximate analytical solution, from which qualitative effects may be extracted explicitly.

$4°$. Cases where an exact analytical solution has been obtained but its representation is inapplicable to practice for specific calculations. An example of this kind is considered in Section 6.1. In such interesting cases, one should look for an alternative approach, which is often the construction of an approximate solution that would be more appropriate for fast computations than the exact analytical solution obtained.

The above situations are not widespread but, when met, are very difficult to cope with efficiently, especially if the researcher does not have sufficient experience in tackling them. This prompted us to conclude each section with a special subsection titled "Helpful Remarks," which may help the reader to build up his or her own less formal conception and allow the creation of a more complete picture of the issue under consideration.

Note that the application of numerical methods in regular cases is well described in the classical literature. For this reason, we only consider numerical methods for some irregular operator problems; see Chapter 9.

To summarize, the main purpose of the present book is to show the close connection between heuristic and rigorous methods in mathematical diffraction theory. We focus on differential and integral equations that can easily be utilized in practical applications.

Such an approach is accounted for by the choice of our potential readers. The book presents clear and elegant methods and is aimed at graduate and post-graduate students, so that they could quickly examine the state of the art in a specific field of interest. At the same time, researchers with considerable expertise in dealing with diffraction theory will hopefully discover that the time of clear explicit solutions in unsolved complex problems has not passed yet—this is demonstrated by the authors' original results in Sections 4.5, 4.6, 5.4–5.7, and 6.3–6.6 as well as in many sections of Chapters 7–9. Furthermore, we hope that an experienced reader will be able to discover for him- or herself new helpful methods, both analytical and numerical.

The reader will see in what follows that we prefer to rely upon classical results of the founders of modern science unlike a rather widespread (mistaken) point of view that only very complicated recent "abstract" theories can provide further progress in contemporary science. We strongly recommend the younger reader to operate with classical mathematical theories, and the present book will demonstrate that the fruitful ideas of Hilbert, Cauchy, Fourier, Abel, Poisson, Weyl, Riemann, Green, Kirchhoff, Rayleigh, Helmholtz, Neumann, and others can guide the reader very efficiently around present-day problems in diffraction theory. It should also be stressed that we tried to avoid too formal presentation, since we believe that wielding thorough knowledge in any mathematical theory implies applying it effectively and successfully to practice rather than operating with the formal apparatus of the theory.

Due to its limited volume, any monograph cannot cover all important questions, and the present book is no exception. For example, the reader will not find here transient problems at all. The presentation is confined to boundary problems for elliptic operators only, and only those with constant coefficients (except Section 3.6). Moreover, the main focus is on methods that provide solutions without too cumbersome mathematical manipulations. For example, the reader will not find the structure of the wave field in the "semi-shadow" zone in diffraction by convex obstacles, and in the method of "edge waves" in diffraction from linear segments, the reader will only find the leading high-frequency asymptotic term, which is constructed by a simple and elegant technique.

The sections, displayed formulas, and figures are enumerated independently within each chapter with the chapter number in front.

The book is intended for the reader familiar with fundamentals of real, complex-valued, and functional analysis within a standard course on calculus in the first three years of any university program of mathematical, physical, or engineering departments.

The style and content of this book have been influenced by the authors' friends, teachers, and colleagues, Alexander Vatulyan (Rostov State University, Russia), Mauro Fabrizio (University of Bologna, Italy), and Dorin Iesan (University of Iasi, Romania).

The authors are grateful to Alexander Manzhirov and Alexei Zhurov for their helpful discussions and comments.

The first author is thankful to his wife, Angela Sumbatyan, and to his daughters, Laura, Carina, and Angelica, who assisted and inspired him in the writing of the book.

M. A. Sumbatyan
A. Scalia

AUTHORS

Mezhlum A. Sumbatyan, Ph.D., D.Sc., is a noted scientist in the field of wave dynamics, diffraction theory, and ill-posed problems.

Mezhlum Sumbatyan graduated from the Faculty of Mechanics and Mathematics, Rostov State University, Russia, in 1969 and received his Ph.D. degree in 1980 at the Institute for Problems in Mechanics, Russian Academy of Sciences, Moscow. His Ph.D. thesis was devoted to asymptotic methods for solving integral equations arising in some problems of mechanics and acoustics with mixed boundary conditions. In 1995, Professor Sumbatyan received his Doctor of Sciences degree; his D.Sc. thesis was dedicated to direct and inverse problems of diffraction theory, with applications to reconstruction of obstacles by ultrasonic techniques.

In 1985–2000, Mezhlum Sumbatyan worked in the Research Institute of Mechanics and Applied Mathematics of the Rostov State University. Since 2001, Professor Sumbatyan has been a member of the staff of the Faculty of Mechanics and Mathematics, Rostov State University.

Professor Sumbatyan has made important contributions to new methods in the theory and analytical methods applied to direct and inverse diffraction problems. He is an author of more than 120 scientific publications, a member of the Russian Acoustical Society, and a member of the Acoustical Society of America.

Address: Faculty of Mechanics and Mathematics   E-mail: sumbat@math.rsu.ru
         Zorge Street 5
         344090 Rostov-on-Don, Russia

Antonio Scalia, Ph.D., D.Sc., is a prominent scientist in the field of mathematical physics.

Antonio Scalia graduated from the Faculty of Mathematics, Physics and Natural Sciences, University of Catania, Italy, in 1972 and received his Ph.D. degree in 1980 at the University of Catania. His thesis was devoted to solvability and uniqueness in mathematical formulation of mechanical and acoustical problems for media with micro-structure.

Since 1989, Professor Scalia has been a member of the staff of the Department of Mathematics and Informatics at the University of Catania as an Associate Professor, and since 2001, as a Full Professor.

Professor Scalia has great expertise in the theory of viscoelasticity, electromagnetism, and acoustics. He has significant achievements in the mixture theory in solids, with applications to wave processes in elastic and acoustic media. Recently, within the theory of porous media, he devised a new approach to the analysis of wave propagation in elastic solids with pores. He explicitly studied a characteristic equation of the porous composite media applied to the layered geometry. Professor Scalia is an author of nearly 100 scientific publications.

Address: Dipartimento di Matematica e Informatica   E-mail: scalia@dmi.unict.it
         Universitá di Catania, Viale A. Doria n. 6
         95125 Catania, Italy

CONTENTS

Chapter 1

Some Preliminaries from Analysis and the Theory of Wave Processes

1.1. Fourier Transform, Line Integrals of Complex-Valued Integrands, and Series in Residues

Let a function $f(x)$ be integrable on the real axis: $f(x) \in L_1(-\infty, \infty)$. Then its Fourier transform $F(s)$ is defined as

$$F(s) = \int_{-\infty}^{\infty} f(x)\, e^{isx}\, dx, \qquad F(s) \in L_1(-\infty, \infty), \tag{1.1}$$

and in the case when $f(x)$ is continuous, the following inversion formula is valid:

$$f(x) = \frac{1}{2\pi} \int_{-\infty}^{\infty} F(s)\, e^{-isx}\, ds = \frac{1}{2\pi} \lim_{a \to \infty} \int_{-a}^{a} F(s)\, e^{-isx}\, ds, \quad x \in (-\infty, \infty). \tag{1.2}$$

The function $f(x)$ will be called an *original* and the function $F(s)$ the Fourier *image* of $f(x)$. The fact that the original and the image are related by formulas (1.1) and (1.2) will be denoted $f(x) \Longrightarrow F(s)$.

Many important and helpful properties of the Fourier transform are well known (e.g., see Titchmarsh, 1948; Bremermann, 1965). We will use only the following two of them:

$1°$. *Fourier image of the derivative.* Let $f(x) \Longrightarrow F(s)$ and $f^{(n)}(x) \in L_1(-\infty, \infty)$, then

$$f^{(n)}(x) \Longrightarrow (-is)^n\, F(s). \tag{1.3}$$

$2°$. *Fourier image of the convolution.* Let $f(x) \in L_1(-\infty, \infty)$, $g(x) \in L_1(-\infty, \infty)$, and $f(x) \Longrightarrow F(s)$, $g(x) \Longrightarrow G(s)$. Then the convolution of $f(x)$ and $g(x)$ is given by

$$h(x) = (f * g)(x) = \int_{-\infty}^{\infty} f(\xi)\, g(x-\xi)\, d\xi \in L_1(-\infty, \infty), \quad \text{and} \quad h(x) \Longrightarrow F(s)\, G(s). \tag{1.4}$$

The first property (1.3) can be obtained by the direct differentiation of Eq. (1.2), and the second property (1.4) follows from a change of variable when applying the Fourier transform to Eq. (1.4).

The Fourier transform can also be defined for functions from the Hilbert space L_2: $f(x) \in L_2(-\infty, \infty)$. The classical *Plancherel theorem* asserts the existence of a Fourier transform $F(s)$ (Wiener, 1934):

$$F(s) = \int_{-\infty}^{\infty} f(x)\, e^{isx}\, ds = \lim_{a \to \infty} \int_{-a}^{a} F(s)\, e^{isx}\, ds, \quad x \in (-\infty, \infty). \tag{1.5}$$

The convergence here is implied in the mean-square sense, i.e., as a convergence in L_2. In this case, $F(s) \in L_2(-\infty, \infty)$, and the inverse Fourier transform is valid in the same sense,

$$f(x) = \frac{1}{2\pi} \int_{-\infty}^{\infty} F(s)\, e^{-isx}\, ds = \frac{1}{2\pi} \lim_{a \to \infty} \int_{-a}^{a} F(s)\, e^{-isx}\, ds, \quad x \in (-\infty, \infty), \tag{1.6}$$

of mean-square convergence. For L_2 functions $f(x), g(x) \in L_2(-\infty, \infty)$ the *Parseval identity* states that if $f(x) \Longrightarrow F(s)$ and $g(x) \Longrightarrow G(s)$, then

$$\int_{-\infty}^{\infty} f(x)\,\overline{g}(x)\,dx = \int_{-\infty}^{\infty} F(s)\,\overline{G}(s)\,ds, \quad \text{in particular,} \quad \int_{-\infty}^{\infty} |f(x)|^2\,dx = \int_{-\infty}^{\infty} |F(s)|^2\,ds,$$

(1.7)

where the bar over a symbol denotes a complex conjugate. The convolution theorem also remains valid in L_2.

Let $H(D)$ denote a set of complex-valued analytic functions $f(z)$ of the complex variable $z = \mathrm{Re}(z) + i\,\mathrm{Im}(z)$ defined over a domain D: $f(z) \in H(D)$, $z \in D$. Recall that this implies that $f(z)$ is analytic and single-valued together with all its derivatives: $f^{(n)}(z) \in H(D)$, $\forall n = 0, 1, 2, \ldots$ (see Markushevich, 1963). Then the Cauchy theorem declares that the value of the line integral

$$I(\Gamma) = \int_{z_A}^{z_B} f(z)\,dz, \qquad z_A, z_B \in D,$$

(1.8)

along a curve $\Gamma \subset D$ of finite length with endpoints z_A, z_B is the same for any Γ, no matter how Γ connects z_A and z_B. This is equivalent to the statement that $I(\Gamma) = 0$ for any closed contour $\Gamma \subset D$ of finite length.

It is clear from the previous consideration that $I(\Gamma)$ in Eq. (1.8) is contour dependent only in the case when $f(z)$ has singular points in D. In the present book, we will consider only *poles* and *branching points* out of the whole variety of singular points.

A point $z_0 \in D$ is a pole of the function $f(z)$ if and only if z_0 is a zero of $g(z) = 1/f(z)$, i.e., $g(z_0) = 0$. The multiplicity n of the zero z_0 of $g(z)$ is, at the same time, the multiplicity of the pole z_0 of $f(z)$. It can be proved that the leading term in the Laurent series of the function $f(z)$ in a neighborhood of z_0 is $(z - z_0)^{-n}$, i.e.,

$$f(z) = \sum_{m=-n}^{\infty} a_m (z - z_0)^m,$$

(1.9)

where the coefficient a_{-1} is called the *residue* of the function $f(z)$ at the pole z_0 and denoted $a_{-1} = \mathrm{Res}\,[f(z), z_0]$. If $n = 1$ in Eq. (1.9), then a_{-1} is the leading coefficient in the Laurent expansion, and such a pole is called a *simple pole*. There is quite a simple way to calculate the residue at a simple pole z_0:

$$\mathrm{Res}\,[f(z),\ z_0] = \frac{h(z_0)}{g'(z_0)} \qquad \text{if} \quad f(z) = \frac{h(z)}{g(z)},$$

(1.10)

which is very efficient with any natural fractional decomposition (as in the case $\tan z = \sin z / \cos z$).

Residues at poles play a key role in the calculation of integrals in the complex plane. This fact is represented by the *Cauchy integral formula* valid for any closed contour $\Gamma \subset D$ traced counterclockwise

$$\int_{\Gamma} f(z)\,dz = 2\pi i \sum_{m} \mathrm{Res}[f(z), z_m],$$

(1.11)

where the residues are taken at all poles z_m (of arbitrary multiplicity) inside Γ.

This formula is very helpful for the calculation of integrals of the type (1.8). Quite often $I(\Gamma)$ in (1.8) may be easily calculated for a certain simple path $I(\Gamma_s)$ passing through the endpoints z_A, z_B. Then the difference between $I(\Gamma)$ and $I(\Gamma_s)$, is equal to the sum of the residues at the poles between Γ and Γ_s, taken with appropriate sign.

This strategy can be applied to integrals written along infinite lines also. In particular, let $f(z)$ in the Fourier transform (1.1) be analytic in a finite-width strip $|\operatorname{Im}(z)| \leq \delta$. Then the integration contour $\Gamma = (-\infty, \infty)$ may be arbitrarily shifted up or down within this strip. Indeed, if the integral (1.1) is finite under the integration along the initial path $(-\infty, \infty)$, this implies that $f(z) \to 0$ as $\operatorname{Re}(z) \to \infty$. Consequently, the integral of the same integrand (1.1) along any closed contour $(-\infty, \infty) \cup (\infty, \infty+i\varepsilon) \cup (\infty+i\varepsilon, -\infty+i\varepsilon) \cup (-\infty+i\varepsilon, -\infty)$, $|\varepsilon| < \delta$, is zero. Since the two integrals over far finite vertical intervals vanish, because $f(z)$ decays in a far-zone, this proves our simple statement. In the case when there is a number of poles between the real axis and the line $\operatorname{Im}(z) = \varepsilon$ parallel to it, it is evident that the same shift of the contour is possible if we add the residues at these poles. Sometimes this technique permits explicit calculation of Fourier transforms.

In order to shift the integration contour Γ more up (or down), outside of a finite-width strip, we need to apply the following

LEMMA (JORDAN). *Let*

$$I_R = \int_{C_R} f(z)\, e^{isz}\, dz, \tag{1.12}$$

where $f(z)$ is analytic everywhere in the upper half-plane $\operatorname{Im}(z) \geq 0$, except perhaps a finite number of poles; $\operatorname{Re}(s) > 0$; $f(z) \to 0$ as $z \to \infty$ uniformly over $0 \leq \arg(z) \leq \pi$; and C_R is an upper semi-circle of radius R: $|z| = R$, $\operatorname{Im}(z) \geq 0$. Then $I_R \to 0$ as $R \to \infty$.

The proof of this lemma is simple and can be found in the classical literature.

Corollary. Under the same conditions, the Fourier transform $F(s)$ of (1.1) can be explicitly expressed as

$$F(s) = \int_{-\infty}^{\infty} f(x)\, e^{isx}\, dx = 2\pi i \sum_{\operatorname{Im}(z_m)>0} \operatorname{Res}\left[f(z),\, z_m\right] e^{isz_m}. \tag{1.13}$$

This result directly follows from the Cauchy integral formula (1.11) if you apply it to the function $f(z)\exp(isz)$ along the contour $\Gamma = (-R, R) \cup C_R$, with $R \to \infty$.

LEMMA (GENERALIZED JORDAN LEMMA). *Let $f(z)$ have a countable set of poles z_m, $m = 1, 2, \ldots$, $\operatorname{Im}(z_m) > 0$, $z_m \to \infty$, $m \to \infty$; and $f(z)$ vanishes uniformly on semi-circles C_{R_m} of radius R_m as $R_m \to \infty$; and each C_{R_m} passes somewhere between z_m and z_{m+1}. Then for any s such that $\operatorname{Re}(s) > 0$, we have*

$$I_{R_m} = \int_{C_{R_m}} f(z)\, e^{isz}\, dz \to 0 \quad as \quad m \to \infty. \tag{1.14}$$

The proof of this less known result repeats the one for the classical Jordan lemma.

Corollary. Under the same conditions, the Fourier transform can be explicitly calculated as an infinite series:

$$F(s) = \int_{-\infty}^{\infty} f(x)\, e^{isx}\, dx = 2\pi i \sum_{m=1}^{\infty} \operatorname{Res}\left[f(z),\, z_m\right] e^{isz_m}. \tag{1.15}$$

This corollary is very helpful when $f(z)$ is *meromorphic*, i.e., is the ratio of two entire functions: $f(z) = h(z)/g(z)$. Recall that *entire* functions are defined as analytic over the whole complex plane, so countable sets of zeros and poles of any meromorphic function $f(z)$ are given by zeros of the entire functions $h(z)$ and $g(z)$, and both the resulting sets are finite if and only if $f(z)$ is rational.

This is clearly demonstrated by the function $f(z) = \tanh(z)/z$, where $h(z) = \sinh(z)/z$ and $g(z) = \cosh(z)$. It is also clear that in this example the set of upper semi-circles C_{R_m}, $m = 1, 2, \ldots$, can be determined from the condition $\tanh(iR_m) = 0 \sim \tan(R_m) = 0 \sim R_m = \pi m$, which causes respective semi-circles to pass through the imaginary points $iR_m = \pi m i$ (see Fig. 1.1).

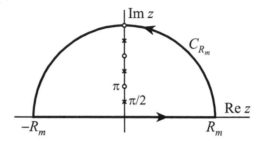

Figure 1.1. Alternate poles and zeros of a meromorphic function

Very often we will encounter below, in diffraction problems, some functions of a complex-valued argument that have branching points and hence are not single-valued. A typical representative here is the root square difference

$$\gamma(z) = \sqrt{z^2 - k^2}, \tag{1.16}$$

with a certain constant positive parameter $k > 0$ (see Mittra and Lee, 1971). Usually, in order to operate with a single-valued function, one has to arrange some cuts that become boundaries between different branches. For the root square difference (1.16) there are two branching points: $z = k$ and $z = -k$, and it is quite natural to make such cuts that allow operating with the arithmetic value of the root square difference, i.e., the branch with $\text{Re}(z) \geq 0$. This can be provided by the cuts shown in Fig. 1.2, one of which passes totally in the upper half-plane $\text{Im}(z) > 0$ and the other in the lower half-plane $\text{Im}(z) < 0$. Note that for real z, $\gamma(z) = \sqrt{z^2 - k^2} \geq 0$ if $|z| \geq k$, and $\gamma(z) = -i\sqrt{k^2 - z^2}$ if $|z| \leq k$.

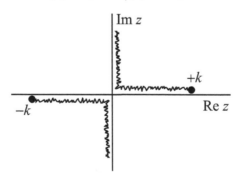

Figure 1.2. Cuts in the complex plane z making the function $\gamma(z) = \sqrt{z^2 - k^2}$ single-valued

For example, with such cuts the integral representation of the Hankel function (Abramowitz and Stegun, 1965)

$$\int_{-\infty}^{\infty} \frac{\exp(ixs)}{\gamma(s)} \, ds = 2 \int_{0}^{\infty} \frac{\cos(xs)}{\gamma(s)} \, ds = \pi i H_0^{(1)}(k|x|) \tag{1.17}$$

implies integration along the real axis, when the integration path lies between the two cuts. Note that the singularities $s = \pm k$ are integrable in the classical sense.

It should be noted that branching functions, like the root square difference (1.16), generally are not analytic. However, some combinations of such functions can yield analytic and even entire functions, as can be seen by the example of the function $\sin[b\gamma(z)]/\gamma(z)$ (b is constant). It is certainly an entire function, since it can be represented by a Taylor series,

$$\frac{\sin[b\gamma(z)]}{\gamma(z)} = \sum_{m=0}^{\infty} \frac{(z^2 - k^2)^m \, b^{2m+1}}{(2m+1)!} (-1)^m, \tag{1.18}$$

that is analytic and convergent for all finite z.

Helpful remarks

$1°$. Interestingly, quite often the "shortest way" between two real points "lies" in the complex plane. To illustrate this, let us consider the following integral over an interval of the real axis and with real integrands:

$$J_1 = \int_0^\infty \frac{\cos(ax)}{x^2 + b^2} \, dx \qquad (a \geq 0, \; b > 0). \tag{1.19}$$

With the help of the Jordan lemma, we obtain

$$J_1 = \frac{1}{2} \int_{-\infty}^\infty \frac{e^{iax} \, dx}{x^2 + b^2} = \pi i \operatorname{Res} \left[\frac{e^{iaz}}{z^2 + b^2}, \; ib \right] = \pi i \frac{e^{-ab}}{2bi} = \frac{\pi}{2b} e^{-ab}, \tag{1.20}$$

where the residue at the simple pole $z = ib$ has been calculated with the method described above.

$2°$. The same approach is applicable to a meromorphic function if you use the generalized Jordan lemma ($a, b > 0$):

$$J_2 = \int_0^\infty \frac{\tanh(bx)}{x} \cos(ax) \, dx = \frac{1}{2} \int_{-\infty}^\infty e^{iax} \frac{\sinh(bx)}{x \cosh(bx)} \, dx$$

$$= \pi i \sum_{m=1}^{\infty} \operatorname{Res} \left[e^{iaz} \frac{\sinh(bz)}{z \cosh(bz)}, \; \frac{\pi i}{b} \left(m - \frac{1}{2} \right) \right] = \pi i \sum_{m=1}^{\infty} \frac{e^{-\pi a(m-1/2)/b}}{\pi (m - 1/2) i} \tag{1.21}$$

$$= \sum_{m=1}^{\infty} \frac{e^{-\pi a(m-1/2)/b}}{m - 1/2} = \ln \frac{1 + e^{-\pi a/2b}}{1 - e^{-\pi a/2b}} = \ln \left[\coth \left(\frac{\pi a}{4b} \right) \right],$$

where, in order to calculate the residues at simple poles, we have put $f(z) = h(z)/g(z)$ with entire functions $h(z) = e^{iaz} \sinh(bz)/z$ and $g(z) = \cosh(bz)$. The following tabulated series has also been taken into account here (Gradshteyn and Ryzhik, 1994):

$$\sum_{m=0}^{\infty} \frac{x^{2m+1}}{2m+1} = \frac{1}{2} \ln \frac{1 + x}{1 - x}, \qquad (|x| < 1). \tag{1.22}$$

$3°$. Another remarkable phenomenon is related to the question why the Fourier transform, which (as a rule) converts real-valued functions to complex-valued, is so helpful when solving real boundary value problems. The answer can be seen from property $1°$ of the Fourier transform, since any derivative of an unknown function is converted to the same image with a factor containing Fourier parameter s. Therefore, if you solve any boundary value problem in a domain where the Cartesian coordinate x varies from $-\infty$ to ∞, then the application of the Fourier transform will allow you to reduce the dimension of the problem by 1. This can reduce ordinary differential equations to algebraic ones, and a partial differential equation in two variables to an ordinary differential equation.

Property $2°$ of the Fourier transform allows you to solve integral equations with convolution kernels explicitly. Both techniques will be demonstrated in detail below.

1.2. Convolution Integral Equations and the Wiener–Hopf Method

Generally, a convolution integral equation has the following form:

$$\alpha\,\varphi(x) + \int_a^b K(x-\xi)\,\varphi(\xi)\,d\xi = f(x), \qquad a < x < b, \tag{1.23}$$

which is evidently an equation of the second kind. In the case $\alpha = 0$ it becomes an equation of the first kind. The function $K(x)$ is a (known) kernel of the equation, and $f(x)$ is a (known) right-hand side. The function $\varphi(x)$ is unknown and is to be determined from Eq. (1.23). There is a special, unique case when equation (1.23) generally admits exact analytical solution. This is the case of $b = \infty$, where we get the Wiener–Hopf equation. In this case, a can be made equal to zero by a linear change of variable and only the first-kind equation ($\alpha = 0$) will be important to us in this case. The solution of this equation is based upon some evident properties of the Fourier transform in the complex plane (see Bremermann, 1965; Mittra and Lee, 1971; Noble, 1958):

$1°$. If $|f(x)| \le A\,e^{\tau_- x}$, $x \to +\infty$, then the function

$$F_+(s) = \int_0^\infty f(x)\,e^{isx}\,dx \tag{1.24}$$

is analytic in the upper half-plane $\operatorname{Im}(s) > \tau_-$.

$2°$. If $|f(x)| \le B\,e^{\tau_+ x}$, $x \to -\infty$, then the function

$$F_-(s) = \int_{-\infty}^0 f(x)\,e^{isx}\,dx \tag{1.25}$$

is analytic in the lower half-plane $\operatorname{Im}(s) < \tau_+$.

$3°$. If both properties $1°$ and $2°$ are satisfied and $\tau_+ > \tau_-$, then the full Fourier transform

$$F(s) = \int_{-\infty}^\infty f(x)\,e^{isx}\,dx \tag{1.26}$$

is analytic in the strip $\tau_- < \operatorname{Im}(s) < \tau_+$, and the inverse Fourier transform may be calculated as follows:

$$f(x) = \frac{1}{2\pi} \int_{-\infty+i\tau}^{\infty+i\tau} F(s)\,e^{-isx}\,ds, \qquad \tau_- < \tau < \tau_+. \tag{1.27}$$

It is obvious from the previous section that you may arbitrarily deform the infinite integration contour Γ in (1.27) within the marked strip, if necessary.

Now we are ready to apply the Wiener–Hopf method to the equation

$$\int_0^\infty K(x-\xi)\,\varphi(\xi)\,d\xi = f(x), \qquad 0 < x < \infty. \tag{1.28}$$

Equation (1.28) is equivalent to

$$\int_{-\infty}^\infty K(x-\xi)\,\varphi_+(\xi)\,d\xi = f_+(x) + f_-(x), \qquad |x| < \infty, \tag{1.29}$$

where

$$\varphi_+(x) = \begin{cases} \varphi(x), & x > 0, \\ 0, & x < 0, \end{cases} \qquad f_+(x) = \begin{cases} f(x), & x > 0, \\ 0, & x < 0, \end{cases} \tag{1.30}$$

and $f_-(x)$ is a certain additional unknown function. Equation (1.29) contains two unknown functions, $\varphi_+(x)$ and $f_-(x)$; however both of them can be determined, as we will see soon, from a single equation. It should be noted that we assume the functions $f_+(x)$ and $f_-(x)$ to satisfy properties 1° and 2°, respectively, and the function $K(x)$ the property 3°, with $\tau_+ > \tau_-$. Then, by applying the Fourier transform to Eq. (1.29) and using the convolution theorem (see the previous section), we arrive at the following relation:

$$L(s)\,\Phi_+(s) = F_+(s) + F_-(s), \qquad \tau_- < \tau < \tau_+, \tag{1.31}$$

where $L(s)$ is the Fourier image of $K(x)$.

The key step of the method is the so-called *factorization* of $L(s)$, i.e., its representation in the form $L(s) = L_+(s)\,L_-(s)$, where $L_+(s)$ is analytic with no singularity and no zero in the upper half-plane $\text{Im}(s) > \tau_-$, and $L_-(s)$ possesses similar properties for $\text{Im}(s) < \tau_+$. Then Eq. (1.31) becomes

$$L_+(s)\,\Phi_+(s) = \frac{F_+(s)}{L_-(s)} + \frac{F_-(s)}{L_-(s)}. \tag{1.32}$$

The next step is *decomposition*, according to which the first fraction on the right-hand side of Eq. (1.28) is represented as

$$\frac{F_+(s)}{L_-(s)} = N_+(s) + N_-(s), \tag{1.33}$$

which yields

$$L_+(s)\,\Phi_+(s) - N_+(s) = N_-(s) + \frac{F_-(s)}{L_-(s)}, \tag{1.34}$$

where the function on the left is analytic for $\text{Im}(s) > \tau_-$ and that on the right is analytic for $\text{Im}(s) < \tau_+$. Since there is a common strip of analyticity, identity (1.34) represents a unique entire function, which coincides with the left-hand side in the upper half-plane and with the right-hand side in the lower half-plane. Typically, it can be proved that this entire function vanishes at infinity, so this implies that it is identically equal to zero. Hence,

$$L_+(s)\,\Phi_+(s) - N_+(s) = 0 \quad \sim \quad \Phi_+(s) = \frac{N_+(s)}{L_+(s)}, \tag{1.35}$$

i.e., the Fourier image of the main unknown function $\varphi(x)$ is defined explicitly, and so is its origin.

Helpful remarks

1°. In many cases factorization and decomposition can be performed in a simple natural way. For instance, some problems have kernels whose Fourier transform $L(s)$ is qualitatively like

$$L(s) = \frac{1}{\sqrt{s^2 + d^2}}\,\frac{P_m(s)}{Q_m(s)}, \tag{1.36}$$

where d is a positive parameter, and $P_m(s)$ and $Q_m(s)$ are some polynomials of the same order. Such a function $L(s)$ admits very clear factorization:

$$L(s) = L_+(s)\,L_-(s), \qquad m^+ + m^- = m,$$

$$L_+(s) = \frac{1}{\sqrt{d - is}}\,\frac{a_p \prod\limits_{k=1}^{m^+}(s - s_{p_k}^-)}{a_q \prod\limits_{k=1}^{m^+}(s - s_{q_k}^-)}, \qquad L_-(s) = \frac{1}{\sqrt{d + is}}\,\frac{\prod\limits_{k=1}^{m^-}(s - s_{p_k}^+)}{\prod\limits_{k=1}^{m^-}(s - s_{q_k}^+)}, \tag{1.37}$$

where a_p and a_q are the leading coefficients of the polynomials P_m and Q_m, $m^\pm$ is the number of zeros of $P_m(s)$ and $Q_m(s)$ with positive (negative) imaginary part (see Fig. 1.3, where all zeros are indicated by crosses), and the cuts to arrange single-valued arithmetic value of branching functions are shown by zigzag lines. Irregular cases, where at least one of the zeros is real, will not be considered here. Let us denote

$$\tau_- = \max\{-d;\ \mathrm{Im}(s_{p_k}^-);\ \mathrm{Im}(s_{q_k}^-)\}, \quad \tau_+ = \min\{d;\ \mathrm{Im}(s_{p_k}^+);\ \mathrm{Im}(s_{q_k}^+)\}. \tag{1.38}$$

Then the function $L_+(s)$ is analytic and has no zeros in the upper half-plane $\mathrm{Im}(s) > \tau_-$, and $L_-(s)$ is analytic and has no zeros in the lower half-plane $\mathrm{Im}(s) < \tau_+$.

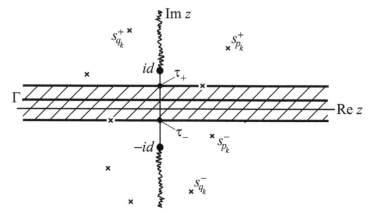

Figure 1.3. The strip of common analyticity for the "+" and "−" functions

Similar idea is applicable to decomposition also. For example, if $L(s) = 1/\sqrt{s^2 + d^2}$, $f(x) = e^{-\beta x}$ $(d,\ \beta > 0)$, then

$$L(s) = \frac{1}{\sqrt{d - is}} \frac{1}{\sqrt{d + is}} = L_+(s)\, L_-(s), \quad F_+(s) = \frac{1}{\beta - is}, \quad \frac{F_+(s)}{L(s)} = \frac{\sqrt{d + is}}{(\beta - is)}. \tag{1.39}$$

The only singular point of this function in the lower half-plane $\mathrm{Im}(s) < d$ is the simple pole $s = -i\beta$, with the factor $h = \sqrt{d + \beta}$. In order to get rid of this pole, you may subtract and add the quantity $h/(\beta - is)$:

$$\frac{F_+(s)}{L_-(s)} = \left(\sqrt{d + is} - \sqrt{d + \beta}\right) \frac{1}{\beta - is} + \frac{\sqrt{d + \beta}}{\beta - is} = N_-(s) + N_+(s). \tag{1.40}$$

Here the first term is analytic in the lower half-plane $\mathrm{Im}(s) < d$ (since it no longer contains the simple pole), and the second term is evidently analytic in the upper half-plane $\mathrm{Im}(s) > -\beta$.

Once we have performed factorization and decomposition, the solution of the Wiener–Hopf equation is given by Eq. (1.35)

$$\Phi_+(s) = \frac{N_+(s)}{L_+(s)} = \frac{\sqrt{d + \beta}\,\sqrt{d - is}}{\beta - is}, \tag{1.41}$$

whose inversion can be found in the tables of inverse Laplace transforms (Bateman and Erdelyi, 1954), since in our problem the Fourier transform is closely connected with the Laplace transform:

$$\Phi_+(s) = \int_0^\infty \varphi(x)\, e^{isx}\, dx = \int_0^\infty \varphi(x)\, e^{-px}\, dx, \quad p = -is, \tag{1.42}$$

so our Fourier image in terms of the Laplace argument p becomes

$$\Phi_+(p) = \frac{\sqrt{d+\beta}\,\sqrt{d+p}}{\beta+p}, \tag{1.43}$$

whose Laplace original is

$$\varphi(x) = \sqrt{d+\beta}\left\{\frac{e^{-dx}}{\sqrt{\pi x}} + \sqrt{d-\beta}\,e^{-\beta x}\,\mathrm{Erf}\left[\sqrt{(d-\beta)x}\right]\right\}, \tag{1.44}$$

where $\mathrm{Erf}(x)$ is the probability integral (error function).

$2°$. If the kernel is too complex to be treated in such a straightforward way as in $1°$, you may apply general formulas of factorization and decomposition, and a good survey of various representations of this kind can be found in (Mittra and Lee, 1971).

But this general approach is hardly reasonable for practical purposes! General formulas are expressed in terms of some line integrals along infinite contours with too complex integrand depending on the variable s. Further, you have to substitute these numerically calculated functions into the infinite integral for the inverse Fourier transform to obtain the originals of your unknown quantities. Implementation of such integrals on a computer is a much more difficult task than your initial convolution integral equation. Instead, it is much more efficient to apply a numerical technique directly to the initial equation. Another possibility is to apply, prior to solving the equation, a uniform approximation to the image of the kernel by a certain function $L(x)$ that admits a simple clear factorization based upon intuition.

1.3. Summation of Divergent Series and Integrals

The concept of divergent series is quite natural in mathematical physics. It also arises in diffraction theory in many boundary value problems. Quite often (and we will see this soon from the forthcoming consideration) the structure of the solution near some boundary lines (like $y = h$ in the example of a layer of constant thickness h) has qualitatively the following behavior:

$$p(x,y) = \sum_{m=1}^{\infty} e^{a_m(y-h)}\, m^\alpha \begin{Bmatrix} \cos \\ \sin \end{Bmatrix}(b_m x); \tag{1.45}$$

$$a_m \sim am, \quad b_m \sim bm, \quad m \to \infty, \qquad (a,b > 0),$$

which in the case $\alpha \geq 0$ yields a series convergent inside the strip $0 < y < h$ only, and divergent on the boundary $y = h$. It is quite natural to treat the series (1.45) also on the boundary line $y = h$, implying that the limit of the sum (1.45) at $y \to h$, if finite, may be admitted as its boundary value. This idea generates a vast theory of generalized summation (see, for example, Hardy, 1956; Rees et al., 1981).

Poisson–Abel summation. If a series $S = \sum_{m=1}^{\infty} a_m$ is divergent in the classical sense and there exists a limit of $S(x) = \sum_{m=1}^{\infty} a_m x^m$ as $x \to 1 - 0$, then the (Poisson–Abel) generalized value of S is $S_{\mathrm{PA}} = \lim_{x \to 1-0} S(x)$.

It follows from the classical Cauchy–Hadamard formula for the convergence radius $R^{-1} = \sup |a_m|^{1/m}$ of a power series (see Smirnov, 1964) in the case where $|a_m|$ grows not faster than any finite power m^α, $m \to \infty$, that $R = 1$. Hence there is a good chance that the generalized value S_{PA} is finite.

It is clear that the Poisson–Abel sum of a series can also be defined alternatively as

$$S_{\mathrm{PA}} = \lim_{x \to 1-0} \sum_{m=1}^{\infty} a_m x^m = \lim_{\varepsilon \to +0} \sum_{m=1}^{\infty} e^{-\varepsilon m} a_m, \quad (x = e^{-\varepsilon}), \tag{1.46}$$

which is in agreement with our heuristic idea discussed above.

A similar technique can be applied to divergent integrals. If the integral

$$S = \int_a^\infty f(x)\,dx \tag{1.47}$$

is divergent at infinity, then its Poisson–Abel generalized value (if exists) is defined as the limit

$$S_{\text{PA}} = \lim_{\varepsilon \to +0} \int_a^\infty e^{-\varepsilon x} f(x)\,dx. \tag{1.48}$$

There are a number of alternative regularization methods for divergent series and integrals (for instance, the Cesaro method is very famous among others), but the Poisson–Abel summation is the most powerful and general.

Examples.

1°. The series

$$S_1 = \sum_{m=1}^\infty m \cos(am) \qquad (0 < a < 2\pi) \tag{1.49}$$

is divergent in the classical sense, since its terms increase without bound as $m \to \infty$. The Poisson–Abel technique gives

$$
\begin{aligned}
S_1 &= \lim_{x \to 1-0} \sum_{m=1}^\infty x^m m \cos(am) = \operatorname{Re} \lim_{x \to 1-0} \sum_{m=1}^\infty m x^m e^{iam} \\
&= \operatorname{Re} \lim_{x \to 1-0} \sum_{m=1}^\infty m \left(x e^{ia} \right)^m = \operatorname{Re} \frac{e^{ia}}{(1 - e^{ia})^2} = -\frac{1}{4\sin^2(a/2)}.
\end{aligned} \tag{1.50}
$$

2°. The integral

$$S_2 = \int_0^\infty \sqrt{x} \cos(ax)\,dx \qquad (a > 0) \tag{1.51}$$

is divergent at infinity. Its Poisson–Abel generalized sum is

$$
\begin{aligned}
S_2 &= \lim_{\varepsilon \to +0} \int_0^\infty e^{-\varepsilon x} \sqrt{x} \cos(ax)\,dx = \operatorname{Re} \lim_{\varepsilon \to +0} \int_0^\infty \sqrt{x}\, e^{-(\varepsilon + ia)x}\,dx \\
&= \operatorname{Re} \lim_{\varepsilon \to +0} \frac{\sqrt{\pi}}{2(\varepsilon + ia)^{3/2}} = \frac{\sqrt{\pi}}{2} \operatorname{Re} \frac{1}{(ia)^{3/2}} = \frac{\sqrt{\pi}}{2a^{3/2}} \operatorname{Re} \left[(e^{\pi i/2})^{-3/2} \right] \\
&= \frac{\sqrt{\pi}}{2a^{3/2}} \cos\left(-\tfrac{3}{4}\pi \right) = -\frac{\sqrt{\pi}}{(2a)^{3/2}}.
\end{aligned} \tag{1.52}
$$

A completely different approach to regularization of divergent series and integrals is based on the theory of generalized functions, or distributions (see Gel'fand and Shilov, 1964; Bremermann, 1965). In these monographs you can find a detailed comparative analysis of methods for generalized summation and regularization of generalized functions. In the ambit of what we are doing in the present section, the following technique gives an appropriate treatment for this sort of series.

Let us consider a series

$$S(\alpha) = \sum_{m=1}^\infty a_m(\alpha), \tag{1.53}$$

where all terms are analytic functions of a complex-valued parameter α in a domain D, $\alpha = \operatorname{Re}(\alpha) + i \operatorname{Im}(\alpha) \in D$, and $S(\alpha)$ is analytic in the same domain D. Then a (regularized) value of the series for $\alpha = \alpha_0 \notin D$ is (if exists) an analytic continuation of the function $S(\alpha)$ from the domain D, where it is analytic, to the value α_0.

The same idea is applicable to divergent integrals. If $f(x, \alpha)$ is analytic with respect to a complex-valued parameter $\alpha \in D$, and the integral,

$$S(\alpha) = \int_a^b f(x, \alpha)\, dx, \qquad (1.54)$$

represents a function analytic in D (any limit in this integral may be equal to $\pm\infty$), then an analytic continuation from the domain D to a value $\alpha = \alpha_0 \notin D$ is called a regularized value of the integral $S(\alpha_0)$.

Let us demonstrate these definitions by examples $1°$ and $2°$ considered above.

$1°$. Let us consider the function

$$S_1(\alpha) = \sum_{m=1}^{\infty} m^\alpha \cos(am), \qquad 0 < a < 2\pi, \qquad (1.55)$$

which is analytic in the infinite domain $D = \{\mathrm{Re}(\alpha) < 0\}$. In this domain the series has an ordinary sum (Gradshteyn and Ryzhik, 1994)

$$S_1(\alpha) = \frac{(2\pi)^{-\alpha}}{4\,\Gamma(-\alpha)\,\cos(\pi\alpha/2)} \left[\zeta\left(1 + \alpha,\, \frac{a}{2\pi}\right) + \zeta\left(1 + \alpha,\, 1 - \frac{a}{2\pi}\right) \right], \qquad (1.56)$$

where $\zeta(s, v) = \sum_{k=0}^{\infty} 1/(k + v)^s$ is a generalized Riemann zeta function. As follows from Eq. (1.56), the function $S_1(\alpha)$ can be continued analytically up to the value $\alpha = 1$. Here we should take into account the limits (Bateman and Erdelyi, 1953)

$$\frac{1}{\Gamma(-\alpha)\,\cos(\pi\alpha/2)} \sim \frac{\alpha - 1}{\cos(\pi\alpha/2)} \sim -\frac{1}{(\pi/2)\sin(\pi\alpha/2)} \to -\frac{2}{\pi}, \qquad \alpha \to 1, \quad (1.57)$$

and the limit value

$$\begin{aligned}
\zeta(2, z) + \zeta(2, 1 - z) &= \sum_{k=0}^{\infty} \frac{1}{(k + z)^2} + \sum_{k=0}^{\infty} \frac{1}{(k + 1 - z)^2} \\
&= \psi'(z) + \psi'(1 - z) = [\psi(z) - \psi(1 - z)]' \\
&= -\pi\, [\cot(\pi z)]' = \frac{\pi^2}{\sin^2(\pi z)},
\end{aligned} \qquad (1.58)$$

where $\psi(z)$ is the logarithmic derivative of Euler's Gamma function $\Gamma(z)$. Collecting formulas (1.55)–(1.58) together, we arrive at the same result (1.50): $S_1(\alpha = 1) = -1/[4\sin^2(a/2)]$. For other values of the parameter a, outside the interval $(0, 2\pi)$, this identity can be continued by periodicity.

$2°$. The function

$$S_2(\alpha) = \int_0^\infty x^\alpha \cos(ax)\, dx \qquad (a > 0) \qquad (1.59)$$

is analytic in the domain $D = \{-1 < \mathrm{Re}(\alpha) < 0\}$, where it is equal to (see Gradshteyn and Ryzhik, 1994)

$$S_2(\alpha) = -\frac{\Gamma(\alpha + 1)}{a^{\alpha+1}} \sin\frac{\pi\alpha}{2}. \qquad (1.60)$$

The analytic continuation of $S_2(\alpha)$ to the point $\alpha = 1/2$ is

$$S_2(1/2) = -\frac{\Gamma(3/2)}{a^{3/2}} \sin\frac{\pi}{4} = -\frac{\sqrt{\pi}}{(2a)^{3/2}}, \tag{1.61}$$

which coincides with (1.52).

Combination of the Fourier transform with the concept of generalized functions permits the application of the Fourier transform to those functions for which it does not exist in the ordinary sense. As a rule, this applies to functions that do not vanish at infinity, and so are not integrable. This automatically implies that the classical Fourier transform for such functions is not finite. A detailed theory of the Fourier transform for generalized functions can be found, for example, in Bremermann (1965) and Zemanian (1969). Here we only cite some rather classical results of this theory, which may be helpful, in one or another way, for further theories. The most important applications are related to the properties of Dirac's *delta function*.

First of all, the Fourier transform of the unit step function is Dirac's delta and some helpful properties of this function and its relations to other known functions are listed below:

$$1(x) \implies 2\pi\delta(s), \quad \frac{d}{dx}|x| = \text{sign}(x),$$

$$\frac{d}{dx}\text{sign}(x) = 2\,\delta(x), \quad \int_{-\infty}^{\infty} f(\xi)\,\delta(x-\xi)\,d\xi = f(x). \tag{1.62}$$

Helpful remarks

If you are an applied mathematician and trust more in concrete results when implemented on your home PC, rather than in abstract theory of generalized functions, then you may want to arrange computations to test correctness of these abstract theories. You may write a code, using an algorithmic language (such as Fortran, C, Pascal, etc.), for computing some formulas like those given by Eq. (1.50). Then, for example, in the case $a = \pi$ you will obtain for the sum of the series $S_1(x) = \sum_{m=1}^{\infty}(-1)^m m x^m$ numbers for various x similar to those obtained by us (the upper limit in the infinite series was taken 10^4 and double-precision Fortran computations were carried out):

x	0.89	0.91	0.93	0.95	0.97	0.99
$S_1(x)$	−0.24915	−0.24944	−0.24967	−0.24983	−0.24994	−0.24999

As $x \to 1-0$, the series approaches the value $S_1(1) = \sum_{m=1}^{\infty}(-1)^m m = -1/[4\sin^2(\pi/2)] = -0.25$.

You may also test the accuracy of formulas (1.52), (1.61). Results of computations for $a = 1/2$ and various ε are shown here:

ε	0.011	0.009	0.007	0.005	0.003	0.001
S_2	−1.710	−1.722	−1.738	−1.742	−1.751	−1.776

These results were calculated with a double-precision Fortran code as an integral with the upper limit 10^4. These values should be compared with $S_2(1/2) = \sqrt{\pi} = -1.772$.

1.4. Asymptotic Estimates of Integrals

It is a rather typical situation that you cannot construct any explicit analytic solution for your complex problem of mathematical physics, but you succeed in finding an approximate solution in some specific range of a certain physical parameter (usually, small or large). The idea of asymptotic estimates is very fruitful in such cases.

We will say that an infinite set of functions $\{g_n(\lambda)\}_{n=0}^{\infty}$ forms an *asymptotic scale* as $\lambda \to \lambda_0$ if $g_{n+1}(\lambda) = o(g_n(\lambda))$, $\lambda \to \lambda_0$. It can easily be proved that if $\{g_n(\lambda)\}$ is an asymptotic scale, then the functions $\{g_n^{\delta}(\lambda)\}$ also form a scale for any $\delta > 0$. In practice we will use only $\lambda_0 = 0$ or $\lambda_0 = \infty$, where the natural scales are formed by powers and logarithmic functions.

In particular, the $\{\lambda^{-\mu_n}\}$, where $\mu_n \to +\infty$ as $n \to \infty$, form a scale as $\lambda \to +\infty$, and so do the functions $\{(\ln \lambda)^{-\mu_n}\}$. Certain combinations of power and logarithmic functions of the type $\{\lambda^{-\delta_n}(\ln \lambda)^{-\mu_n}\}$ can also form asymptotic scales as $\lambda \to +\infty$. By analogy, both sets $\{\lambda^{\mu_n}\}$ and $\{[\ln(1/\lambda)]^{-\mu_n}\}$, as well as some of their combinations, form asymptotic scales as $\lambda \to +0$.

If $\{g_n(\lambda)\}$ forms an asymptotic scale as $\lambda \to \lambda_0$, then the formal expansion $f(\lambda) \sim \sum_{n=0}^{\infty} a_n g_n(\lambda)$ is called a full *asymptotic expansion* of the function $f(x)$ at $\lambda \to x_0$ if for any $N = 0, 1, 2, \ldots$

$$f(\lambda) = \sum_{n=0}^{N} a_n g_n(\lambda) + o(g_N(\lambda)), \qquad \lambda \to \lambda_0. \tag{1.63}$$

In the case where relation (1.63) is valid only for some N, it is an asymptotic expansion of order N. As a rule, in actual complex problems only the leading asymptotic term ($N = 0$) can be constructed in a direct elegant way. Further terms usually require too cumbersome transformations.

The most widespread type of integrals admitting natural asymptotic estimates contains a large asymptotic parameter in the argument of exponential function. In the case where the phase function $S(x)$ is real-valued, we arrive at the Laplace integral (Erdelyi, 1956)

$$I(\lambda) = \int_a^b f(x) e^{\lambda S(x)}\, dx, \qquad (a < b \le \infty). \tag{1.64}$$

If the density $f(x)$ and the phase $S(x)$ are smooth—more precisely if $f(x), S(x) \in C^{\infty}[a, b]$—and $\max_{x \in [a,b]} S(x) = S(x_0)$, where x_0 is an internal point of the interval $[a, b]$, then the full power asymptotic expansion for $I(\lambda)$ is expressed as follows (Erdelyi, 1956; Fedorjuk, 1977):

$$I(\lambda) = \int_a^b f(x) e^{\lambda S(x)}\, dx \sim e^{\lambda S(x_0)} \sum_{n=0}^{\infty} c_n \lambda^{-n-1/2}, \qquad \lambda \to +\infty, \qquad \text{where}$$

$$c_n = \frac{\Gamma(n + 1/2)}{(2n-1)!} \lim_{x \to x_0} \left(\frac{d}{dx}\right)^n \left\{ f(x) \left[\frac{S(x_0) - S(x)}{(x - x_0)^2}\right]^{-n-1/2} \right\}. \tag{1.65}$$

Heuristically, the principal asymptotic contribution to the integral $I(\lambda)$ is given by a small neighborhood of the point x_0 (where $S'(x_0) = 0$), and the relative contribution of all other points is exponentially small.

It is obvious from Eq. (1.65) that the leading asymptotic term is

$$I(\lambda) \sim \sqrt{-\frac{2\pi}{\lambda S''(x_0)}}\, f(x_0) e^{\lambda S(x_0)} + O\left(\lambda^{-3/2}\right), \qquad \lambda \to +\infty, \tag{1.66}$$

where, of course, we have $S''(x_0) < 0$ at the point x_0 of (local) maximum. The estimate (1.66) is also valid for $f(x) \in C_1[a, b]$.

In the case where the point x_0 of global maximum coincides with a boundary point ($x_0 = a$ or $x_0 = b$), the considered integral (1.64) usually looks like a Laplace-transform integral:

$$I(\lambda) = \int_0^a x^{\beta-1} f(x) \exp(-\lambda x^\alpha)\, dx \qquad (\alpha, \beta > 0, \; 0 < a \leq \infty). \qquad (1.67)$$

If $f(x) \in C^\infty[0, a]$ here, then the asymptotic expansion is given by the following Watson lemma (see Erdelyi, 1956; Fedorjuk, 1977).

LEMMA (WATSON).

$$\begin{aligned} I(\lambda) &= \int_0^a x^{\beta-1} f(x)\, e^{-\lambda x^\alpha}\, dx \\ &\sim \frac{1}{\alpha} \sum_{n=0}^{\infty} \Gamma\left(\frac{n+\beta}{\alpha}\right) \frac{f^{(n)}(0)}{n!} \lambda^{-(n+\beta)/\alpha}, \quad \lambda \to +\infty. \end{aligned} \qquad (1.68)$$

Estimates like (1.65) and (1.68) can be derived, after some appropriate preliminary treatment, either with the help of integration by parts or using Taylor series expansion for the density $f(x)$.

For a general analytic phase function the Laplace transform (1.67) admits the same integration by parts, which yields the following asymptotic expansion:

$$I(\lambda) = \int_0^a f(x)\, e^{-\lambda S(x)}\, dx \sim e^{-\lambda S(0)} \sum_{n=0}^{\infty} \frac{c_n}{\lambda^{n+1}}, \quad \lambda \to +\infty \quad (0 < a \leq \infty),$$

$$c_n = \left(\frac{1}{S'(x)} \frac{d}{dx}\right)^n \left(\frac{f(x)}{S'(x)}\right)\Bigg|_{x=0}, \qquad f(x),\, S(x) \in C^\infty[0, a], \qquad (1.69)$$

if $x_0 = 0$ is a point of global maximum of the phase $S(x)$.

Example.
This example demonstrates an asymptotic expansion with logarithmic scale. The integral

$$J(\lambda) = \int_0^\infty e^{-\lambda x} \frac{dx}{x \ln^2 x} \qquad (1.70)$$

is finite in the classical sense, since the singularity at the origin ($x \to 0$) is integrable, which can be proved by integration by parts:

$$J(\lambda) = -e^{-\lambda x} \frac{1}{\ln x}\Bigg|_{x=0}^{\infty} - \lambda \int_0^\infty \frac{e^{-\lambda x}}{\ln x}\, dx = -\lambda \int_0^\infty \frac{e^{-\lambda x}}{\ln x}\, dx. \qquad (1.71)$$

If we are interested in an asymptotic estimate of $J(\lambda)$ for large λ, then we can use a Taylor expansion:

$$\begin{aligned} J(\lambda) &= -\int_0^\infty \frac{e^{-x}\, dx}{\ln(x/\lambda)} = \int_0^\infty \frac{e^{-x}\, dx}{\ln \lambda - \ln x} = \frac{1}{\ln \lambda} \int_0^\infty \frac{e^{-x}\, dx}{1 - (\ln x/\ln \lambda)} \\ &\sim \sum_{n=0}^{\infty} \frac{1}{(\ln \lambda)^{n+1}} \int_0^\infty e^{-x} \ln^n(x)\, dx = \sum_{n=0}^{\infty} \frac{\Gamma^{(n)}(1)}{(\ln \lambda)^{n+1}}, \end{aligned} \qquad (1.72)$$

where the value of the last integral has been taken from the tables (see Gradshteyn and Ryzhik, 1994).

More often, in diffraction theory, the exponential kernel in integrals is not a positive (as in Eqs. (1.64), (1.67)) but an oscillating function. The following lemma is analogous to the Watson lemma for integrals with positive kernels (Erdelyi, 1956; Fedorjuk, 1977).

LEMMA (ERDELYI). *Let* $a \leq \infty$; $\alpha, \beta > 0$; $f(x) \in C^\infty[0, a]$; *and* $f^{(n)}(a) = 0$ $(\forall n = 0, 1, 2, \ldots)$. *Then*

$$\int_0^a x^{\beta-1} f(x) e^{i\lambda x^\alpha} \, dx \sim \frac{1}{\alpha} \sum_{n=0}^{\infty} \Gamma\left(\frac{n+\beta}{\alpha}\right) \frac{f^{(n)}(0)}{n!} (-i\lambda)^{-(n+\beta)/\alpha}, \quad \lambda \to +\infty. \quad (1.73)$$

The more general Fourier integral

$$I(\lambda) = \int_a^b f(x) e^{i\lambda S(x)} \, dx, \qquad S(x) \neq \text{const}, \quad (1.74)$$

where the phase function $S(x)$ is real-valued, can also be estimated. As the parameter λ increases the exponential function in (1.74) becomes rapidly oscillating; however the integral itself is asymptotically small, since the contribution of its positive and negative parts almost cancel each other out. The only unclear question is how small it is.

If $f(x), S(x) \in C^\infty[a, b]$ and $S'(x) \neq 0$, $x \in [a, b]$ $(a < b < \infty)$, then

$$I(\lambda) = \int_a^b f(x) e^{i\lambda S(x)} \, dx$$

$$\sim e^{i\lambda S(a)} \sum_{n=0}^{\infty} \frac{d_n}{(i\lambda)^{n+1}} - e^{i\lambda S(b)} \sum_{n=0}^{\infty} \frac{e_n}{(i\lambda)^{n+1}}, \quad \lambda \to +\infty, \quad (1.75)$$

$$d_n = \left(\frac{1}{S'(x)} \frac{d}{dx}\right)^n \left(\frac{f(x)}{S'(x)}\right)\Bigg|_{x=a}, \quad e_n = \left(\frac{1}{S'(x)} \frac{d}{dx}\right)^n \left(\frac{f(x)}{S'(x)}\right)\Bigg|_{x=b}.$$

It follows from the last expansion that, with exponentially small error, the principal contribution to the asymptotics of $I(\lambda)$, $\lambda \to +\infty$ is made by small neighborhoods of the boundary points a and b, and the leading term has the order of $O(1/\lambda)$, $\lambda \to +\infty$. If $f(x) \in C_1[a, b]$ only, then the estimate (1.75) remains valid, containing only the leading contributions from the ends. In the case where the density $f(x)$ vanishes at the ends of the interval $[a, b]$ together with all its derivatives, $f^{(n)}(a) = f^{(n)}(b) = 0$, $\forall n$, the value of $I(\lambda)$ is exponentially small.

It should be noted that the formal structures of Eqs. (1.69) and (1.75) are quite similar. This is because both expansions have been derived with the help of integration by parts. However, the intrinsic essence of these expansions is totally different. The former is a consequence of the fact that the integrand decays exponentially far away from the endpoint $x_0 = 0$. By contrast, the latter is determined by mutual cancellation of the positive and negative contributions, far away from the endpoints.

DEFINITION. *A point* $x_0 \in (a, b)$ *where* $S'(x_0) = 0$ *is called a* stationary point.

Intuitively, since the stationary point is at the same time a point of local extremum, the phase function $S(x)$ is almost constant in a small neighborhood of any stationary point. We thus expect that its contribution to the integral $I(\lambda)$ of (1.75) is asymptotically more significant than those from the endpoints. A precise result here is given by the following statement, which represents a *stationary phase method*.

Let $f(x), S(x) \in C^\infty[a, b]$ and let there exist only one internal stationary point x_0: $a < x_0 < b$, $S'(x_0) = 0$, $S''(x_0) \neq 0$. Then if $f^{(n)}(a) = f^{(n)}(b) = 0$, $\forall n = 0, 1, 2, \ldots$, we have

$$I(\lambda) = \int_a^b f(x) e^{i\lambda S(x)} \, dx \sim \sum_{n=0}^{\infty} \frac{a_n}{\lambda^{n+1/2}}, \quad \lambda \to +\infty, \quad (1.76)$$

where the leading asymptotic term is

$$I(\lambda) \sim \sqrt{\frac{2\pi}{\lambda |S''(x_0)|}} \, e^{i[\lambda S(x_0) + \pi \delta/4]} \left[f(x_0) + O\left(\frac{1}{\lambda}\right) \right], \quad \lambda \to +\infty, \quad \delta = \text{sign}[S''(x_0)].$$

(1.77)

We can see that the contribution of the stationary point is of the order of $O(\lambda^{-1/2})$ (compare with the contribution given by an endpoint, which is $O(1/\lambda)$). Note that if there are several stationary points on the interval $x \in (a, b)$, then the complete asymptotics is given by adding all the contributions together.

This stationary phase method can be extended to multiple integrals (Fedorjuk, 1962; 1977). Let $\Omega \subset \mathbb{R}^N$ be a bounded domain and let the density $f(x) \in C^\infty(\Omega)$ be compactly supported (i.e., it vanishes when approaching the boundary of Ω, together with all its partial derivatives). Let the real-valued phase $S(x)$, $x \in \Omega$, have a unique stationary point $x_0 \in \Omega$, such that the gradient $S'(x_0) = 0$ and $\det S''(x_0) \neq 0$, where $S''(x)$ is the Hessian of the function $S(x)$, i.e., the matrix of its second-order partial derivatives $\partial^2 S / \partial x_n \partial x_k$. Then the multiple stationary-phase method determines the contribution of the stationary point x_0 as follows:

$$I(\lambda) = \int_\Omega f(x) e^{i\lambda S(x)} \, dx \sim e^{i\lambda S(x_0)} \sum_{n=0}^\infty \frac{a_n}{\lambda^{n+N/2}}, \quad \lambda \to +\infty,$$

(1.78)

with the leading asymptotic term being

$$I(\lambda) \sim \left(\frac{2\pi}{\lambda}\right)^{N/2} e^{i[\lambda S(x_0) + \pi i \delta/4]} \frac{f(x_0)}{\sqrt{|\det S''(x_0)|}}, \quad \lambda \to +\infty,$$

(1.79)

where $\delta = \nu_+[S''(x_0)] - \nu_-[S''(x_0)]$ is the difference between the number of positive and negative eigenvalues of the Hessian $S''(x)$ at the stationary point x_0.

Sometimes it is very important to know the behavior of the Laplace integral (1.67) not only for large but also for small λ. Unexpectedly, this turns out to be a harder task. An important result is stated as follows (Handelsman and Lew, 1970; Riekstinsh, 1977).

If

$$f(x) \sim \sum_{n=0}^\infty \frac{c_n}{x^{\mu_n}}, \quad x \to +\infty, \quad n < \mu_n < n+1 \quad (c_0 \neq 0),$$

(1.80)

then

$$I(\lambda) = \int_0^\infty e^{-\lambda x} f(x) \, dx \sim \sum_{n=0}^\infty c_n \Gamma(1 - \mu_n) \lambda^{\mu_n - 1} + \sum_{n=0}^\infty \frac{(-1)^n}{n!} d_n \lambda^n, \quad \lambda \to +0, \quad (1.81)$$

where

$$d_n = \int_0^\infty x^n \left[f(x) - \sum_{j=0}^n \frac{c_j}{x^{\mu_j}} \right] dx.$$

Note that the μ_n in Eqs. (1.80), (1.81) are noninteger. Otherwise, in the case of integer μ_n we have: if

$$f(x) \sim \sum_{n=1}^\infty \frac{c_n}{x^n}, \quad x \to +\infty \quad (c_1 \neq 0),$$

(1.82)

then

$$I(\lambda) \sim \ln \lambda \sum_{n=0}^\infty \frac{(-1)^n}{n!} c_{n+1} \lambda^n + \sum_{n=0}^\infty \frac{(-1)^n}{n!} e_n \lambda^n, \quad \lambda \to +0,$$

(1.83)

where

$$e_n = \int_0^1 x^n \left[f(x) - \sum_{j=1}^{n} c_j x^{n-j} \right] dx + \int_1^{n} x^n \left[f(x) - \sum_{j=1}^{n+1} \frac{c_j}{x^j} \right] dx + c_{n+1} \left[\Gamma'(1) + \sum_{j=1}^{n} \frac{1}{j} \right].$$

These results can be extended to Fourier-type integrals. If the behavior of the function $f(x)$ at infinity is given by Eq. (1.80), then

$$I(\lambda) = \int_0^{\infty} e^{i\lambda x} f(x)\, dx$$

$$\sim \sum_{n=0}^{\infty} c_n \Gamma(1 + n - \mu_n)(-1)^n (-i\lambda)^{\mu_n - 1} + \sum_{n=0}^{\infty} \frac{d_n}{n!} (i\lambda)^n, \quad \lambda \to +0, \tag{1.84}$$

where d_n is the same as in Eq. (1.81).

In the case where the μ_n are positive integers, i.e., if $f(x)$ possesses for large x asymptotics of the type (1.82), we have (Grosjean, 1965; Riekstinsh, 1981)

$$I(\lambda) \sim -\ln \lambda \sum_{n=0}^{\infty} \frac{c_{n+1}}{n!} (i\lambda)^n + \sum_{n=0}^{\infty} \frac{h_n}{n!} (i\lambda)^n, \quad \lambda \to +0, \qquad h_n = e_n + \frac{\pi i}{2n!}, \tag{1.85}$$

where e_n is given by Eq. (1.83).

Helpful remarks

$1°$. If you compare the properties of asymptotic series with those of usual convergent series, you may find that these properties, according to their definitions, are absolutely different. If we treat the expansion (1.63) as an infinite series (with $N = \infty$), then the convergence of an ordinary series implies that the difference between the left and the right hand sides in (1.63) vanishes as $N \to \infty$. By contrast, the expansion (1.63) treated in the asymptotic sense means that this discrepancy vanishes as $\lambda \to \lambda_0$. We can conclude from these arguments that typically convergent series do not represent any asymptotic series, and vice versa, many asymptotic series, which are very helpful in efficient calculations with λ around λ_0, do not converge in the classical sense.

$2°$. If you compare the asymptotic behavior of the Laplace transform

$$F(\lambda) = \int_0^{\infty} e^{-\lambda x} f(x)\, dx \tag{1.86}$$

for large (Eq. (1.68)) and small (Eq. (1.81)) values of the parameter λ in the case when the principal term of the density is a noninteger power, you will be surprised that the leading asymptotic terms coincide in the sense that ($A = \text{const}$)

$$\text{if} \quad f(x) \sim A x^{\mu} \quad (-1 < \mu < 0), \quad x \to \begin{Bmatrix} +0 \\ +\infty \end{Bmatrix},$$

$$\text{then} \quad F(\lambda) \sim A \Gamma(1 + \mu) \lambda^{-\mu - 1}, \quad \lambda \to \begin{Bmatrix} +0 \\ +\infty \end{Bmatrix}. \tag{1.87}$$

Remarkably, the respective leading terms of the Fourier integrals have the same form (see Eqs. (1.73), (1.83) with $n = 0$) if you formally substitute $-i\lambda$ for λ in the Laplace integral.

$3°$. We considered above only real-valued and imaginary phases. In some problems the phase $S(x)$ appears complex-valued, which leads to the so-called *steepest descent method*, described in detail in the literature (e.g., see Fedorjuk, 1962). This method is thus designed to deal with more general cases than Laplace or oscillating integrals. Amazingly, after some refined manipulation, almost always such problems can be reduced to a combination of Laplace-type and Fourier-type integrals.

For example, the integral

$$J = \int_{-\infty}^{\infty} F(\alpha)\, e^{-i\alpha x} e^{-\gamma(\alpha) y}\, d\alpha, \qquad \gamma(s) = \sqrt{\alpha^2 - k^2} \qquad (k > 0), \qquad (1.88)$$

where $y > 0$ and we assume the function $F(s)$ to be even for simplicity (so J is even with respect to x), is usually studied for $r = \sqrt{x^2 + y^2} \to \infty$ by the steepest descent method (Mittra and Lee, 1971) with the help of a certain transformation of the integration contour in the complex plane (recall the analytic properties of the branching function $\gamma(\alpha) = \sqrt{\alpha^2 - k^2}$, Section 1.1). Let us show that the integral (1.88) can be correctly estimated just over the real ray $\mathrm{Im}(\alpha) = 0$, $\mathrm{Re}(\alpha) \geq 0$:

$$\begin{aligned} J &= \int_0^{\infty} F(\alpha)\big(e^{i\alpha|x|} + e^{-i\alpha|x|}\big) e^{-\gamma(\alpha) y}\, d\alpha \\ &= \int_0^k F(\alpha)\big(e^{i\alpha|x|} + e^{-i\alpha|x|}\big) e^{i\sqrt{k^2-\alpha^2}\, y}\, d\alpha + \int_k^{\infty} F(\alpha)\big(e^{i\alpha|x|} + e^{-i\alpha|x|}\big) e^{-\sqrt{\alpha^2-k^2}\, y}\, d\alpha \quad (1.89) \\ &= J_1 + J_2. \end{aligned}$$

The second integral in (1.89) admits estimation of the leading term from (1.67), where $\alpha = 1/2$, $\beta = 1$, $\lambda = y$, $n = 0$:

$$\begin{aligned} |J_2| &= \left| \int_0^{\infty} F(\alpha + k)\left[e^{i(\alpha+k)|x|} + e^{-i(\alpha+k)|x|} \right] e^{-\sqrt{\alpha+2k}\, y}\, e^{-\sqrt{\alpha}\, y}\, d\alpha \right| \\ &\leq \int_0^{\infty} |F(\alpha + k)|\, e^{-\sqrt{\alpha}\, y}\, d\alpha \sim O\left(\frac{1}{y^2}\right), \qquad y \to +\infty. \end{aligned} \qquad (1.90)$$

This relation holds uniformly over $|x| < \infty$. For any fixed y, J_2 can be estimated for $x \to \infty$ by (1.75) as $O(1/|x|)$. Therefore, $J_2 \sim O(1/r)$ as $r = \sqrt{x^2 + y^2} \to \infty$.

The first integral J_1 can be treated by the stationary phase method. Let $|x| = r \sin \varphi$, $y = r \cos \varphi$, $r \to \infty$. Then in the integral J_1,

$$J_1 = \int_0^k F(\alpha)\left[e^{ir(\sqrt{k^2-\alpha^2}\, \cos \varphi - \alpha \sin \varphi)} + e^{ir(\sqrt{k^2-\alpha^2}\, \cos \varphi + \alpha \sin \varphi)} \right] d\alpha, \qquad (1.91)$$

the phase $S(\alpha) = \sqrt{k^2 - \alpha^2} \cos \varphi - \alpha \sin \varphi$ of the first term has no stationary points ($S'(\alpha) \neq 0$, $0 \leq \alpha \leq k$), so its behavior is again of the order of $O(1/r)$. The other term has the phase $S(\alpha) = \sqrt{k^2 - \alpha^2} \cos \varphi + \alpha \sin \varphi$ with the stationary point $\alpha_0 = k \sin \varphi$, $S'(\alpha_0) = 0$. Taking into account that $S''(\alpha_0) = -1/(k \cos^2 \varphi)$, $\delta = -1$, $S(\alpha_0) = k$, we thus arrive at the final result, in accordance with Eq. (1.77):

$$J \sim \sqrt{\frac{2\pi k}{r}}\, \cos \varphi\, e^{i(kr - \pi/4)}\, f(k \sin \varphi) + O\left(\frac{1}{r}\right), \qquad r \to \infty. \qquad (1.92)$$

$4°$. In practice, you may always try to extract the leading asymptotic term by letting λ tend to infinity. Such an approach can be, as a rule, proved strictly as a theorem. Thus, the following integral:

$$\int_0^{\infty} \frac{J_0(\lambda x)\, dx}{\sqrt{x^2 + a^2}} = \int_0^{\infty} \frac{J_0(t)\, dt}{\sqrt{t^2 + (a\lambda)^2}}, \qquad (1.93)$$

after the change of variable $\lambda x = t$, can be estimated by letting $\lambda \sim +\infty$, which implies the first term under the square root to be negligibly small compared with $(a\lambda)$. This yields the leading asymptotic term in the form

$$\int_0^\infty \frac{J_0(\lambda x)\,dx}{\sqrt{x^2 + a^2}} \sim \frac{1}{a\lambda} \int_0^\infty J_0(t)\,dt = \frac{1}{a\lambda}, \quad \lambda \to +\infty \quad \left(\int_0^\infty J_0(t)\,dt = 1 \right), \quad (1.94)$$

where we have used the value of a tabulated integral (Gradshteyn and Ryzhik, 1994).

In the cases where you do not succeed in applying such a simple and clear procedure, it is quite probable that your asymptotic expansion involves logarithmic or fractional power functions rather than integer powers. For example, the integral

$$\int_0^\infty \frac{J_0^2(\lambda x)\,dx}{\sqrt{x^2 + a^2}} = \int_0^\infty \frac{J_0^2(t)\,dt}{\sqrt{t^2 + (a\lambda)^2}}, \quad (1.95)$$

which is allied to (1.93), cannot be estimated for large λ in the same way because the integral $\int_0^\infty J_0^2(t)\,dt$ diverges at infinity. This is a sure sign for a logarithmic scale to appear. The same feature becomes apparent for both integrals (1.93), (1.95) as $\lambda \to 0$. Indeed, with $\lambda \to +0$, all arising integrals,

$$\int_0^\infty \frac{dx}{\sqrt{x^2 + a^2}} \quad (J_0(0) = 1), \qquad \int_0^\infty \frac{J_0(t)\,dt}{t}, \qquad \int_0^\infty \frac{J_0^2(t)\,dt}{t}, \quad (1.96)$$

diverge, the first one at infinity and the others at the origin.

5°. The phenomenon indicated in 4° is related to a very difficult case where the integrand does not contain any logarithmic function but, at the same time, the asymptotic expansion yields a logarithmic term in the scale. In this case it is advised that you reduce your integral to any combination of standard special functions, which have been well studied over the last 100 years and whose asymptotics are described in detail.

For example, the integral (1.93) is

$$\int_0^\infty \frac{J_0(\lambda x)\,dx}{\sqrt{x^2 + a^2}} = I_0\left(\frac{a\lambda}{2}\right) K_0\left(\frac{a\lambda}{2}\right), \quad (1.97)$$

where I_0 is the modified Bessel function of order 0 and K_0 is the McDonald function of the same order. First of all, this leads to a full asymptotics as $\lambda \to +\infty$ (Abramowitz and Stegun, 1965):

$$\int_0^\infty \frac{J_0(\lambda x)\,dx}{\sqrt{x^2 + a^2}} \sim \frac{1}{a\lambda} + \sum_{n=1}^\infty (-1)^n \frac{(2n-1)!!}{(2n)!!\,(a\lambda)^{2n+1}} \prod_{j=1}^n (2j-1)^2, \quad \lambda \to +\infty, \quad (1.98)$$

which, in particular, gives our leading term (1.94).

Representation (1.97) permits also a small-λ expansion, which contains logarithmic functions. It can be obtained from the series:

$$I_0(z) = \sum_{n=0}^\infty \frac{(z/2)^{2n}}{(n!)^2}, \quad K_0(z) = -[\ln(z/2) + \gamma]\,I_0(z) + \sum_{n=1}^\infty \frac{(z/2)^{2n}}{(n!)^2}\left(\sum_{j=1}^n \frac{1}{j}\right), \quad (1.99)$$

where $\gamma = 0.577216$ is the Euler constant. In particular, the first two leading terms are:

$$\int_0^\infty \frac{J_0(\lambda x)\,dx}{\sqrt{x^2 + a^2}} \sim -\ln\lambda + [\ln(4/a) - \gamma] + O(\lambda^2 \ln\lambda), \quad \lambda \to +0. \quad (1.100)$$

The same idea can be applied to the integral (1.95). To this end, we use the integral representation (Abramowitz and Stegun, 1965)

$$J_0^2(t) = \frac{2}{\pi} \int_0^{\pi/2} J_0(2t \cos \psi)\, d\psi,$$ (1.101)

which, with the help of (1.97), gives

$$\int_0^\infty \frac{J_0^2(\lambda x)\, dx}{\sqrt{x^2 + a^2}} = \frac{2}{\pi} \int_0^{\pi/2} I_0(a\lambda \cos \psi) K_0(a\lambda \cos \psi)\, d\psi,$$ (1.102)

and, by using series (1.99), for small λ we finally obtain

$$\int_0^\infty \frac{J_0^2(\lambda x)\, dx}{\sqrt{x^2 + a^2}} \sim -\ln \lambda + [\ln(4/a) - 2\gamma] + O(\lambda^2 \ln \lambda), \quad \lambda \to +0,$$

$$\text{since} \quad \int_0^{\pi/2} \ln(\cos \psi)\, d\psi = \pi(\gamma - \ln 2)/2$$ (1.103)

(see Gradshteyn and Ryzhik, 1994).

The most difficult problem is to construct the asymptotics of the integral (1.80) as $\lambda \to +\infty$, which requires a more refined approach. With the help of tabulated integrals (Gradshteyn and Ryzhik, 1994) one can observe that

$$\begin{aligned}
\int_0^\infty \frac{J_0^2(\lambda x)\, dx}{\sqrt{x^2 + a^2}} &= \frac{2}{\pi} \int_0^\infty J_0^2(x)\, dx \int_0^\infty \cos(\xi x)\, K_0(|\alpha|\xi)\, d\xi \\
&= \frac{2}{\pi} \int_0^\infty K_0(|\alpha|\xi) \left\{ \begin{array}{ll} 0, & \xi > 2 \\ \frac{1}{2} P_{-1/2}(\frac{1}{2}\xi^2 - 1), & 0 < \xi < 2 \end{array} \right\} d\xi \\
&= \frac{1}{\pi|\alpha|} \int_0^{2|\alpha|} K_0(\xi)\, P_{-1/2}\left(\frac{\xi^2}{2\alpha^2} - 1 \right) d\xi,
\end{aligned}$$ (1.104)

where $P_\nu(x)$ is the Legendre function, whose asymptotic behavior near the singular point $x = -1$ is (Bateman and Erdelyi, 1953)

$$P_\nu(x) \sim \frac{\sin(\pi\nu)}{\pi} \left[\ln \frac{1+x}{2} + \gamma + 2\psi(\nu+1) + \pi \cot(\pi\nu) \right] + o(1).$$ (1.105)

Now asymptotic results of the theorem directly follow from the exponential decay of the McDonald function at infinity, which yields the first two leading asymptotic terms in the following form:

$$\begin{aligned}
\int_0^\infty \frac{J_0^2(x)\, dx}{\sqrt{\alpha^2 + x^2}} &\sim -\frac{1}{\pi^2 |\alpha|} \int_0^\infty K_0(\xi) \left[\ln \frac{\xi^2}{4\alpha^2} + \gamma + 2\psi\left(\frac{1}{2}\right) \right] d\xi \\
&= \frac{1}{\pi|\alpha|} \left[\ln|\alpha| - \psi\left(\frac{1}{2}\right) + 2\ln 2 + \frac{\gamma}{2} \right] \\
&= \frac{1}{\pi|\alpha|} \left[\ln|\alpha| + 4\ln 2 + \frac{3}{2}\gamma \right], \quad \alpha \to \infty,
\end{aligned}$$ (1.106)

since $\psi(1/2) = -(\gamma + 2\ln 2)$. Here the following tabulated integrals have been taken into account (Gradshteyn and Ryzhik, 1994):

$$\int_0^\infty K_0(\xi)\, d\xi = \frac{\pi}{2}, \quad \int_0^\infty K_0(\xi) \ln \xi\, d\xi = -\frac{\pi}{2}(\ln 2 + \gamma).$$ (1.107)

1.5. Fredholm Theory for Integral Equations of the Second Kind

The theory of integral equations of the second kind

$$u(x) - \mu \int_S G(x, \xi) \, u(\xi) \, ds(\xi) = f(x), \qquad x \in S, \tag{1.108}$$

was devised by Fredholm and further developed by Riesz (see, for example, Riesz and Sz.-Nagy, 1972). Here the right-hand side $f(x)$ and the kernel $G(x, \xi)$ are some known functions, μ is a known parameter, S is a piecewise-smooth surface (or line), and $u(x)$ is an unknown function to be determined from Eq. (1.108). Note that, in order to specify the variable (x or ξ) with respect to which the integral operator is applied, we explicitly indicate that the element ds depends on the variable ξ.

The following results of the Fredholm theory can be regarded as classical. Let us rewrite Eq. (1.108) as a linear functional equation

$$(I - \mu G) \, u = f, \qquad (Gu)(x) = \int_S G(x, \xi) \, u(\xi) \, ds(\xi). \tag{1.109}$$

We will consider it in the normal space of continuous functions $X = C(S)$ with the norm $\|\varphi\|_{C(S)} = \max_{x \in S} |\varphi(x)|$. The set $U \subset X$ is called *compact* if an arbitrary sequence $\{u_n\} \subset U$ contains a convergent subsequence.

THEOREM 1 (ARZELA–ASCOLI). *A set $U \subset X$ of functions $u(x)$, $x \in S$ is compact in $C(S)$ if and only if:*

1) there exists a constant A such that $|u(x)| \leq A$ for all $x \in S$ and all $u(x) \in U$;

2) for any $\varepsilon > 0$ there exists a constant $\delta > 0$ such that for all $x, y \in S : |x - y| < \delta$ and all $u(x) \in U$ the inequality $|u(x) - u(y)| < \varepsilon$ holds.

The first property means that the set U is uniformly bounded in $C(S)$, and the second one means that the functions from U are equicontinuous.

An operator $G : X \to X$ is called *compact* if it maps any bounded set to a compact one. It is well known that any linear compact operator is continuous (i.e., bounded). Below in this section we will consider only compact operators G.

A simple theorem states that if the kernel of Eq. (1.108) is continuous with respect to both its variables, $G(x, \xi) \in C(S \times S)$, then the operator G in (1.109) is compact. The proof directly follows from the Arzela–Ascoli theorem. The same result holds if the operator G is weakly singular, i.e., if the kernel $G(x, \xi)$ is continuous for $x \neq \xi$ and $G(x, \xi) \leq B|x - \xi|^{\alpha - n}$ ($0 < \alpha < 1$) (n is the dimension of S, i.e., $n = 2$ when S is a surface, and $n = 1$ when S is a line). The kernel $G(x, \xi)$ with weak singularity is integrable over S. Note that the logarithmic singularity, when $G(x, \xi) \sim A \ln |x - \xi|$, $\xi \to x$, is also weak. It should also be noted that the same properties of compactness, as in $C(S)$, take place in the Hilbert space $L_2(S)$, too. Two operators $A, B : X \to X$ are called *conjugate* to each other if for all u, v we have $(Au, v) = (u, Bv)$. It is evident that if $G(x, \xi)$ in (2) is continuous or weakly singular, then the integral operator G^* with kernel $G^*(x, \xi) = \bar{G}(\xi, x)$ is conjugate to the operator G. Thus, any symmetric real-valued kernel yields some *self-conjugate* (*self-adjoint*) operator.

THEOREM 2 (FIRST FREDHOLM THEOREM). *The subspaces of solutions to the homogeneous equations*

$$(I - \mu G) \, u = 0, \qquad (I - \mu G^*) \, v = 0, \tag{1.110}$$

have equal finite dimension $n < \infty$ in X. In particular, these subspaces may be empty (i.e., $n = 0$).

THEOREM 3 (SECOND FREDHOLM THEOREM). *Nonhomogeneous equations*

$$(I - \mu G)\, u = f, \qquad (I - \mu G^*)\, v = g, \tag{1.111}$$

have solutions if and only if $(f, v_j) = 0$, $(u_j, g) = 0$ *for all solutions of the conjugate homogeneous equations*

$$(I - \mu G)\, u_j = 0, \qquad (I - \mu G^*)\, v_j = 0. \tag{1.112}$$

From these two theorems you can directly come to the conclusion stated by the following

THEOREM 4 (FREDHOLM'S ALTERNATIVE, OR THIRD FREDHOLM THEOREM). *If the homogeneous equations (1.110) have (simultaneously) only the trivial solution* $u = 0$, $v = 0$, *then the nonhomogeneous equations (1.111) have unique solutions for all* $f, g \in C(S)$. *In this case there exist continuous inverse operators* $(I - \mu G)^{-1}$, $(I - \mu G^*)^{-1}$. *In the case where equations (1.110) have (simultaneously) nontrivial solutions, the nonhomogeneous equations (1.111) have a solution if and only if relations (1.112) hold.*

DEFINITION. *The set of irregular values of the parameter* λ *related to the second case in Fredholm's alternative, when* $G u_j = \lambda u_j$, $u_j \neq 0$, *is called the* spectrum *of the operator* G, *or the set of its* eigenvalues. *The corresponding values* $\mu = 1/\lambda$ *are called characteristic values of* G.

THEOREM 5 (FOURTH FREDHOLM THEOREM). *The set of characteristic values of integral operators with continuous or weakly singular kernels consists of a (possibly complex-valued) countable discrete array* μ_n, $n = 1, 2, \ldots$, *which has no finite limit point.*

The theory briefly outlined above provides us with a powerful method for solving integral equations of the second kind. First of all, it is clear that for sufficiently small μ the first case of two different possibilities in Fredholm's alternative always holds, and so equations (1.111) are uniquely solvable for all $f, g \in C(S)$. To show this, let us start with proof by contradiction. If there is a nontrivial solution u_j to Eq. (1.110)$_1$, then $u_j = \mu\, G u_j$, which implies $\|u_j\| = |\mu|\, \|G u_j\| \leq |\mu|\, \|G\|\, \|u_j\| \sim |\mu| \geq 1/\|G\|$. The last inequality is wrong, since for a compact operator, we have $G : \|G\| \leq A$, and hence $\|\mu\| \geq 1/A$, which contradicts the assumption that μ is small enough.

The last conclusion in particular means that for small μ the inverse operators $(I - \mu G)^{-1}$ and $(I - \mu G^*)^{-1}$ exist and, besides, are continuous (bounded). Moreover, they may be explicitly calculated as Neumann series in iterated kernels: $(I - \mu G)^{-1} = \sum_{j=0}^{\infty} \mu^j G^j$ (where $G^j = G \ldots G$ with the j-times repeated product of G), and the parameter μ is small enough.

Another important practical conclusion from Fredholm's alternative is that it predetermines, in the case of invertibility, continuous dependence on the right-hand sides f, g, since in this case $u = (I - \mu G)^{-1} f$, $v = (I - \mu G^*)^{-1} g$ with both continuous operators, which guarantees stability of a solution with respect to small perturbations in f and g. This property is very important in the context of what is discussed in Chapter 8, devoted to the theory of ill-posed problems, which covers also operator equations of the first kind.

A more precise estimate for a maximum value of the parameter μ, for which the operator $I - \mu G$ is surely invertible, follows from the well-known theorem: if the norm of the operator $K : X \to X$ is $\|K\| < 1$, then the operator $(I - K)$ is invertible. From this statement we can conclude that Eq. (1.108) or (1.109) is surely solvable and there exists a continuous inverse operator $(I - \mu G)^{-1}$ if $|\mu| < 1/\|G\|$. Of course, this does not mean that for large μ Eq. (1.108) is unsolvable.

The Fredholm theory for equations of the second kind gives you also a powerful practical tool for solving these equations numerically. Let us assume that we replace the operator G by some approximate operator G_1 that admits a simple and direct numerical treatment, so that it is much easier to solve the equation $(I - \mu G_1) u_1 = f$ instead of Eq. (1.109). Then if $(I - \mu G)^{-1}$ is bounded and

$$\|G - G_1\| \le \varepsilon / (\mu \|(I - \mu G)^{-1}\|), \tag{1.113}$$

with $0 < \varepsilon < 1$ being a small quantity, the operator $(I - \mu G_1)^{-1}$ is bounded too, and $\|u - u_1\| \le \varepsilon \|u\| \|(I - \mu G_1)^{-1}\| / (I - \mu G)^{-1}\|$, hence $\|u - u_1\| \to 0$ as $\varepsilon \to +0$.

In order to prove this, let us consider the operator $(I - \mu G_1) = (I - \mu G) + \mu(G - G_1)$. Note that $(I - \mu G)^{-1}(I - \mu G_1) = (I - \mu G)^{-1} [(I - \mu G) + \mu(G - G_1)] = I + \mu(I - \mu G)^{-1}(G - G_1)$. Then, according to the last remark of the previous paragraph and due to condition (1.113), it turns out that the operator $\mu(I - \mu G)^{-1}(G - G_1)$ has a finite norm that is less than unit (more precisely, less than $\varepsilon < 1$), and hence the operator $(I - \mu G)^{-1}(I - \mu G_1)$ is invertible. This implies the invertibility of $(I - \mu G_1)$. Now we have

$$\begin{aligned} u - u_1 &= u - (I - \mu G_1)^{-1} f = u - (I - \mu G_1)^{-1}(I - \mu G)u \\ &= u - (I - \mu G_1)^{-1} [(I - \mu G_1) + \mu(G_1 - G)]u \\ &= -\mu(I - \mu G_1)^{-1}(G_1 - G)u, \end{aligned} \tag{1.114}$$

and consequently

$$\|u - u_1\| \le \varepsilon \|u\| \|(I - \mu G_1)^{-1}\| / (I - \mu G)^{-1}\|, \tag{1.115}$$

as was to be proved.

The last result allows you to construct efficient algorithms for solving equation (1.108). This is well described in the literature (see, for example, Banerjee and Butterfield, 1981; Hackbusch, 1995), and we recommend that you use the most natural and direct algorithm, the so-called *collocation method*, which involves splitting the total domain S into N small subregions and then applying a simple quadrature formula with a constant integrand over each small subregion S_j of measure ΔS_j, $j = 1, 2, \ldots, N$:

$$Gu = \int_S G(x, \xi) u(\xi) \, d\xi \approx \sum_{j=1}^{N} G(x, \xi_j) \Delta S_j u_j \approx G_1 u, \quad u_j = u(\xi_j), \quad \xi_j \in S_j. \tag{1.116}$$

It is quite obvious that with $N \gg 1$ the operator G_1 represents a good approximation to G in the space $X = C(S)$.

If you change the integral operator G in Eq. (1.108) by its approximation (1.116) and let the variable x run through the same set of nodes $\{x_i\}_{i=1}^{N} = \{\xi_j\}_{j=1}^{N}$; $x_i = \xi_i \in S_i$, $i = 1, 2, \ldots, N$, then you arrive at a linear algebraic system of dimension $N \times N$ with respect to the unknowns $u_i = u(x_i)$, $i = 1, 2, \ldots, N$:

$$u_i - \mu \sum_{j=1}^{N} G(x_i, \xi_j) \Delta S_j u_j = f(x_i), \quad i = 1, 2, \ldots, N. \tag{1.117}$$

Helpful remarks

$1°$. An excellent survey of existing methods for the investigation of various types of integral equations can be found in (Polyanin and Manzhirov, 1998); and in Hackbusch (1995). Description of some efficient classical numerical methods for constructing solutions of integral equations is presented, for example, in Hackbusch (1995).

$2°$. It is rather unexpected that this elegant Fredholm theory is absolutely inapplicable for equations of the first kind. Furthermore, as we will see below, equations of the first kind possess qualitative properties opposite to equations of the second kind.

1.6. Fredholm Integral Equations of the First Kind

Let us start with the attempt to extract from the Fredholm alternative something helpful for equations of the first kind. We could see in the previous section that if G is a linear continuous compact operator, then the operator $(\mu G - I)$ is surely invertible if $|\mu|$ is small enough, which is equivalent to invertibility of $(G - \lambda I)$ for sufficiently large $|\lambda|$ ($\lambda = 1/\mu$). Now, when studying an operator equation of the first kind

$$Gu = f, \tag{1.118}$$

we should estimate how small λ can be, provided the operator $(G - \lambda I)^{-1}$ exists and is bounded. It is clear that: if λ here can be arbitrarily small, then it is very probable that the operator G is invertible.

Recall that the irregular values of the parameter λ pertaining to the second case in the Fredholm alternative, when the homogeneous equation has a nontrivial solution $u_j \neq 0$, $Gu_j = \lambda u_j$, form the spectrum $\{\lambda_n\}$ of the operator G or its eigenvalues. Evidently, for these $\lambda = \lambda_n$ the operator $(G - \lambda I)^{-1}$ does not exist, since in this case there is a nontrivial solution to the corresponding homogeneous equation. Let us consider for simplicity the case of self-adjoint operator: $G^* = G$. Then the set of eigenvalues can be completely described in the case of positive definite operator.

DEFINITION. *Let $X = H$ be a Hilbert space. Then a linear continuous operator $G : H \to H$ is called* positive definite *if $(Gu, u) \geq \gamma^2(u, u) \sim (Gu, u) \geq \gamma^2 \|u\|^2$ for all $u \in H$, with some positive constant $\gamma > 0$.*

It is simply proved (see Mikhlin, 1964; Kantorovich and Akilov, 1980) that all eigenvalues of a self-adjoint and positive definite operator G are positive and displaced on the real axis between γ and the norm of the operator $M = \|G\|$: $\gamma \leq \lambda_n \leq M$.

Perhaps, the most impressive result of the theory of linear operators is stated by the following theorem (Kantorovich and Akilov, 1980).

THEOREM 1. *A linear continuous (not necessarily compact) operator $G : X \to Y$ acting from a normed space X to a normed space Y is invertible if and only if the homogeneous equation $Gu = 0$ has only the trivial solution $u = 0$.*

Corollary. Any self-adjoint positive definite linear continuous operator in Hilbert space is invertible.

Indeed, if there is a nontrivial solution $u \neq 0$ to the homogeneous equation $Gu = 0$, this means that the point $\lambda_0 = 0$ belongs to the spectral set of the operator, which is impossible when G is self-adjoint and positive definite.

Unfortunately, all these results become useless when we consider compact operators, because of the following theorem.

THEOREM 2. *If $G : X \to Y$ (X and Y are arbitrary normed spaces) is a bounded compact linear operation, then G^{-1} cannot exist.*

Indeed, if G^{-1} is continuous, then $I = G^{-1}G : X \to X$, as a composition of the compact (G) and the continuous (G^{-1}) operators, maps any bounded set $U \subset X$ onto a compact set $V \subset X$, which for the identity operator I means $U = V$. Thus, in this case any bounded set in X would be simultaneously compact, which is impossible if the space X is not finite-dimensional.

This result is certainly disappointing, since all integral operators with regular and even weakly singular kernels are compact (see the previous section). So, the operator of the Fredholm integral equation (1.118) cannot be invertible. However, this does not mean that Eq. (1.118) cannot have solutions for special classes of right-hand sides f from a set F of a space Y: $F \subset Y$.

Besides, it follows from Theorem 2 and the corollary of Theorem 1 that a positive compact operator cannot be positive definite. Generally, there is no constructive result that would guarantee that Eq. (1.118) possesses a solution in one or another case. However, an interesting theory for Fredholm integral operators with convolution kernels has been developed. It is genetically connected with the general theory of weak (or generalized) solutions of operator equations.

Let us consider an integral equation of the first kind, which we assume, for simplicity, to hold on a finite interval of the real axis. By a trivial change of variable the equation can be made symmetric with respect to the origin:

$$Gu = f \sim \int_{-a}^{a} G(x - \xi) \, u(\xi) \, d\xi = f(x), \qquad |x| < a. \tag{1.119}$$

We assume that the kernel $G(x)$ is given for all arguments, $|x| < \infty$. It is very important, for further consideration, to know qualitative properties of the Fourier image $L(s)$ of the kernel:

$$L(s) = \int_{-\infty}^{\infty} G(x) \, e^{isx} \, dx, \qquad G(x) = \frac{1}{2\pi} \int_{-\infty}^{\infty} L(s) \, e^{-isx} \, ds. \tag{1.120}$$

We assume that the Fourier image $L(s)$ is continuous, positive, even and bounded, $0 < L(s) \le L_0$. Note that the evenness of $L(s)$ guarantees that the kernel $G(x - \xi)$ is real and symmetric, and the operator G is self-conjugate in the real Hilbert space $H = L_2(-a, a)$. Indeed,

$$G(x - \xi) = \int_{-\infty}^{\infty} L(s) \, e^{is(x-\xi)} \, ds = 2 \int_{0}^{\infty} L(s) \, \cos[s(x - \xi)] \, ds =$$

$$= 2 \int_{0}^{\infty} L(s) \, \cos[s(\xi - x)] \, ds = \int_{-\infty}^{\infty} L(s) \, e^{is(\xi-x)} \, ds = G(\xi - x). \tag{1.121}$$

Now, there are various possibilities:

1°. The lower bound of $L(s)$ is positive: $0 < l_0 \le L(s) \le L_0$. Let us prove that in this case the operator G is positive definite in $L_2(-a, a)$, and hence, according to the corollary of Theorem 1, is invertible in this Hilbert space. Indeed,

$$(Gu, u) = \int_{-a}^{a} \int_{a}^{a} G(x - \xi) \, u(\xi) \, u(x) \, dx \, d\xi = \frac{1}{2\pi} \int_{-\infty}^{\infty} L(s) \, |U(s)|^2 \, ds$$

$$\ge \frac{l_0}{2\pi} \int_{-\infty}^{\infty} |U(s)|^2 \, ds = l_0 \int_{-a}^{a} u^2(x) \, dx = l_0 \, (u, u), \tag{1.122}$$

where we have used the Parseval identity (1.7) and the property (1.4). Note that to use these properties, we assume that the function $u(x)$ is extended to the entire axis $x \in (-\infty, \infty)$, outside from the interval $x \in [-a, a]$, by a zero value: $u(x) = 0$, $|x| > a$.

Thus, in this case, G is positive definite and Eq. (1.119) has a unique solution for arbitrary $f(x) \in L_2(-a, a)$.

It should be noted that in this case, Eq. (1.118)–(1.119) is not a standard Fredholm equation of the first kind. Actually, if this was a Fredholm-type equation with a compact operator G, then G would not be invertible. In this sense, the Fourier transform of the kernel, $L(s)$, considered here determines an operator that is much like a second-kind Fredholm operator from the previous section rather than the first-kind one.

In order to explain this in more detail, let us consider a rather widespread case where

$$L(s) = l_0 + \widetilde{L}(s), \qquad \widetilde{L}(s) = O(s^{-\alpha}), \quad \alpha > 1, \quad s \to \infty. \tag{1.123}$$

Then the kernel becomes as follows ($\delta(x)$ is the Dirac delta, see Eq. (1.62)):

$$G(x) = \frac{l_0}{2\pi} \int_{-\infty}^{\infty} e^{-isx}\, ds + \frac{1}{2\pi} \int_{-\infty}^{\infty} \widetilde{L}(s) e^{-isx}\, ds = l_0\, \delta(x) + \widetilde{G}(x), \quad |\widetilde{L}(s)| \le \frac{A}{(|s| + 1)^{\alpha}}, \quad (1.124)$$

since $\widetilde{L}(s)$ is bounded for finite s, with the asymptotic behavior given by Eq. (1.123).

Now it follows from Eq. (1.124) that

$$|\widetilde{G}(x)| \le \frac{1}{2\pi} \int_{-\infty}^{\infty} |\widetilde{L}(s)|\, ds \le \frac{A}{2\pi} \int_0^{\infty} \frac{ds}{(s+1)^{\alpha}} = \frac{A}{\pi(1-\alpha)}, \qquad (1.125)$$

so the kernel $\widetilde{G}(x)$ is continuous and bounded, generating a compact operator in $L_2(-a, a)$. Now for the full operator we have (cf. Eq. (1.62))

$$\begin{aligned}(Gu)(x) &= \int_{-a}^{a} G(x - \xi)u(\xi)\, d\xi = l_0 \int_{-a}^{a} \delta(x - \xi)u(\xi)\, d\xi + \int_{-a}^{a} \widetilde{G}(x - \xi)u(\xi)\, d\xi \\ &= l_0 u(x) + (\widetilde{G}u)(x),\end{aligned} \qquad (1.126)$$

with compact operator $\widetilde{G}$. We thus have arrived at a standard Fredholm integral equation of the second-kind, and the result obtained here offers an alternative treatment for second-kind Fredholm equations. Specifically, if one considers an integral equation with a bounded continuous convolution kernel $\widetilde{G}(x - \xi)$,

$$(I + \widetilde{G})\, u = f, \qquad (1.127)$$

such that the Fourier transform of the full kernel is positive,

$$1 + \widetilde{L}(s) > 0, \quad \text{and} \quad \widetilde{L}(s) = O(|s|^{-\alpha}), \quad \alpha > 1, \quad s \to \infty, \qquad (1.128)$$

then equation (1.126) is uniquely solvable in $L_2(-a, a)$.

2°. The lower bound of $L(s)$ is equal to zero. The most typical case is when $L(s)$ vanishes at infinity ($L(s) \to 0$, $s \to \infty$) as some power: $L(s) = O(|s|^{-\alpha})$, $s \to \infty$, $\alpha > 0$. It follows from general properties of the Fourier transform (see Sections 1.1 and 1.4) that $G(x - \xi)$ is not worse than a weakly singular kernel. Hence, G is a compact operator and so cannot be invertible. But ideas of the standard theory of weak (generalized) solutions of operator equations in energetic spaces may be applied to study this case.

Let $X = H = L_2(-a, a)$ and let us study equation (1.119) in the Hilbert space. We take the scalar product of Eq. (1.119) with a function $u(x) \in L_2(-a, a)$ to obtain

$$(Gu, u) = (f, u). \qquad (1.129)$$

DEFINITION. *Any function $u \in L_2(-a, a)$ that satisfies identity (1.129) is called a weak, or generalized solution of equation (1.119).*

Let us write out the expressions in Eq. (1.129) in detail, by using the Parseval identity and the convolution theorem for the Fourier transform (cf. Eq. (1.122)) (the bar denotes a complex conjugate):

$$\int_{-\infty}^{\infty} L(s)\, |U(s)|^2\, ds = \int_{-\infty}^{\infty} \overline{F}(s)\, U(s)\, ds, \qquad (1.130)$$

where $F(s)$ is the Fourier transform of the right-hand side: $f(x) \Longrightarrow F(s)$. Now we introduce the new (*energetic*) Hilbert space with the scalar product

$$(u, v)_e = \int_{-\infty}^{\infty} L(s) U(s) V(s) \, ds. \tag{1.131}$$

By standard methods (see Mikhlin, 1964) it can directly be proved that this expression possesses all required properties of a scalar product in a Hilbert functional space. We thus have constructed a new functional linear space and the only missed property is completeness. To complete this space, let us supplement the space by all its limit points. The thus constructed space H_e becomes a standard Hilbert space, which is a complete subspace of $L_2(-a, a)$. The variational identity (1.129)–(1.130) can be rewritten now as

$$(u, u)_e = (f, u). \tag{1.132}$$

Let us study the solvability of our operator equation written in the form (1.132), by using the following estimate:

$$|(f, u)| = \left| \int_{-\infty}^{\infty} \bar{F}(s) U(s) \, ds \right| = \left| \int_{-\infty}^{\infty} \sqrt{L(s)} \, U(s) \frac{\bar{F}(s)}{\sqrt{L(s)}} \, ds \right|$$

$$\leq B \left[\int_{-\infty}^{\infty} L(s) |U(s)|^2 \, ds \right]^{1/2} = B \|u\|_e, \quad B = \left[\int_{-\infty}^{\infty} \frac{|F(s)|^2}{L(s)} \, ds \right]^{1/2}. \tag{1.133}$$

It is now evident that the linear functional (f, u) on the right-hand side is bounded (continuous) if the constant B is finite. It certainly depends on the asymptotic behavior of $F(s)$ and $L(s)$ at infinity, because these functions provide the regularity of the integrand in the constant B at any finite point, including $s = 0$, if we assume that the constant $L(0)$ is finite and positive:

$$0 < L(0) < \infty. \tag{1.134}$$

The most typical behavior of the symbolic function $L(s)$ at infinity is that of a power-law function:

$$L(s) = O(|s|^{-\alpha}), \quad \alpha > 0, \tag{1.135}$$

as follows from results of Section 1.4 (see formula (1.73) in the case $S(x) \equiv x$, $\lambda = s$).

We will distinguish between three different cases.

2.1. $0 < \alpha < 1$. Here the constant B is finite, $B < \infty$ at least if $f(x) \in C_1[-a, a]$, since in this case $F(s) = O(1/s)$, see Eq. (1.73), and so

$$\frac{|F(s)|^2}{L(s)} = O(|s|^{\alpha - 2}), \tag{1.136}$$

which is integrable over the line $-\infty < s < \infty$.

2.2. $1 \leq \alpha < 3$. Here a necessary condition for the constant B to be finite is that the asymptotics of $F(s)$, $s \to \infty$, must be like $o(1/s)$, which requires that two leading asymptotic terms in Eq. (1.73) vanish. A sufficient condition for this is satisfied if at least $f(x) \in C_2[-a, a]$, and $f(-a) = f(a) = 0$. Indeed, under such conditions $F(s) = O(s^{-2})$, and hence

$$\frac{|F(s)|^2}{L(s)} = O(|s|^{\alpha - 4}), \tag{1.137}$$

which together with $\alpha < 3$ provides that the constant B in Eq. (1.133) is finite.

2.3. $\alpha \geq 3$. This case is of no practical significance, so we do not pay any attention to it, although it admits quite a similar analysis as in 2.1 and 2.2.

In all these cases, where $B < \infty$, the functional (f, u) is continuous in H_e. Consequently, according to the classical Riesz theorem (Mikhlin, 1964; Riesz and Sz.-Nagy, 1972), there exists a unique element $u_0 \in H_e$ such that $(f, u) = (u_0, u)_e$, and this allows us to rewrite the energetic relation (1.132) in the equivalent form

$$(u, u)_e = (u_0, u)_e. \tag{1.138}$$

Hence, it is clear that in all these cases an integral equation of the first kind has a unique weak (or generalized, or energetic) solution $u = u_0 \in H_e$.

Helpful remarks

$1°$. Of course, the existence of a weak solution does not imply the existence of a classical one. However, it is obvious that if there is a classical solution to Eq. (1.119), it is at the same time a weak solution.

$2°$. The uniqueness of the solution is guaranteed by the positiveness of the symbolic function $L(s)$, both in the classical and weak sense, since if there are two solutions, $Gu_1 = f$, $Gu_2 = f$, then it follows from $G(u_1 - u_2) = 0$ that (see Eq. (1.122))

$$(G(u_1 - u_2), (u_1 - u_2)) = \frac{1}{2\pi} \int_{-\infty}^{\infty} L(s)|U_1(s) - U_2(s)|^2 \, ds = 0,$$

$$\text{hence} \quad U_1(s) \equiv U_2(s) \quad \sim \quad u_1(x) \equiv u_2(x). \tag{1.139}$$

$3°$. In a sense, Fredholm equations of the second and first kinds possess opposite properties. The former admits a clear description by the Fredholm theory, which in the case of regular kernel gives explicit-form conditions of unique solvability of the equation. Thus, the better the kernel the better qualitative properties of the equation. By contrast, equations of the first kind with regular kernels are, as a rule, unsolvable. Therefore we are faced here with an astonishing phenomenon—the worse the kernel the better qualitative properties of the equation.

$4°$. The most interesting and important for applications in diffraction theory is the case when

$$L(s) = \frac{D}{|s|} \left[1 + O\left(\frac{1}{s}\right) \right], \quad s \to \infty. \tag{1.140}$$

Unfortunately, the theory discussed here states a (weak) solvability of Eq. (1.119) only for those right-hand sides that vanish at the ends of the interval: $f(\pm a) = 0$, $f(x) \in C_1[-a, a]$. However, in this case formulas (1.82)–(1.85) explicitly describe the behavior of the kernel as follows:

$$G(x) \in C[-2a, 2a] \setminus \{0\}; \tag{1.141}$$

$$G(x) \sim -D \ln |x| \, [1 + O(x)], \quad x \to 0, \tag{1.142}$$

i.e., we arrive at equations with logarithmic-type kernels. The theory of integral equations with such kernels is closely connected with Cauchy-type integral equations and admits an absolutely different and much more advanced study based on the Riemann boundary value problem and the principal value of the Cauchy-type integrals. This theory is presented in the next section.

$5°$. Note that the described poor theory of equations of the first kind does not give you any appropriate algorithm how to treat these equations numerically.

$6°$. It is very interesting to understand in more detail what the real structure of the kernel in case 2.1 is, when a generalized solution exists being unique for arbitrary smooth right-hand side. If we assume that

$$L(s) = \frac{D}{|s|^\alpha} \left[1 + O\left(\frac{1}{s}\right) \right], \quad s \to \infty, \tag{1.143}$$

then formulas (1.82)–(1.83) determine the behavior of the kernel,

$$G(x) \sim \frac{D\Gamma(1-\alpha)}{|x|^{1-\alpha}} \left[1 + O(x)\right], \quad x \to 0, \tag{1.144}$$

which is a classical Fredholm operator with weak singularity in the kernel.

1.7. Singular Integral Equations with a Cauchy-Type Singularity in the Kernel

Integral operators of the type

$$(Gu)(x) = \int_{-a}^{a} \frac{u(\xi)\,d\xi}{x - \xi}, \quad x \in (-a, a), \tag{1.145}$$

do not fall into any class among those considered in the previous two sections. If we calculate the Fourier image of the kernel (which is certainly a convolution-type kernel), then we obtain

$$L(s) = \int_{-\infty}^{\infty} \frac{e^{isx}}{x}\,dx = 2 \int_{0}^{\infty} \frac{\sin sx}{x}\,dx = \pi \operatorname{sign}(s), \tag{1.146}$$

which does not decrease at infinity, i.e., as $s \to \infty$. Another problem is related to the fact that the integral (1.146) does not exist in any classical sense.

If the density $u(x)$ is at least from the Lipschitz class, $u(x) \in \text{Lip}[-a, a]$, i.e.,

$$|u(x_1) - u(x_2)| \le A\,|x_1 - x_2|, \quad \forall x_1, x_2 \in [-a, a], \tag{1.147}$$

with a certain constant $A > 0$, then

$$(Gu)(x) = \int_{-a}^{a} \frac{u(\xi) - u(x)}{x - \xi}\,d\xi + u(x) \int_{-a}^{a} \frac{d\xi}{x - \xi}, \tag{1.148}$$

and we can reduce the integral (1.145) to the case with constant density, because the first integrand in Eq. (1.148) is bounded and hence gives a finite integral. It should be noted that all the results below in this section are valid even in a more general case when $u(x)$ is from the Hölder class H^δ with some positive exponent δ: $|u(x_1) - u(x_2)| \le A|x_1 - x_2|^\delta$ ($\delta > 0$).

Let us try to treat the last integral in the generalized sense (see Eq. (1.54)):

$$
\begin{aligned}
\int_{-a}^{a} \frac{d\xi}{x - \xi} &= \lim_{\alpha \to +0} \int_{-a}^{a} \frac{|x - \xi|^\alpha\,d\xi}{x - \xi} \\
&= \lim_{\alpha \to +0} \left[\int_{-a}^{x} (x - \xi)^{\alpha-1}\,d\xi - \int_{x}^{a} (\xi - x)^{\alpha-1}\,d\xi \right] \\
&= \lim_{\alpha \to +0} \frac{(x+a)^\alpha - (a-x)^\alpha}{\alpha} \\
&= \lim_{\alpha \to +0} \left[(x+a)^\alpha \ln(x+a) - (a-x)^\alpha \ln(a-x) \right] = \\
&= \ln \frac{a+x}{a-x}, \quad x \in (-a, a),
\end{aligned}
\tag{1.149}
$$

where we have used the L'Hospital rule. Occasionally, it happens that the final result coincides with that trivially obtained when neglecting for a moment that the integrand is not integrable:

$$\int_{-a}^{a} \frac{d\xi}{x - \xi} = -\ln|\xi - x|\Big|_{\xi=-a}^{a} = \ln\left|\frac{a+x}{a-x}\right| = \ln\frac{a+x}{a-x} \quad \text{if} \quad x \in (-a, a). \tag{1.150}$$

In order to clear up completely the treatment of this sort of integrals, Cauchy proposed to consider the so-called *principal value*.

DEFINITION. *If the limit*

$$\lim_{\varepsilon \to +0} \left(\int_{-a}^{x-\varepsilon} \frac{u(\xi)\, d\xi}{x - \xi} + \int_{x+\varepsilon}^{a} \frac{u(\xi)\, d\xi}{x - \xi} \right) \tag{1.151}$$

is finite, then it is called the principal value *of the integral (1.145).*

The definition for Cauchy-type integrals over arbitrary contour is similar. The key step consists in deleting a small symmetric ε-neighborhood in Eq. (1.151).

It is directly seen that if $u(x) \in \text{Lip}[-a, a]$, then the principal value (1.151) exists and coincides with expressions (1.149)–(1.150).

Cauchy-type integrals, which will also be called here *singular integrals*, possess very interesting unique properties. The most interesting properties of singular integrals are: 1) their analytic properties in the complex plane; 2) explicit inversion of the characteristic singular operator; 3) relation between the Fredholm theory and the theory of singular integral equations.

First of all, let us consider an arbitrary closed contour $\Gamma \subset D$ of finite length in some domain D of the complex plain $z = \{\text{Re}\, z, \text{Im}\, z\}$. Then the theory of analytic functions states that

$$\frac{1}{2\pi i} \int_{\Gamma} \frac{u(\zeta)\, d\zeta}{\zeta - z} = \begin{cases} u(z), & z \text{ inside } \Gamma, \\ 0, & z \text{ outside } \Gamma, \end{cases} \tag{1.152}$$

provided that the function $u(z)$ is analytic in D and the contour Γ is traced counterclockwise. This result directly follows from the residue theorem (see formula (1.11)), since $\zeta = z$ is a simple pole of the integrand. But what happens at the contour, $z \in \Gamma$?

The *Sokhotsky–Plemelj* formula gives the answer to this question even if $u(\zeta)$ is not analytic but only $u(\zeta) \in \text{Lip}(\Gamma)$:

$$\lim_{z \to z_0 \in \Gamma} \frac{1}{2\pi i} \int_{\Gamma} \frac{u(\zeta)\, d\zeta}{\zeta - z} = \pm \frac{u(z)}{2} + \frac{1}{2\pi i} \int_{\Gamma} \frac{u(\zeta)\, d\zeta}{\zeta - z_0}, \quad z \in \Gamma, \tag{1.153}$$

where the plus sign corresponds to the interior limit and the minus sign to the exterior limit, and the last integral is treated as a singular one. Note that in the case where $u(\zeta)$ is analytic, the integral is equal to zero. The proof is very simple:

$$\int_{\Gamma} \frac{u(\zeta)\, d\zeta}{\zeta - z_0} = \int_{\Gamma} \frac{u(\zeta) - u(z_0)}{\zeta - z_0}\, d\zeta + u(z_0) \int_{\Gamma} \frac{d\zeta}{\zeta - z_0} = 0. \tag{1.154}$$

The integrand of the first integral here is analytic and hence is equal to zero owing to the Cauchy integral theorem, but the last one is calculated directly to be trivial. We thus come to the statement that if $u(z)$ is analytic in a domain containing a closed contour Γ, then

$$\frac{1}{2\pi i} \int_{\Gamma} \frac{u(\zeta)\, d\zeta}{\zeta - z} = \begin{cases} u(z), & z \text{ inside } \Gamma, \\ \frac{1}{2}u(z), & z \text{ on } \Gamma, \\ 0, & z \text{ outside } \Gamma. \end{cases} \tag{1.155}$$

Therefore, we can finally conclude that the analytic properties of a Cauchy-type integral are such that it takes a "jump" every time the point z crosses the integration contour Γ. Moreover, if z passes from outside to inside of Γ, then this jump is equal to the residue at the point z; but if z, in its passage from outside to inside of Γ, stays just on the contour, then the jump is equal to half of the residue (taken with appropriate sign). The same properties are valid for a singular integral along an arbitrary nonclosed contour of finite length

$$\int_{z_A}^{z_B} \frac{u(\zeta)\,d\zeta}{\zeta - z}, \tag{1.156}$$

i.e., it has discontinuities every time when z crosses the contour (z_A, z_B), and the "full jump" when passing across the contour is the residue of the integrand (with appropriate sign), while it is half of the residue if the point z "stops" on the contour.

Let us however come back to the characteristic singular integral equation holding over the interval $(-a, a)$

$$\int_{-a}^{a} \frac{u(\xi)\,d\xi}{x - \xi} = f(x), \qquad |x| < a, \qquad f(x) \in \mathrm{Lip}[-a, a], \tag{1.157}$$

which is the main subject of our investigation. It is proved (see, for example, Gakhov, 1966) that the general integrable solution to this equation has the following form:

$$u(x) = \frac{1}{\pi^2 \sqrt{a^2 - x^2}} \left[C - \int_{-a}^{a} \frac{\sqrt{a^2 - \xi^2}\, f(\xi)}{x - \xi}\, d\xi \right], \qquad |x| < a, \tag{1.158}$$

where C is an arbitrary constant. The inversion formula (1.158) is a conclusion from a very thorough theory of the Riemann boundary value problem for analytic functions.

Therefore, in the case of characteristic Cauchy-type kernel the considered singular equation admits an explicit-form exact solution by quadrature. But what can be said about the solvability of the full equation

$$(Gu)(x) = \int_{-a}^{a} \left[\frac{1}{x - \xi} + K(x, \xi) \right] u(\xi)\,d\xi = f(x), \qquad |x| < a, \tag{1.159}$$

if the kernel $K(x, \xi)$ is regular? All results below are valid when at least $K(x, \xi) \in C_1[(-a, a) \times (-a, a)]$ and $f(x) \in \mathrm{Lip}[-a, a]$.

The theory of singular equations (1.159)—and also more general equations of the second kind, when the unknown function $u(x)$ is present also outside of the integral—was developed by Noether, and it is now known in the literature as the Noether theory (see Gakhov, 1966). As applied to an equation written in the special form (1.159) it can be represented by the theorem that follows below, similar in a sense to the Fredholm theorems.

Let us consider the conjugate equation

$$(G^*u)(x) = \int_{-a}^{a} \left[\frac{1}{x - \xi} + K^*(x, \xi) \right] v(\xi)\,d\xi = g(x), \qquad |x| < a. \tag{1.160}$$

The Noether theory requires that the conjugate of Eq. (1.160) be considered in a conjugate class of functions as follows. If we seek a solution of the main equation (1.159) in the class h of functions unbounded at both endpoints $x = \pm a$, then the conjugate equation (1.160) should be considered in a conjugate class h' of functions bounded at one of two edges $x = a$ or $x = -a$.

Let us also introduce an index of the problem χ; in the considered first-kind case (1.159) it is simply the number of the intervals where the equation holds, i.e., $\chi = 1$ here. Then the Noether theory states the following (compare with the Fredholm theory).

THEOREM 1 (FIRST NOETHER THEOREM). *The subspaces of solutions to the homogeneous equations*

$$Gu = 0, \qquad G^*v = 0 \qquad (1.161)$$

have finite dimensions n *and* n', *respectively, where* $n - n' = \chi$, *i.e.,* $n - n' = 1$ *in our problem.*

THEOREM 2 (SECOND NOETHER THEOREM). *A nonhomogeneous equation*

$$Gu = f \qquad (1.162)$$

has a solution of the class h *if and only if* $(f, v_j) = 0$ *for all solutions* v_j *of the conjugate equation* $G^*v = 0$ *in the class* h'.

Let us verify the correctness of these results by the example of the characteristic equation (1.157). As we can see, $n = 1$ in this example, as it follows from Eq. (1.158). At the same time, the conjugate operator G^* here coincides with the original one G, with the same solution (1.158). It is obvious that the latter has only the trivial solution of the class h' for the trivial right-hand side. Thus, we have $n = 1$, $n' = 0 \implies n - n' = 1$.

A union of so powerful mathematical instruments as generalized treatment of integrals and operation with singular integrals can be very fruitful for many applications in pure mathematics and mathematical physics. For example, let us give an explicit solution of the Wiener–Hopf equation (1.28) in the case of the above considered kernel with the Fourier image $L(s) = 1/\sqrt{s^2 + d^2}$. As we could see from (1.32), (1.39), in our case this Wiener–Hopf equation in Fourier images has the form

$$\frac{\Phi_+(s)}{\sqrt{d - is}} = \left[\sqrt{d + is}\right]_- F_+(s) + \left[\sqrt{d + is}\right]_- F_-(s). \qquad (1.163)$$

Further, the decomposition of the function

$$\left[\sqrt{d + isz},\right]_- F_+(s) = N_+(s) + N_-(s), \qquad (1.164)$$

according to the Sokhotsky–Plemelj formulas (Gakhov, 1966; Muskhelishvili, 1965), can be written as follows:

$$N_+(s) = \frac{1}{2}(I + S)\sqrt{d + is}\, F_+(s), \qquad (1.165)$$

where I is the identity operator and S is a singular integral operator with a Cauchy kernel.

Equation (1.163) then becomes

$$\frac{\Phi_+(s)}{\sqrt{d - is}} - N_+(s) = N_-(s) + \sqrt{d + is}\, F_-(s) \equiv 0, \qquad (1.166)$$

and consequently

$$\Phi_+(s) = \sqrt{d - is}\, N_+(s) = \frac{1}{2}\sqrt{d - is}\,\sqrt{d + is}\, F_+(s) + \frac{1}{2}\sqrt{d - is}\, S\left[\sqrt{d + is}\, F_+(s)\right]. \qquad (1.167)$$

In order to perform the Fourier inversion in the last identity, the following tabulated operational relations may be used (Bateman and Erdelyi, 1954):

$$\sqrt{d - is} \impliedby -\frac{1}{2\sqrt{\pi}}\left(\frac{e^{-d|x|}}{|x|^{3/2}}\right)_+, \qquad \sqrt{d + is} \impliedby -\frac{1}{2\sqrt{\pi}}\left(\frac{e^{-d|x|}}{|x|^{3/2}}\right)_-, \qquad (1.168)$$

where the plus sign means that the support of the corresponding function is the positive semi-axis, and the minus sign means the same, with the negative semi-axis. Of course, integrals arising in these relations are understood in the generalized sense (see Section 1.3).

Now, returning to origins gives

$$\varphi(x) = \frac{1}{8\pi} \left\langle \left(\frac{e^{-d|x|}}{|x|^{3/2}} \right)_+ * \left\{ \left(\frac{e^{-d|x|}}{|x|^{3/2}} \right)_- * f(x) \right\} \right. \\ \left. + \left(\frac{e^{-d|x|}}{|x|^{3/2}} \right)_+ * \left\{ \text{sign}(x) \left(\frac{e^{-d|x|}}{|x|^{3/2}} \right)_- * f(x) \right\} \right\rangle . \tag{1.169}$$

Here we have used the following relation:

$$\text{sign}(x)\, g(x) \Longrightarrow (SG)(s) \quad \text{if} \quad g(x) \Longrightarrow G(s), \tag{1.170}$$

which directly follows from the convolution theorem and the formula (1.146) for the Fourier transform of the Cauchy-type singular kernel studied in this section. Recall that the sign $*$ stands for the convolution of two functions.

Identity (1.169) can be written out more explicitly:

$$\varphi(x) = \frac{1}{8\pi} \left\{ \int_{-\infty}^{0} \frac{e^{-d(x-y)}\, dy}{(x-y)^{3/2}} \int_0^{\infty} \frac{e^{d(y-t)} f(t)}{(t-y)^{3/2}}\, dt \right. \\ + \int_0^{x} \frac{e^{-d(x-y)}\, dy}{(x-y)^{3/2}} \int_y^{\infty} \frac{e^{d(y-t)} f(t)}{(t-y)^{3/2}}\, dt \\ - \int_{-\infty}^{0} \frac{e^{-d(x-y)}\, dy}{(x-y)^{3/2}} \int_0^{\infty} \frac{e^{d(y-t)} f(t)}{(t-y)^{3/2}}\, dt \\ \left. + \int_0^{x} \frac{e^{-d(x-y)}\, dy}{(x-y)^{3/2}} \int_y^{\infty} \frac{e^{d(y-t)} f(t)}{(t-y)^{3/2}}\, dt \right\}, \tag{1.171}$$

or

$$\varphi(x) = \frac{1}{4\pi} \int_0^{x} \frac{e^{-d(x-y)}\, dy}{(x-y)^{3/2}} \int_y^{\infty} \frac{e^{d(y-t)} f(t)}{(t-y)^{3/2}}\, dt, \quad x > 0. \tag{1.172}$$

Both integrals here are treated in the generalized sense.

Let us make sure that in the case $f(x) = e^{-\beta x}$ this result coincides with (1.44). To this end, we ought to calculate the internal integral (Gradshteyn and Ryzhik, 1994), which is again treated as a generalized value (see Section 1.3)

$$\int_y^{\infty} \frac{e^{d(y-t)}\, e^{-\beta t}}{(t-y)^{3/2}}\, dt = e^{-\beta y} \int_0^{\infty} \frac{e^{-(d+\beta)t}}{t^{3/2}}\, dt = \Gamma\left(-\frac{1}{2} \right) \sqrt{d+\beta}\, e^{-\beta y}, \tag{1.173}$$

and the generalized value of the next integral can be calculated with the help of integration by parts

$$\int_0^{x} \frac{e^{-d(x-y)}\, e^{-\beta y}}{(x-y)^{3/2}}\, dy = e^{-\beta x} \int_0^{x} \frac{e^{-(d-\beta)y}}{y^{3/2}}\, dy \\ = e^{-\beta x} \left[-\frac{2}{\sqrt{x}}\, e^{-(d-\beta)x} - 2(d-\beta) \int_0^{x} e^{-(d-\beta)y}\, \frac{dy}{\sqrt{y}} \right] \tag{1.174} \\ = -\frac{2}{\sqrt{x}}\, e^{-dx} - 2\sqrt{\pi(d-\beta)}\, e^{-\beta x}\, \text{Erf}\left[\sqrt{(d-\beta)}\, x \right],$$

so the solution we seek is

$$\varphi(x) = \sqrt{d+\beta} \left\{ \frac{e^{-dx}}{\sqrt{\pi x}} + \sqrt{d-\beta}\, e^{-\beta x}\, \mathrm{Erf}\left[\sqrt{(d-\beta)x}\right] \right\}, \qquad (1.175)$$

which coincides with (1.44).

The described theory of Cauchy singular integrals gives also an efficient tool for solving convolution equations with kernels containing logarithmic singularity, which could not be constructed completely by methods of first-kind Fredholm operators (see remark 4° in the previous section). First of all, it is clear that the characteristic equation

$$\int_{-a}^{a} \ln|x-\xi|\, u(\xi)\, d\xi = f(x), \qquad |x| \leq a, \qquad (1.176)$$

is reduced, after differentiation with respect to x, to Eq. (1.157), which admits explicit inversion. The unknown constant C arising during these transformations can be uniquely determined by substituting an expression of the type (1.158) (with $f'(x)$ substituted for $f(x)$) into the initial equation (1.176) with a certain value of x, for example, at $x = 0$. This procedure was first performed by Carleman (Carleman, 1922), who gave the following representation for the (unique) integrable solution:

$$u(x) = \frac{1}{\pi^2\sqrt{a^2-x^2}} \left[\int_{-a}^{a} \frac{\sqrt{a^2-\xi^2}\, f'(\xi)}{\xi-x}\, d\xi + \frac{1}{\ln(a/2)} \int_{-a}^{a} \frac{f(\xi)\, d\xi}{\sqrt{a^2-\xi^2}} \right], \qquad |x| < a, \quad (1.177)$$

in the case $a \neq 2$, $f'(x) \in \mathrm{Lip}[-a, a]$. In the case $a = 2$ the solution has a slightly different representation.

Then, if we study a full equation (of the first or second kind) with a logarithmic-type kernel, considerations here follow the strategy applied for full singular equations with the Cauchy-type kernels. You can apply explicit inversion of the characteristic logarithmic kernel, reducing the problem to a Fredholm equation of the second kind, whose theory is covered by the Fredholm theorems (see Section 1.5).

Helpful remarks

1°. Recall remark 3° in the previous section. Here we continue to trace the wonderful feature of integral equations of the first kind—worse kernels yield wider classes of solutions. Singular kernels in this sense are even more advanced than kernels with weak singularities (see Section 1.6), since they yield a wide one-parameter class of solutions with arbitrary constant C rather than a unique solution. Thus, singular integral equations are closer to first-order ordinary differential equations than to classical Fredholm-type integral equations.

2°. It is very interesting to note that the Noether theory is quite similar to that of Fredholm. The only difference lies in the fact that dimensions of the nuclear subspaces are different for the main and the conjugate operators.

3°. The Noether theory remains valid for an arbitrary singular integral equation of the second kind containing a Cauchy-type operator and a regular part of the kernel. The only difference from the described first-kind case (which is of key importance for further context) lies in a more specific definition of the index χ.

So, there is essentially no difference between the first- and the second-kind operators in the theory of singular integral equations in contrast to Fredholm-type integral equations. The solvability problem for the first- and the second-kind operators is described by the same Noether theory.

1.8. Hyper-Singular Integrals and Integral Equations

DEFINITION. *A convolution integral of the form*

$$(Gu)(x) = \int_{-a}^{a} \frac{u(\xi)\, d\xi}{(x-\xi)^2}, \qquad |x| < a, \tag{1.178}$$

is called a hyper-singular integral.

First of all, we should clarify in which sense hyper-singular integrals as given by Eq. (1.178) may be treated, since they exist neither as improper integrals of the first kind nor as Cauchy-type singular integrals. At least three different definitions of hyper-singular integrals are known (see Samko, 2000):

1. The integral is a derivative of the Cauchy principal value:

$$\int_{-a}^{a} \frac{u(\xi)\, d\xi}{(x-\xi)^2} = -\frac{d}{dx} \int_{-a}^{a} \frac{u(\xi)}{x-\xi}\, d\xi. \tag{1.179}$$

2. The integral is treated as a Hadamard principal value (see Belotserkovsky and Lifanov, 1993):

$$\int_{-a}^{a} \frac{u(\xi)\, d\xi}{(x-\xi)^2} = \lim_{\varepsilon \to +0} \left[\left(\int_{-a}^{x-\varepsilon} + \int_{x+\varepsilon}^{a} \right) \frac{u(\xi)\, d\xi}{(x-\xi)^2} - \frac{2u(x)}{\varepsilon} \right]. \tag{1.180}$$

3. The integral is a residue value, in the sense of generalized functions (see Section 1.3), which is an analytic continuation of the integral

$$\int_{-a}^{a} |x-\xi|^{\alpha}\, u(\xi)\, d\xi, \tag{1.181}$$

from a domain where it exists in the classical sense to the value $\alpha = -2$.

When $u(x) \equiv 1$, the three different approaches give the same result:

$$\int_{-a}^{a} \frac{d\xi}{(x-\xi)^2} = -\frac{d}{dx} \int_{-a}^{a} \frac{d\xi}{x-\xi} = \frac{d}{dx} \ln\left(\frac{a-x}{a+x}\right) = -\frac{2a}{a^2-x^2}, \tag{1.182}$$

$$\int_{-a}^{a} \frac{d\xi}{(x-\xi)^2} = \lim_{\varepsilon \to +0} \left[\left(-\frac{1}{a+x} + \frac{1}{\varepsilon} + \frac{1}{\varepsilon} + \frac{1}{x-a} \right) - \frac{2}{\varepsilon} \right] = -\frac{2a}{a^2-x^2}, \tag{1.183}$$

$$\int_{-a}^{a} |x-\xi|^{\alpha}\, d\xi = \int_{x}^{a} (\xi-x)^{\alpha}\, d\xi = \frac{(a+x)^{\alpha+1}}{\alpha+1} + \frac{(a-x)^{\alpha+1}}{\alpha+1}, \tag{1.184}$$

when $\mathrm{Re}(\alpha) > -1$. The analytic continuation of (1.184) from the half-plane $\mathrm{Re}(\alpha) > -1$ to the value $\alpha = -2$ gives again

$$\int_{-a}^{a} \frac{d\xi}{(x-\xi)^2} = \lim_{\alpha \to -2} \left[\frac{(a+x)^{\alpha+1}}{\alpha+1} + \frac{(a-x)^{\alpha+1}}{\alpha+1} \right] = -\frac{2a}{a^2-x^2}. \tag{1.185}$$

All three definitions are also equivalent to each other if the density $u(x) \in \mathrm{Lip}[-a, a]$, that is $u'(x) \in \mathrm{Lip}^{1}[-a, a]$. Indeed, let $\mathrm{Re}(\alpha) > -1$ then, according to the definition of the

Cauchy principal value of singular integrals, we have

$$-\frac{d}{dx}\int_{-a}^{a}\frac{u(\xi)\,d\xi}{x-\xi} = -\lim_{\varepsilon\to+0}\frac{d}{dx}\left(\int_{-a}^{x-\varepsilon}+\int_{x+\varepsilon}^{a}\right)\frac{u(\xi)\,d\xi}{x-\xi}$$

$$= \lim_{\varepsilon\to+0}\left[\left(\int_{-a}^{x-\varepsilon}+\int_{x+\varepsilon}^{a}\right)\frac{u(\xi)\,d\xi}{(x-\xi)^2}-\frac{2\varphi(x)}{\varepsilon}\right],$$

$$\int_{-a}^{a}|x-\xi|^{\alpha}\,u(\xi)\,d\xi = \lim_{\varepsilon\to+0}\left(\int_{-a}^{x-\varepsilon}+\int_{x+\varepsilon}^{a}\right)|x-\xi|^{\alpha}\,u(\xi)\,d\xi \qquad (1.186)$$

$$= \lim_{\varepsilon\to+0}\left[\int_{-a}^{x-\varepsilon}(x-\xi)^{\alpha}\,u(\xi)\,d\xi+\int_{x+\varepsilon}^{a}(\xi-x)^{\alpha}\,u(\xi)\,d\xi\right]$$

$$= \frac{d}{dx}\lim_{\varepsilon\to+0}\left[-\int_{-a}^{x-\varepsilon}\frac{(x-\xi)^{\alpha+1}}{\alpha+1}\,u(\xi)\,d\xi+\int_{x+\varepsilon}^{a}\frac{(\xi-x)^{\alpha+1}}{\alpha+1}\,u(\xi)\,d\xi\right],$$

where we have used the rule of differentiation of integrals containing outer variable both in the limit of integration and in the integrand. Now, by applying the analytic continuation to the last relation, one can see (with the help of the standard (ε,δ) formalism) that the right-hand side results in (1.179). Therefore, equivalence of definitions 1, 2 and 3 (in case all these integrals are finite) is evident. Let us prove that if $u(x)\in\mathrm{Lip}^1[-a,a]$, then a finite value of the limit in expression (1.180) exists, and hence for $x\in(a,b)$ the integral (1.178) is finite in any sense. Indeed, the expression in square brackets in Eq. (1.180) is

$$\left(\int_{-a}^{x-\varepsilon}+\int_{x+\varepsilon}^{a}\right)\frac{u(\xi)-u(x)-u'(x)(\xi-x)}{(x-\xi)^2}\,d\xi - u(x)\frac{2a}{a^2-x^2}+u'(x)\ln\frac{a-x}{a+x}, \quad (1.187)$$

which has a finite limit at $\varepsilon\to+0$.

Let us proceed however to the investigation of the hyper-singular equation with the characteristic kernel

$$\int_{-a}^{a}\frac{u(\xi)\,d\xi}{(x-\xi)^2}=f'(x), \qquad |x|<a, \qquad f(x)\in\mathrm{Lip}^1[-a,a], \qquad (1.188)$$

and construct its bounded solution with the help of relation (1.179) and the inversion of the characteristic singular integral operator (see Section 1.7). Indeed, by definition (1.179), equation (1.188) is equivalent to

$$\frac{d}{dx}\int_{-a}^{a}\frac{u(\xi)\,d\xi}{x-\xi}=-f'(x) \quad\sim\quad \int_{-a}^{a}\frac{u(\xi)\,d\xi}{x-\xi}=-f(x)+C_1, \qquad |x|<a, \qquad (1.189)$$

where C_1 is an arbitrary constant. Now, the inversion formula for the Cauchy characteristic integral operator (see the previous section) determines the function $u(x)$ in a form containing another arbitrary constant C_2, in addition to C_1. This pair of constants should be chosen so as to provide finite values of the solution at both endpoints $x=\pm a$. This can directly be made, resulting in the following

THEOREM. *If $f(x)\in\mathrm{Lip}^1[-a,a]$, then the bounded solution of Eq. (188) is unique and given by the following expression:*

$$u(x)=\frac{\sqrt{a^2-x^2}}{\pi^2}\int_{-a}^{a}\frac{f(\xi)\,d\xi}{\sqrt{a^2-\xi^2}\,(x-\xi)}, \qquad |x|<a. \qquad (1.190)$$

You can also directly verify that the above holds true for

$$C_1=\frac{1}{\pi^2}\int_{-a}^{a}\frac{f(\xi)\,d\xi}{\sqrt{a^2-\xi^2}}. \qquad (1.191)$$

It is interesting to note that the bounded solution of Eq. (1.188) vanishes as $x \to \pm a$. It is obvious that the full hyper-singular equation

$$(Gu)(x) = \int_{-a}^{a} \left[\frac{1}{(x-\xi)^2} + K(x,\xi) \right] u(\xi)\, d\xi = f(x), \qquad |x| < a, \qquad (1.192)$$

where the kernel $K(x,\xi)$ is regular, can be studied by explicit inversion of its characteristic part. Investigation of this problem is similar to that for full Cauchy-type or logarithmic-type equations.

In the final part of this section we study analytic properties of hyper-singular integrals

$$\int_{\Gamma} \frac{u(\zeta)\, d\zeta}{(\zeta - z)^2}, \qquad (1.193)$$

as functions of a complex-valued argument z with density $u(z)$ analytic in some domain D containing a simple closed contour Γ of finite length. Here we would like to give a comparison with analytic properties of the Cauchy-type singular integral (1.155).

Let us start with the case of a constant density and prove that the integral

$$\int_{\Gamma} \frac{d\zeta}{(\zeta - z)^2} \equiv 0 \qquad (1.194)$$

is identically equal to zero. When z lies inside the contour, this result is evident because the residue of the integrand (i.e., the factor in front of the first term in the Laurent series expansion in negative powers of $(\zeta - z)$) is equal to zero. When z lies outside Γ this is evident due to analyticity of the integrand inside Γ. If $z \in \Gamma$, it follows immediately from the first two of three above techniques that this is trivial indeed. Now, for arbitrary analytic density we have

$$\int_{\Gamma} \frac{u(\zeta)\, d\zeta}{(\zeta - z)^2} = \int_{\Gamma} \frac{u(\zeta) - u(z)}{(\zeta - z)^2}\, d\zeta = \int_{\Gamma} \frac{v(\zeta)\, d\zeta}{\zeta - z}, \qquad v(\zeta) = \frac{u(\zeta) - u(z)}{\zeta - z}, \qquad (1.195)$$

where $v(\zeta)$ is now analytic, and we can use appropriate results from the Cauchy-type integral theory. It is obvious that $v(\zeta)|_{\zeta = z} = v'(z)$, so formula (1.155) yields

$$\frac{1}{2\pi i} \int_{\Gamma} \frac{u(\zeta)\, d\zeta}{\zeta - z} = \begin{cases} 0, & z \text{ inside } \Gamma, \\ \frac{1}{2} u'(z), & z \text{ on } \Gamma, \\ 0, & z \text{ outside } \Gamma. \end{cases} \qquad (1.196)$$

Helpful remarks

Let us verify formula (1.196) by the example of the integral

$$I(t) = \int_{-\infty}^{\infty} \frac{e^{ibs}\, ds}{(s-t)^2}, \qquad b > 0, \qquad -\infty < t < \infty. \qquad (1.197)$$

Since the function e^{ibs} with $b > 0$ exponentially decays in the upper half-plane $\text{Im}(s) > 0$, we can close the integration contour by a semi-circle of infinite radius. Then the value of this integral, according to Eq. (1.196), is $I(t) = -\pi b\, e^{ibt}$. At the same time, direct calculations show that

$$I(t) = \frac{d}{dt} \int_{-\infty}^{\infty} \frac{e^{ibs}\, ds}{s - t} = \frac{d}{dt} \int_{-\infty}^{\infty} \frac{e^{ib(s+t)}\, ds}{s} = \frac{d}{dt} \left[e^{ibt} \int_{-\infty}^{\infty} \frac{e^{ibs}\, ds}{s} \right]$$
$$= (ib\, e^{ibt})\, 2i \int_{0}^{\infty} \frac{\sin(bs)}{s}\, ds = (ib\, e^{ibt})(\pi i) = -\pi b\, e^{ibt}, \qquad (1.198)$$

which is exactly the same result.

By the way, extracting separately the real and imaginary parts of this integral, you can evaluate the following real-valued integrals:

$$\int_{-\infty}^{\infty} \frac{\cos(bs)}{(s-t)^2}\, ds = -\pi|b|\cos(bt), \qquad \int_{-\infty}^{\infty} \frac{\sin(bs)}{(s-t)^2}\, ds = -\pi|b|\sin(bt), \qquad (1.199)$$

valid for arbitrary real b.

1.9. Governing Equations of Hydroaeroacoustics, Electromagnetic Theory, and Dynamic Elasticity

Linear Hydroaeroacoustics

The governing equations of linear acoustics are essentially linearized equations of general nonlinear fluid and gas dynamics with small perturbation velocities.

Let a fluid or a gas have under the at-rest conditions a constant density ρ_0, and a uniform pressure p_0, and zero velocity $\mathbf{v}_0 = \mathbf{0}$. Then the total pressure is $p = p_0 + p'$, the total density $\rho = \rho_0 + \rho'$, and the net velocity vector $\mathbf{v} = \mathbf{v}'$, where all quantities marked with a prime are small, and we should perform linearization with respect to these small parameters. Then Euler's equations of motion

$$\rho \frac{\partial \mathbf{v}}{\partial t} + \rho(\mathbf{v} \cdot \nabla)\mathbf{v} + \operatorname{grad} p = 0, \qquad (1.200)$$

in the case of small perturbations (linearized equations), have the following form:

$$\rho_0 \frac{\partial \mathbf{v}'}{\partial t} + \operatorname{grad} p' = 0. \qquad (1.201)$$

Let us assume that the fluid is barotropic. Then the constitutive equation $p = p(\rho) \sim \rho = \rho(p)$ in the linearized form is expressed as

$$p = p_0 + \left.\frac{\partial p}{\partial \rho}\right|_{\rho=\rho_0} (\rho - \rho_0) \quad \sim \quad p' = c^2 \rho' + \text{const}, \qquad (1.202)$$

where the quantity c,

$$c^2 = \left.\frac{\partial p}{\partial \rho}\right|_{\rho=\rho_0}, \qquad (1.203)$$

is called *the wave speed* in the medium. It is easily seen that the dimension of the constant physical parameter c is m/s, which justifies its name. Besides, it is clear for physical reasons that for real media the dependence of density upon pressure is monotonically increasing, so $\partial p/\partial \rho > 0$, which yields $c^2 > 0$. Therefore, the wave speed is always real and positive.

The continuity equation

$$\frac{\partial \rho}{\partial t} + \operatorname{div}(\rho\mathbf{v}) = 0, \qquad (1.204)$$

in the linearized form is

$$\frac{\partial \rho'}{\partial t} + \rho_0 \operatorname{div}(\mathbf{v}') = 0. \qquad (1.205)$$

Let us derive from Eqs. (1.201), (1.202), (1.205) the wave equation (the primes are omitted from now on). To this end, let us differentiate Eq. (1.205) with respect to time, and then substitute the expression $\rho_0 \, \partial \mathbf{v}/\partial t$, taken from Eq. (1.201), to the obtained equation (Δ is the Laplacian operator):

$$\frac{\partial^2 \rho}{\partial t^2} = \operatorname{div}(\operatorname{grad} p) \quad \sim \quad \frac{\partial^2 \rho}{\partial t^2} = \Delta p. \tag{1.206}$$

Then from the last equation and Eq. (1.202), we obtain the *wave equation*

$$\frac{\partial^2 p}{\partial t^2} = c^2 \Delta p. \tag{1.207}$$

In the case when the wave process is harmonic in time, $p(x, y, z, t) = \operatorname{Re}\{e^{-i\omega t}\widetilde{p}(x, y, z)\}$, where ω is the *angular frequency* and the new function $\widetilde{p}$ is independent of time, we obtain the *Helmholtz equation* (the signs of real part and the tildes are further omitted)

$$\Delta p + k^2 p = 0, \qquad k = \frac{\omega}{c}, \tag{1.208}$$

where k is called the *wave number*. Note that in the forthcoming consideration the time-dependent factor $e^{-i\omega t}$ will be omitted every time we study a harmonic problem. In this harmonic case the velocity vector can be expressed in terms of pressure from Eq. (1.201) as follows:

$$\mathbf{v} = \frac{1}{i\omega\rho_0} \operatorname{grad} p. \tag{1.209}$$

Let $\{x_i\}$, $i = 1, 2, 3$, be a fixed system of rectangular Cartesian coordinates. In the harmonic regime of oscillations with an angular frequency ω, when the complex amplitude of pressure (or, simply the *pressure function*) satisfies Eq. (9), we introduce the term *plane wave* as a solution to the Helmholtz equation (1.208) of the following form:

$$\begin{aligned} p &= f(\mathbf{n}\cdot\mathbf{r}) = f(n_1 x_1 + n_2 x_2 + n_3 x_3), \quad \mathbf{r} = \{x_1, \, x_2, \, x_3\}, \\ \Delta p &= (n_1^2 + n_2^2 + n_3^2)f'' = f'', \quad \text{hence} \quad f'' + k^2 f = 0 \ \sim \ f = A_1 e^{ik\,\mathbf{n}\cdot\mathbf{r}} + A_2 e^{-ik\,\mathbf{n}\cdot\mathbf{r}}, \end{aligned} \tag{1.210}$$

where $\mathbf{n}$ is an arbitrary fixed unit vector in the space, so that the full structure of the solution corresponding to the plane wave is

$$p = A_1 e^{i(k\mathbf{n}\cdot\mathbf{r}-\omega t)} + A_2 e^{-i(k\mathbf{n}\cdot\mathbf{r}+\omega t)}, \tag{1.211}$$

where the argument of an oscillating exponential function is called the *phase* of the corresponding wave.

DEFINITION. *For any moment t, a set in the three-dimensional (3D) space that corresponds to a constant value of the phase is called a* wave front *set. The velocity of propagation of this wave front in time is called the* phase velocity.

Let us show that the first term in Eq. (1.211) represents a wave going away to infinity, and the second term a wave arriving from infinity. Indeed, the wave front of the two terms is a set, where

$$k\,\mathbf{n}\cdot\mathbf{r} \mp \omega t = \text{const} \quad \sim \quad \mathbf{n}\cdot\mathbf{r} = \text{const} \pm ct, \tag{1.212}$$

since $\omega/k = c$. It is known from analytic geometry that if a unit direction vector is $\mathbf{n} = \{\cos\alpha, \cos\beta, \cos\gamma\}$, then the equation $x_1 \cos\alpha + x_2 \cos\beta + x_3 \cos\gamma - q = 0$ defines a plane with normal $\mathbf{n}$ and distance from the origin equal to q. In our case $\mathbf{n} = \{n_1, n_2, n_3\}$, and the distance from the origin is equal to $q(t) = \text{const} \pm ct$. Thus, the wave front represents a plane, which justifies the name of the wave, and the direction **n** determines a unit normal to this plane front. Moreover, the velocity of propagation of this front is

$$\dot{q}(t) = \pm c, \tag{1.213}$$

which defines the velocity of the wave front propagation. It is no accident that it coincides with the wave speed in the medium. Now it becomes clear that the upper sign is related to the wave that travels away from the origin to infinity, and the lower sign, to that arriving from infinity. A special discussion in Chapter 3 will be devoted to the important question what is physically and mathematically the intrinsic difference between these two types of waves. Here we only mention that throughout the text, diffracted (scattered, reflected) waves must satisfy the so-called *radiation condition*, whose strict formulation will be given below. In the simplest intuitive sense, the radiation condition reads: among the number of possible solutions in diffraction problems, only diffracted waves going away to infinity are physically realizable, and the waves arriving from infinity are fictitious. Thus, out of two plane waves in Eq. (1.211), only the first one satisfies the radiation condition.

The *wavelength* λ is determined as the minimum distance in the direction **n** for which the phases of oscillations (i.e., the arguments of the exponential functions in (1.211)) coincide at the points defined by the radius vector **r** and the radius vector **r** + λ**n**. This implies $\lambda k = 2\pi$, or

$$\lambda = \frac{2\pi}{k} = \frac{2\pi c}{\omega} = \frac{c}{f}, \qquad \text{where} \quad \omega = 2\pi f. \tag{1.214}$$

Here ω is the angular frequency, which is measured in rad/s, and f is called the *cyclic frequency* and is measured in Hz = $1/s$.

In diffraction problems where an incident wave falls onto an obstacle, complete mathematical formulation implies some boundary conditions on the boundary surface (3D case) or boundary line (2D case) of the obstacle. Typically, there are two types of boundary conditions: 1) acoustically hard boundary and 2) acoustically soft boundary. In the first case the medium particles cannot penetrate through the boundary, so the normal component of the velocity vanishes, $v_n = 0$, which with the help of Eq. (1.209) means

$$\left.\frac{\partial p}{\partial n}\right|_S = 0. \tag{1.215}$$

In the second case it is assumed that the total pressure on the boundary is equal to zero:

$$p|_S = 0. \tag{1.216}$$

Electromagnetic wave theory

Here the governing equations for an isotropic medium are given by the system of Maxwell's equations

$$\begin{cases} \operatorname{rot} \mathbf{E} = -\dfrac{\mu}{c}\dfrac{\partial \mathbf{H}}{\partial t}, & \operatorname{div} \mathbf{E} = 0, \\[2mm] \operatorname{rot} \mathbf{H} = \dfrac{\sigma \mathbf{E}}{c} + \dfrac{\varepsilon}{c}\dfrac{\partial \mathbf{E}}{\partial t}, & \operatorname{div} \mathbf{H} = 0, \end{cases} \tag{1.217}$$

where **E** is the electric field strength, **H** is the magnetic field strength, ε, μ, σ are some positive constant physical parameters, and c is the wave speed.

If we apply the operator rot to the first equation (1.217), and eliminate the magnetic strength, then we arrive at the following equation with respect to **E**:

$$\Delta \mathbf{E} = \frac{\mu\varepsilon}{c^2}\frac{\partial^2 \mathbf{E}}{\partial t^2} + \frac{\mu\sigma}{c^2}\frac{\partial \mathbf{E}}{\partial t}, \tag{1.218}$$

where we have taken into account that $\operatorname{rot}\operatorname{rot}\mathbf{E} = \operatorname{grad}\operatorname{div}\mathbf{E} - \Delta\mathbf{E}$ and the additional condition $\operatorname{div}\mathbf{E} = 0$.

If the wave process is harmonic in time, with the usual time-dependent factor $e^{-i\omega t}$, then Eq. (1.218) becomes the Helmholtz partial differential equation

$$\Delta\mathbf{E} + k^2\mathbf{E} = 0, \quad k = \frac{\omega}{c}\sqrt{\mu\varepsilon + i\frac{\sigma}{\omega}} \qquad (\operatorname{Re} k > 0, \quad \operatorname{Im} k \geq 0), \qquad (1.219)$$

with some (generally) complex-valued wave number k, which in the case of vacuum medium ($\sigma = 0$) becomes real-valued: $k = (\omega/c)\sqrt{\mu\varepsilon}$. The magnetic strength $\mathbf{H}$ is expressed in terms of $\mathbf{E}$ as follows:

$$\mathbf{H} = \frac{c}{i\omega\mu}\operatorname{rot}\mathbf{E}. \qquad (1.220)$$

The boundary conditions for Eq. (1.219) usually correspond to the surface of a perfect conductor S, where the tangential component of $\mathbf{E}$ vanishes:

$$\mathbf{n}\times\mathbf{E}|_S = 0. \qquad (1.221)$$

Linear dynamic elasticity

If a linear isotropic elastic medium occupies some domain in the 3D space, then at every point of the medium there are three components of the displacement vector $\mathbf{u} = \{u_1, u_2, u_3\}$ and six components of the stress tensor. The latter can be represented as a 3×3 symmetric matrix $\{\sigma_{ij}\}$, $i, j = 1, 2, 3$, $\sigma_{ij} = \sigma_{ji}$ whose components σ_{ij} determine the jth stress component arising on a small area element crossing the given point and possessing external normal parallel to x_i.

The constitutive equations of the linear isotropic elastic medium are

$$\sigma_{ij} = \mu\left(\frac{\partial u_i}{\partial x_j} + \frac{\partial u_j}{\partial x_i}\right) + \lambda\,\delta_{ij}\sum_{k=1}^{3}\frac{\partial u_k}{\partial x_k}, \qquad (1.222)$$

and the equations of motion are expressed as

$$\mu\Delta\mathbf{u} + (\lambda + \mu)\operatorname{grad}\operatorname{div}\mathbf{u} = \rho\frac{\partial^2\mathbf{u}}{\partial t^2}, \qquad (1.223)$$

where μ and λ are the Lamé elastic moduli, and ρ is the mass density of the material.

It is proved that, unlike hydroaeroacoustics and electromagnetism, where physically only one wave speed c can exist, two independent elastic wave velocities,

$$c_p^2 = \frac{\mu}{\rho}, \qquad c_s^2 = \frac{\lambda + 2\mu}{\rho}, \qquad (1.224)$$

can be introduced in dynamic elasticity. These are the so-called *longitudinal* and *transverse* wave speeds, respectively.

Then, according to classical results (see, for example, Achenbach, 1973), the following Lamé potential representation gives a general solution to system (1.223):

$$\mathbf{u} = \operatorname{grad}\varphi + \operatorname{rot}\psi \qquad (\operatorname{div}\psi = 0), \qquad (1.225)$$

where φ is a longitudinal scalar potential and ψ is a transverse vector potential, which satisfy the following wave equations, respectively:

$$\Delta\varphi = \frac{1}{c_p^2}\frac{\partial^2\varphi}{\partial t^2}, \qquad \Delta\psi = \frac{1}{c_s^2}\frac{\partial^2\psi}{\partial t^2}. \qquad (1.226)$$

Of course, in the case of a harmonic process, with a factor $e^{-i\omega t}$, the wave equations (1.226) become the Helmholtz equations

$$\Delta\varphi + k_p^2\,\varphi = 0, \quad \Delta\psi + k_s^2\,\psi = \mathbf{0}, \qquad k_p = \frac{\omega}{c_p}, \quad k_s = \frac{\omega}{c_s}, \qquad (1.227)$$

with two different wave numbers k_p and k_s.

The most typical boundary conditions in diffraction problems for elastic waves correspond to the case of boundary surfaces free of load. This means trivial components of the stress over the boundary surface S. Since the stress vector on the elementary area is expressed in terms of components of the stress tensor, this implies

$$P_i\big|_S = \sum_{j=1}^{3} \sigma_{ij} n_j = 0, \qquad i = 1, 2, 3. \qquad (1.228)$$

The last condition can be reformulated in terms of the Lamé potentials, by using relations (1.222) and (1.225).

Helpful remarks

$1°$. It is interesting to note that the principal governing equation of all three theories of absolutely different physical nature are covered by identical wave or (in a monochromatic process) Helmholtz equations. The difference in the governing equations is that in acoustics the wave equation is scalar, and the total theory can be formulated in terms of a scalar pressure function. In electromagnetic theory the governing equations are vector, but all the components of the vector fields are described by the same wave equation with a single wave speed. The governing equations of dynamic elasticity are tensor, and generally they are described by a pair of wave equations with two different wave speeds.

$2°$. The mathematical uniformity of wave processes in so physically different media explains why they are usually studied as sections of a unified theory. This even results in a mutual influence of terminologies, heuristic ideas, and other things. Thus, in hydroacoustics and elastic wave theory there is a very widespread idea about *light* and *shadow* zones in geometrical diffraction theory, though there are no light rays or any other matters that could produce light or shadow. Then, harmonic processes in scalar and elastic theories are often called *monochromatic* (*single-colored* in Greek), a terminology certainly generated by the electromagnetic theory, since an electromagnetic wave of a single fixed frequency leads to a certain color in the rainbow. Such examples of genetic uniformity can be cited in various aspects and in many problems.

$3°$. It suffices for many applications to study the so-called (2D) *in-plane* problem, in Cartesian coordinates (x, y). In scalar acoustics its solution is determined as a solution of the wave equation which is sought (if exists) in the form $p = p(x, y, t)$ in transient case and $p = p(x, y)$ in harmonic case. Then the corresponding wave and Helmholtz equations become

$$\frac{\partial^2 p}{\partial t^2} = c^2\left(\frac{\partial^2 p}{\partial x^2} + \frac{\partial^2 p}{\partial y^2}\right), \qquad \frac{\partial^2 p}{\partial x^2} + \frac{\partial^2 p}{\partial y^2} + k^2 p = 0 \quad (k = \omega/c). \qquad (1.229)$$

In electromagnetism the plane problem can be separated into two independent, *E*-polarized and *H*-polarized, wave fields. We will write out respective governing equations only for the harmonic case. The former type of polarization corresponds to

$$\mathbf{E} = \{0,\, 0,\, E(x, y)\}, \qquad \frac{\partial^2 E}{\partial x^2} + \frac{\partial^2 E}{\partial y^2} + k^2 E = 0,$$

$$\mathbf{H} = \{H_x(x, y),\, H_y(x, y),\, 0\}, \qquad H_x = \frac{c}{i\omega\mu}\frac{\partial E}{\partial y}, \qquad H_y = -\frac{c}{i\omega\mu}\frac{\partial E}{\partial x}. \qquad (1.230)$$

The latter type of polarization is described by the following governing equations:

$$\mathbf{H} = \{0,\, 0,\, H(x,y)\}, \qquad \frac{\partial^2 H}{\partial x^2} + \frac{\partial^2 H}{\partial y^2} + k^2 H = 0,$$

$$\mathbf{E} = \{E_x(x,y),\, E_y(x,y),\, 0\}, \qquad E_x = -\frac{c}{i\omega\varepsilon}\frac{\partial H}{\partial y}, \quad E_y = \frac{c}{i\omega\varepsilon}\frac{\partial H}{\partial x}. \tag{1.231}$$

In dynamic elasticity the in-plane problem implies that the displacement vector has only two first components of three, with both of them depending on (x,y) only. It can directly be proved that in this case the vector Lamé potential ψ can be taken scalar with the only third nontrivial component depending on (x,y), which automatically makes the condition in brackets in Eq. (1.225) valid. The complete system of governing equations here is given by

$$\mathbf{u} = \{u(x,y),\, v(x,y),\, 0\}, \quad u = \frac{\partial\varphi}{\partial x} + \frac{\partial\psi}{\partial y}, \quad v = \frac{\partial\varphi}{\partial y} - \frac{\partial\psi}{\partial x},$$

$$\varphi = \varphi(x,y), \quad \psi = \psi(x,y), \quad \frac{\partial^2\varphi}{\partial x^2} + \frac{\partial^2\varphi}{\partial y^2} + k_p^2\,\varphi = 0, \quad \frac{\partial^2\psi}{\partial x^2} + \frac{\partial^2\psi}{\partial y^2} + k_s^2\psi = 0. \tag{1.232}$$

The components of the stress tensor can be found from the Hooke law (1.222).

4°. In dynamic theory of elasticity there is the so-called *anti-plane*, or *SH*-wave process; this is the case when the problem can be reduced to a single scalar equation with only one (transverse) wave number k_s. This is characterized by the main property that the wave is polarized in the z-direction but propagates in the (x,y) plane. Mathematically, this is covered by the equations

$$\mathbf{u} = \{0,\, 0,\, w(x,y)\}, \quad \frac{\partial^2 w}{\partial x^2} + \frac{\partial^2 w}{\partial y^2} + k_s^2\,w = 0, \quad \sigma_{xz} = \mu\frac{\partial w}{\partial x}, \quad \sigma_{yz} = \mu\frac{\partial w}{\partial y}, \tag{1.233}$$

with all other components of the stress tensor being trivial.

5°. Due to the outlined unity of wave processes in the three mentioned classes of problems, we will operate, as a rule, with scalar acoustic wave fields in the forthcoming consideration of various theoretical aspects. Only sometimes, we will treat some elastic wave problems in order to demonstrate that the developed methods are efficient in the most complex tensor (elastic) case, too.

Chapter 2

Integral Equations of Diffraction Theory for Obstacles in Unbounded Medium

2.1. Properties of the Potentials of Single and Double Layers

As noted in Section 1.9, the main subject of our consideration is related to harmonic wave processes in scalar media, so we will mainly deal with the scalar Helmholtz partial differential equation

$$\Delta p + k^2 p = 0, \qquad k = \frac{\omega}{c}, \tag{2.1}$$

where ω is the angular frequency, c is the wave speed, and k is the wave number. This equation is to be solved in some (3D) domain D bounded by a simple surface S. If the problem is two-dimensional (2D), then we study Eq. (2.1) in some plane domain D with a simple boundary line l. As mentioned in Section 1.9, the standard boundary conditions may be one of the following types: (1) acoustically hard boundary,

$$v_n|_S = 0 \qquad \sim \qquad \frac{\partial p}{\partial n}\bigg|_S = 0, \tag{2.2}$$

or (2) acoustically soft boundary,

$$p|_S = 0. \tag{2.3}$$

DEFINITION. *A fundamental solution (or a Green's function) is a solution $\Phi(r)$ to the Helmholtz equation that depends only on the distance r between the origin and the current point.*

In order to construct a Green's function, we are obliged to treat the 2D and 3D cases separately.

3D case. The Laplace operator in the spherical coordinate system (r, φ, θ) has the form

$$\Delta = \frac{1}{r^2} \frac{\partial}{\partial r}\left(r^2 \frac{\partial}{\partial r}\right) + \frac{1}{r^2 \sin\theta} \frac{\partial}{\partial \theta}\left(\sin\theta \frac{\partial}{\partial \theta}\right) + \frac{1}{r^2 \sin^2\theta} \frac{\partial^2}{\partial \varphi^2}. \tag{2.4}$$

Consequently, in the case where the only nontrivial derivatives are those applied with respect to the r-coordinate, a Green's function can be obtained from the equation

$$\left[\frac{1}{r^2} \frac{d}{dr}\left(r^2 \frac{d}{dr}\right) + k^2\right]\Phi(r) = 0, \tag{2.5}$$

which is evidently equivalent to

$$\frac{d^2}{dr^2}(r\Phi) + k^2(r\Phi) = 0. \tag{2.6}$$

Equation (2.6) is an ordinary differential equation with constant coefficients for $r\Phi(r)$. It admits exact analytical solution by constructing a characteristic polynomial (Smirnov, 1964), which gives $r\Phi = A\,e^{\pm ikr}$ with a certain constant A, so that the general solution is given by

$$\Phi(r) = A_1 \frac{e^{ikr}}{r} + A_2 \frac{e^{-ikr}}{r}. \tag{2.7}$$

2D case. Here the Helmholtz equation (2.1) becomes again an ordinary differential equation but with nonconstant coefficients,

$$\frac{d^2\Phi}{dr^2} + \frac{1}{r}\frac{d\Phi}{dr} + k^2\Phi = 0, \tag{2.8}$$

which is a Bessel equation (Abramowitz and Stegun, 1965). Its general solution is expressed as

$$\Phi(r) = A_1 J_0(kr) + A_2 Y_0(kr), \tag{2.9}$$

where J_0 and Y_0 are the Bessel functions of the first and second kinds (also called the Bessel and Neumann functions, respectively) of order 0.

In this chapter we will consider only diffraction by finite obstacles occupying a domain D and placed in an unbounded acoustic medium, i.e., only exterior problems. For a correct selection of a unique solution, we should adopt a correct condition at infinity, which again is the radiation condition (see Section 1.9).

Let us recall that the full structure of the obtained solution (2.7), as a function of the space coordinate r and time, is as follows:

$$\Phi(r,t) = A_1 \frac{e^{i(kr-\omega t)}}{r} + A_2 \frac{e^{-i(kr+\omega t)}}{r}. \tag{2.10}$$

It is clear, by analogy with what was written in Section 1.9, that the wave front corresponding to these two waves is defined by $kr \mp \omega t = \text{const}$ for any chosen moment t, i.e., $r = \pm(\omega/k)\,t + \text{const} = \pm ct + \text{const}$. We thus can see that a Green's function represents a *spherical* wave, since for any fixed t the obtained wave front equation represents a spherical surface. The phase velocity is $\dot{r} = \pm c$, i.e., it again coincides with the wave speed in the medium. Only the first term in Eq. (2.1) satisfies the radiation condition in the considered 3D case, which gives a Green's function in the form

$$\Phi(r) = \frac{e^{ikr}}{4\pi r}, \tag{2.11}$$

where the harmonic (in time) factor $\exp(-i\omega t)$ is omitted, and the constant A is taken equal to $1/(4\pi)$ for convenience, which will become clear very soon. It is very important to stress that the Green's function (2.11) is analytic and satisfies the Helmholtz equation only outside a small neighborhood of the origin.

In order to make a correct choice between the two functions in the 2D case, let us quote the far-field asymptotics of Bessel functions (Abramowitz and Stegun, 1965):

$$J_0(kr) \sim \sqrt{\frac{2}{\pi kr}}\,\cos\left(kr - \frac{\pi}{4}\right) + O(r^{-3/2}),$$
$$Y_0(kr) \sim \sqrt{\frac{2}{\pi kr}}\,\sin\left(kr - \frac{\pi}{4}\right) + O(r^{-3/2}), \qquad r \to \infty. \tag{2.12}$$

It is now evident that the only linear combination of these functions that generates a wave propagating from the origin to infinity, in accordance with the radiation condition, is

$$\Phi(r) = A\left[J_0(kr) + iY_0(kr)\right] = AH_0^{(1)}(kr)$$

$$\sim \sqrt{\frac{2}{\pi kr}}\,\exp\left[i\left(kr - \frac{\pi}{4}\right)\right] + O(r^{-3/2}), \qquad r \to +\infty. \tag{2.13}$$

The expression in the first square brackets here is called the Hankel function of the first kind of order 0 (see Abramowitz and Stegun, 1965). Only for convenience, we specify the constant A and take the 2D Green's function in the form

$$\Phi(r) = \frac{i}{4}H_0^{(1)}(kr). \tag{2.14}$$

Note that the Green's function of the 2D theory (2.14) has again a singularity at the origin.

DEFINITION. *The integral*

$$(G_s u)(x) = \int_{\partial D}\Phi(|x-y|)\,u(y)\,ds_y, \tag{2.15}$$

taken over the boundary ∂D of the domain D, is called a single-layer potential. Here $|x-y|$ is the distance between the points $x = (x_1, x_2, x_3)$ and $y = (y_1, y_2, y_3)$ in the 3D space $\mathbb{R}^3$, or between the points $x = (x_1, x_2)$ and $y = (y_1, y_2)$ in $\mathbb{R}^2$. The subscript y indicates that the integration is applied with respect to y.

It is evident from Eqs. (2.11), (2.14) and the properties of the Hankel function for small arguments (Abramowitz and Stegun, 1965),

$$H_0^{(1)}(kr) \sim \frac{2i}{\pi}\left[\ln(kr) + \gamma - \ln 2\right] + 1 + O\big((kr)^2\ln(kr)\big), \qquad (kr) \to +0, \tag{2.16}$$

where γ is the Euler constant, that the integral operator G_s in (2.15) has a weakly singular kernel (see Section 1.5). It follows that G_s is a compact operator in the Banach space of continuous functions $u(y) \in C(\partial D)$. For further consideration we will also assume that the boundary ∂D is piecewise smooth.

THEOREM 1. *If the density $u(y) \in C(\partial D)$, then the single-layer potential (2.15) represents a continuous function in $\mathbb{R}^3$ ($\mathbb{R}^2$).*

Proof. We give a proof only for the 3D case, and the 2D case can be studied by analogy. Since the kernel of the integral operator (2.15) is continuous for all $x, y \in \mathbb{R}^3$ excluding a small neighborhood of the origin (when $|x-y| \to 0$), it is evident that the integral (2.15) is certainly a continuous function of x far away from the boundary $S = \partial D$. Now, let $x \in S$; then we will show that the difference $|(G_s u)(x') - (G_s u)(x)|$ can be made arbitrarily small if the difference $|x' - x|$ is small enough.

Let $\varepsilon > 0$ be an arbitrary small positive parameter and $\delta = a\varepsilon^3$ ($a > 0$). Let us denote a part of the surface S that represents a small neighborhood of the point x with radius $r_\varepsilon = b\varepsilon$ ($b > 0$) by S_ε. The constants a and b do not depend on ε. The rest of S is $S \setminus S_\varepsilon$. Now suppose $|x' - x| \le \delta$ and estimate the following difference:

$$|(G_s u)(x') - (G_s u)(x)| \le M\left(\int_{S_\varepsilon} + \int_{S\setminus S_\varepsilon}\right)|\Phi(|x'-y|) - \Phi(|x-y|)|\,ds_y,$$

$$M = \max_{y\in S}|u(y)|. \tag{2.17}$$

The first integral is

$$\int_{S_\varepsilon} |\Phi(|x'-y|) - \Phi(|x-y|)| \, ds_y \le \int_{S_\varepsilon} |\Phi(|x'-y|)| + |\Phi(|x-y|)| \, ds_y$$

$$= \int_{S_\varepsilon} \left(\frac{1}{|x-y|} + \frac{1}{|x'-y|} \right) ds_y \qquad (2.18)$$

$$\le 2 \int_{S_\varepsilon} \frac{ds_y}{|x-y|} = \int_0^{r_\varepsilon} \int_0^{2\pi} \frac{\rho \, d\rho \, d\psi}{\rho} = 2\pi r_\varepsilon = 2\pi b\varepsilon,$$

where ψ is the polar angle of the point y in the polar coordinate system arranged in a tangential plane with center x. Here we have taken into account that if the point x' approaches x along the normal direction, then $|x'-y| \ge |x-y|$.

The second integral in (2.17) can be estimated as follows:

$$\int_{S \setminus S_\varepsilon} |\Phi(|x'-y|) - \Phi(|x-y|)| \, ds_y$$

$$= \int_{S \setminus S_\varepsilon} \left| \frac{e^{ik|x-y|} - e^{ik|x'-y|}}{|x-y|} + \frac{e^{ik|x'-y|} (|x'-y| - |x-y|)}{|x-y| \, |x'-y|} \right| ds_y \qquad (2.19)$$

$$\le \int_{S \setminus S_\varepsilon} \frac{|e^{ik|x-y|} - e^{ik|x'-y|}|}{|x-y|} \, ds_y + \int_{S \setminus S_\varepsilon} \frac{||x-y| - |x'-y||}{|x-y| \, |x'-y|} \, ds_y.$$

Further, since

$$\left| e^{ik|x-y|} - e^{ik|x'-y|} \right| = 2 \left| \sin \frac{|x-y| - |x'-y|}{2} \right| \le ||x-y| - |x'-y||, \qquad (2.20)$$

equation (2.19) is reduced to

$$\int_{S \setminus S_\varepsilon} |\Phi(|x'-y|) - \Phi(|x-y|)| \, ds_y \le \int_{S \setminus S_\varepsilon} \frac{||x-y| - |x'-y||}{|x-y|} \left(1 + \frac{1}{|x'-y|} \right) ds_y. \quad (2.21)$$

The last step is to use an inequality evident from the geometry of triangles: $|\bar{z}_1 - \bar{z}_2| \ge ||\bar{z}_1| - |\bar{z}_2||$, which in our case may be applied in the form $||x-y| - |x'-y|| \le |x-x'|$, hence

$$\int_{S \setminus S_\varepsilon} |\Phi(|x'-y|) - \Phi(|x-y|)| \, ds_y \le |x-x'| \int_{S \setminus S_\varepsilon} \frac{1}{|x-y|} \left(1 + \frac{1}{|x'-y|} \right) ds_y$$

$$\le 2 |x-x'| \int_{S \setminus S_\varepsilon} \frac{ds_y}{|x-y| \, |x'-y|} \qquad (2.22)$$

$$\le \frac{2\delta}{r_\varepsilon^2} S^* = \frac{2aS^*\varepsilon}{b^2}$$

(S^* is the square of the surface S), because for $y \in S \setminus S_\varepsilon$ we have $|x-y| \ge r_\varepsilon$, $|x'-y| \ge r_\varepsilon$ (recall that x' lies on the normal to the point x).

By collecting together Eqs. (2.17), (2.18), (2.22), and setting $a = 1/(64\pi^2 M^3 S^*)$ and $b = 1/(4\pi M)$, we arrive at the final estimate: for arbitrary small $\varepsilon > 0$ there exists $\delta = \varepsilon^3/(64\pi^2 M^3 S^*)$ such that

$$|(G_s u)(x') - (G_s u)(x)| \le \frac{\varepsilon}{2} + \frac{\varepsilon}{2} = \varepsilon \qquad (2.23)$$

if $|x' - x| \leq \delta$. This is indeed the definition of continuity of any function, which was to be proved.

The last integral was estimated over a plane small disk of the radius r_ε rather than the corresponding nonplane small area S_ε, which is the same, provided that the small quantities of the order of $O(\varepsilon^2)$ are neglected.

Note that the presented proof is valid if x lies on a smooth part of the surface S. If x hits an edge or any other sharp ledge of the piecewise boundary the proof is quite similar, with some minor modifications.

DEFINITION. *Let* $\mathbf{n}_y$ *($y \in \partial D$) be an outward unit normal. Then the integral*

$$(G_d\, u)(x) = \int_{\partial D} \frac{\partial \Phi(|x - y|)}{\partial n_y}\, u(y)\, ds_y \qquad (2.24)$$

is called the double-layer *potential.*

For the same reason as for the single-layer potential this integral is evidently continuous outside of a neighborhood of the boundary ∂D. The following theorem states that it is discontinuous when passing across the boundary ∂D.

THEOREM 2. *If* $u(y) \in C(\partial D)$, *then at* $x \to y_0 \in \partial D$ *the double-layer potential (2.24) can be continuously extended both from outside of the boundary ∂D and from inside, with their limit values being, respectively,*

$$(G_d\, u)_\pm(y_0) = \int_{\partial D} \frac{\partial \Phi(|y_0 - y|)}{\partial n_y}\, u(y)\, ds_y \pm \frac{u(y_0)}{2}. \qquad (2.25)$$

Proof. To be more specific, we will demonstrate the proof for the case of a 2D problem with a boundary contour $l = \partial D$, and the limit $x \to y_0 \in l$ from outside of the contour. The other cases can be proved likewise.

Let us represent the integral (2.24) as follows:

$$\begin{aligned}(G_d\, u)(x) = &\int_l \left[\frac{\partial \Phi(|x - y|)}{\partial n_y} - \frac{\partial \Phi_0(|x - y|)}{\partial n_y} \right] u(y)\, dl_y \\ &+ \int_l \frac{\partial \Phi_0(|x - y|)}{\partial n_y} [u(y) - u(x)]\, dl_y + u(x) \int_l \frac{\partial \Phi_0(|x - y|)}{\partial n_y}\, dl_y,\end{aligned} \qquad (2.26)$$

where $\Phi_0 = \Phi|_{k=0} = -(1/2\pi) \ln|x-y|$ is the low-frequency Green's function, so that $(\Phi - \Phi_0) \in C_1(\mathbb{R}^2)$ (cf. Eq. (2.16)). It can easily be shown that the first two integrals here are continuous in $\mathbb{R}^2$, and major effort should be directed at the third one.

Let us take the two functions

$$u_1(y) \equiv 1, \qquad u_2(y) = \Phi_0(|x - y|). \qquad (2.27)$$

Obviously, $\Phi_0(|x - y|)$ as a function of the argument y satisfies the Laplace equation $\Delta_y \Phi_0(|x-y|) = 0$ throughout $\mathbb{R}^2$, except a small neighborhood of the point x. Therefore, if x lies in the exterior of l, then both functions $u_1(y)$ and $u_2(y)$ satisfy the Laplace equation inside l. Then, by using Green's formula, we can write out successively:

$$\begin{aligned}0 &= \iint_D \left[u_1(y)\Delta_y u_2(y) - u_2(y)\Delta_y u_1(y) \right] dy_1\, dy_2 \\ &= \int_l \left[u_1(y)\frac{\partial u_2(y)}{\partial n_y} - u_2(y)\frac{\partial u_1(y)}{\partial n_y} \right] dl_y = \int_l \frac{\partial \Phi_0(|x - y|)}{\partial n_y}\, dl_y \\ &\sim \int_l \frac{\partial \Phi_0(|x - y|)}{\partial n_y}\, dl_y = 0, \quad x \in \mathbb{R}^2 \setminus D,\end{aligned} \qquad (2.28)$$

hence the outer limit here is:

$$\lim_{x \to y_0 \in l} \int_l \frac{\partial \Phi_0(|x - y|)}{\partial n_y} \, dl_y = 0. \tag{2.29}$$

We thus can evaluate this limit value of the full expression (2.26) as follows:

$$\lim_{x \to y_0 \in l} (G_d u)(x) = \int_l \left[\frac{\partial \Phi(|y_0 - y|)}{\partial n_y} - \frac{\partial \Phi_0(|y_0 - y|)}{\partial n_y} \right] u(y) \, dl_y + \int_l \frac{\partial \Phi(|y_0 - y|)}{\partial n_y} u(y) \, dl_y$$

$$= \int_l \frac{\partial \Phi(|y_0 - y|)}{\partial n_y} u(y) \, dl_y - u(y_0) \int_l \frac{\partial \Phi(|y_0 - y|)}{\partial n_y} \, dl_y. \tag{2.30}$$

In order to calculate the last integral, let us operate again with Green's formula for the same two functions $u_1(y) \equiv 1$ and $\Phi_0(|y_0 - y|)$ and apply it over the domain $\widetilde{D}$ contained between the contour l and a semi-circle $\Gamma = \Gamma_1 \cup \Gamma_2$ of a large radius R whose diameter Γ_1 is tangent to l at the point y_0, so that y_0 is the center of the semi-circle (see Fig. 2.1).

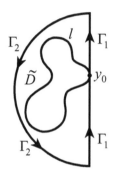

Figure 2.1. Contour $\Gamma = \Gamma_1 \cup \Gamma_2$ is tangent to the boundary line l

It can easily be proved that, despite a singularity of the function $\Phi_0(|y_0 - y|)$, considered as a function of the argument y, near the point y_0, the application of Green's formula is correct for two reasons. Firstly, let us show that the integral

$$I = \iint_{D_\varepsilon} \Delta_y \Phi_0(|y_0 - y|) \, dy_1 \, dy_2 \tag{2.31}$$

over a small neighborhood of this singular point of radius $\varepsilon > 0$ is small as $\varepsilon \to 0$, and then the integral over the domain $\widetilde{D}$ is finite. Indeed, if the contour l has a certain radius of curvature ρ at the point y_0, then simple observations, with the help of some classical results from differential geometry, show that the part of this ε-neighborhood belonging to the domain $\widetilde{D}$ has the square $\varepsilon^3/(6\rho^2) + O(\varepsilon^5)$. But the singularity of the quantity $\Delta_y \Phi_0(|y_0 - y|)$ is of the order of $O(1/\varepsilon^2)$, so the integral over this small domain D_ε is of the order of $O(\varepsilon)$, which is evidently small.

Secondly, let us prove that $\partial \Phi_0(|y_0 - y|)/\partial n_y = O(1)$ in a small neighborhood of y_0, both over Γ_1 and l. Indeed,

$$\frac{\partial \Phi_0(|y_0 - y|)}{\partial n_y} = \left(\mathbf{n}_y, \, \mathrm{grad}_y \, \Phi_0 \right), \qquad y = (y_1, y_2), \qquad y_0 = (y_{0_1}, y_{0_2}),$$

$$\mathrm{grad}_y \, \Phi_0(|y_0 - y|) = -\frac{1}{2\pi|y_0 - y|} \left\{ \frac{y_1 - y_{0_1}}{|y_0 - y|}, \, \frac{y_2 - y_{0_2}}{|y_0 - y|} \right\}, \qquad \text{hence} \tag{2.32}$$

$$\left(\mathbf{n}_y, \, \mathrm{grad}_y \, \Phi_0 \right) = -\frac{(\mathbf{n}_y, \, y - y_0)}{2\pi|y_0 - y|^2}, \qquad y, y_0 \in l, \, \Gamma_1.$$

It is clear that if $y, y_0 \in \Gamma_1$, then $(\mathbf{n}_y, y - y_0) = 0$, because these two vectors are orthogonal to each other. If $y, y_0 \in l$, then

$$(\mathbf{n}_y, y - y_0) = |y - y_0| \cos(\mathbf{n}_y, y - y_0) = -|y - y_0| \left[\frac{|y - y_0|}{2\rho} + O\left(|y - y_0|^2\right) \right], \qquad (2.33)$$

where ρ is the curvature radius of the contour l at the point y_0. Consequently,

$$\frac{\partial \Phi_0(|y_0 - y|)}{\partial n_y} = -\frac{1}{4\pi\rho} + O(|y - y_0|), \qquad y, y_0 \in l, \qquad (2.34)$$

i.e., both integrals along the boundary lines l and Γ are finite, since their integrands are continuous and bounded.

Now, the application of Green's formula over the domain $\widetilde{D}$ gives

$$\left(\int_\Gamma - \int_l \right) \frac{\partial \Phi_0(|y_0 - y|)}{\partial n_y} \, dl_y = 0 \sim \int_l \frac{\partial \Phi_0(|y_0 - y|)}{\partial n_y} \, dl_y$$
$$= \left(\int_{\Gamma_1} + \int_{\Gamma_2} \right) \frac{\partial \Phi_0(|y_0 - y|)}{\partial n_y} \, dl_y. \qquad (2.35)$$

The first integral over Γ_1 here is trivial, since $\mathbf{n}_y \perp (y_0 - y)$. The second integral is

$$\int_{\Gamma_2} \frac{\partial \Phi_0(|y_0 - y|)}{\partial n_y} \, dl_y = -\frac{1}{2\pi} \int_0^\pi \frac{\partial}{\partial R} (\ln R) \, R \, d\varphi = -\frac{1}{2}. \qquad (2.36)$$

It finally follows from Eqs. (2.30), (2.36) that

$$\lim_{x \to y_0 \in l} (G_d u)(x) = \int_l \frac{\partial \Phi(|y_0 - y|)}{\partial n_y} u(y) \, dl_y + \frac{u(y_0)}{2}. \qquad (2.37)$$

At the concluding part of the present section we cite without proof two results related to the normal derivatives of the single and double layer potentials (see, for example, Colton and Kress, 1983).

THEOREM 3. *If $u(y) \in C(\partial D)$, then at $x \to y_0 \in \partial D$ the limit values of the normal derivative of single-layer potential are determined as follows:*

$$\left[\frac{\partial (G_s u)(x)}{\partial n_x} \right]_{x=y_0}^\pm = \int_{\partial D} \frac{\partial \Phi(|y_0 - y|)}{\partial n_{y_0}} u(y) \, ds_y \mp \frac{u(y_0)}{2}. \qquad (2.38)$$

THEOREM 4. *If $u(y) \in C(\partial D)$, then at $x \to y_0 \in \partial D$ the normal derivative of the double-layer potential represents a function, which is continuous in $\mathbb{R}^3$ ($\mathbb{R}^2$):*

$$\left[\frac{\partial (G_s u)(x)}{\partial n_x} \right]_{x=y_0}^+ = \left[\frac{\partial (G_s u)(x)}{\partial n_x} \right]_{x=y_0}^-. \qquad (2.39)$$

Helpful remarks

1°. Interestingly, Theorem 2 asserts that the double-layer potential (2.24) is continuous up to the boundary line l, both from outside and inside. The key feature of the two limit boundary values $(G_d u)_\pm$ is that they are different, and the *jump* at the boundary line is equal to the value of the density at the boundary point: $u(y_0)$. In this sense, properties of the double-layer potential are congeneric with those of the Cauchy-type singular integral.

2°. Theorem 2 was proved above for the case when $y_0 \in l$ lies on the smooth portion of the piecewise smooth boundary line. If y_0 comes to a break of smoothness, of the inner angle α, then the term $u(y_0)/2$ in Theorem 2 must be changed by $\alpha u(y_0)/(2\pi)$. Theorem 1 also remains valid in this case.

3°. The kernel of the integral operator (2.25) is piecewise continuous for piecewise smooth contour, since due to Eq. (2.34) we can observe that

$$\frac{\partial \Phi(|y_0 - y|)}{\partial n_y} \sim \frac{\partial \Phi_0(|y_0 - y|)}{\partial n_y} \sim -\frac{1}{4\pi\rho} + O(|y - y_0|), \qquad y \to y_0 \quad (y, y_0 \in l). \quad (2.40)$$

4°. A more detailed and more formal treatment of the modern potential theory, with its application to diffraction problems, can be found in Colton and Kress (1983).

5°. Physicists prefer an alternative definition of a Green's function, as a solution to the equation

$$\Delta_y \Phi + k^2 \Phi = -\delta(y - x), \quad (2.41)$$

where the presence of Dirac's delta function means that a point source is applied at the point x. However, it can be directly tested, by using properties of the Fourier transform (1.3) and the delta function (1.62), that such an approach leads to the same representations (2.11) and (2.14) with $r = |x - y|$.

2.2. Basic Integral Equations of the Diffraction Theory

Let us study the exterior boundary value problem for the Helmholtz equation in the scalar case:

$$\Delta u + k^2 u = 0, \qquad k = \frac{\omega}{c}, \quad (2.42)$$

which holds for acoustic pressure $u(y)$ outside a simple closed piecewise smooth boundary of finite measure (see Fig. 2.2).

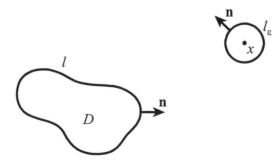

Figure 2.2. Boundary contour l and a small ε-neighborhood of the observation point x

THEOREM. *Let $x \in \mathbb{R}^n \setminus D$ ($n = 2, 3$). If $u(y)$ is a solution of Eq. (2.42) satisfying the radiation condition, then the following Kirchhoff–Helmholtz integral formula is valid:*

$$u(x) = \int_{\partial D} \left[u(y) \frac{\partial \Phi(|x-y|)}{\partial n_y} - \Phi(|x-y|) \frac{\partial u(y)}{\partial n_y} \right] ds_y. \tag{2.43}$$

Here $\mathbf{n}_y$ is an outer normal to the domain D.

Proof. To be more concrete, we give again a proof in the 2D case, because the 3D case is studied in very much the same way.

Let us consider a domain $\widetilde{D}$ contained inside a disk D_R of infinitely large radius R, with the region D and an ε-neighborhood D_ε of the point x removed (see Fig. 2.2). Then both functions $u(y)$ and $\Phi(|x-y|)$, the latter treated as a function of the argument y, satisfy the Helmholtz equation in $\widetilde{D}$. Therefore, Green's formula applied to this pair of functions leads to the following succession of relations:

$$0 = \iint_{\widetilde{D}} \left[u(y) \Delta_y \Phi(|x-y|) - \Phi(|x-y|) \Delta_y u(y) \right] dy_1\, dy_2 =$$

$$= \int_{\partial \widetilde{D}} \left[u(y) \frac{\partial \Phi(|x-y|)}{\partial n_y} - \Phi(|x-y|) \frac{\partial u(y)}{\partial n_y} \right] dl_y, \quad \partial \widetilde{D} = l \cup l_R \cup l_\varepsilon, \tag{2.44}$$

where $\mathbf{n}_y$ points outwards of $\widetilde{D}$.

First of all, we will prove that, since both functions $u(y)$ and $\Phi(|x-y|)$ satisfy the radiation condition, the integral along a far-zone circle l_R vanishes as $R \to \infty$. Indeed, all functions satisfying the radiation condition in unbounded acoustic medium have far-field asymptotics of the following form:

$$u(y) \sim u_0(\varphi) \frac{e^{ikR}}{\sqrt{R}}, \qquad \Phi(|x-y|) = \Phi^*(x, \varphi) \frac{e^{ikR}}{\sqrt{R}}, \qquad R \to \infty, \tag{2.45}$$

where $x = (R\cos\varphi,\ R\sin\varphi)$ are polar coordinates of the point x. This fact can be strictly proved for an arbitrary function satisfying the radiation condition, but we demonstrate it here only for the Green's function.

In the Cartesian coordinate system, let $x = (x_1, x_2)$ and $y = (y_1, y_2)$. Then

$$|x-y| = \sqrt{(y_1 - x_1)^2 + (y_2 - x_2)^2} = \sqrt{|y|^2 - 2(y,x) + |x|^2}$$

$$= |y| - \frac{(y,x)}{|y|} + O\left(\frac{1}{|y|}\right) = R - |x|\cos(\alpha - \varphi) + O\left(\frac{1}{|y|}\right), \tag{2.46}$$

$$x = (|x|\cos\alpha,\ |x|\sin\alpha), \quad |y| \to \infty.$$

Now let us recall the asymptotic behavior of the Green's function $\Phi(|x-y|) = (i/4) \times H_0^{(1)}(k|x-y|)$ at infinity (see Eq. (2.13)):

$$H_0^{(1)}(k|x-y|) \sim \sqrt{\frac{2}{\pi k R}}\, e^{i(k|x-y|-\pi/4)} + O\left(\frac{1}{R^{3/2}}\right), \quad |y| = R \to \infty. \tag{2.47}$$

So, with the help of the far-field asymptotics (2.46), we have

$$H_0^{(1)}(k|x-y|) \sim \sqrt{\frac{2}{\pi k R}}\, e^{ikR}\, e^{i[k|x|\cos(\alpha-\varphi)+\pi/4]}$$

$$= \Phi^*(x, \varphi) \frac{e^{ikR}}{\sqrt{R}} + O\left(\frac{1}{R^{3/2}}\right), \quad |y| = R \to \infty, \tag{2.48}$$

for fixed x.

Let us in the meantime come back to the integral along l_R, where $\partial/\partial n_y = \partial/\partial R$ if $y \in l_R$. It can trivially be seen from Eq. (2.45) that

$$\left[u(y) \frac{\partial \Phi(|x-y|)}{\partial n_y} - \Phi(|x-y|) \frac{\partial u(y)}{\partial n_y} \right] = O\left(\frac{1}{R^{3/2}} \right), \quad R \to \infty, \quad \text{hence}$$

$$\int_{l_R} \left[u(y) \frac{\partial \Phi(|x-y|)}{\partial n_y} - \Phi(|x-y|) \frac{\partial u(y)}{\partial n_y} \right] dl_y = 2\pi R \, O\left(\frac{1}{R^2} \right) = O\left(\frac{1}{R} \right). \tag{2.49}$$

The next step is to estimate a similar integral over l_ε (see Fig. 2.2). We have for $\varepsilon \to +0$ (i.e., $y \to x$):

$$\int_{l_\varepsilon} \left[u(y) \frac{\partial \Phi(|x-y|)}{\partial n_y} - \Phi(|x-y|) \frac{\partial u(y)}{\partial n_y} \right] dl_y$$

$$= u(x) \int_{l_\varepsilon} \frac{\partial \Phi(|x-y|)}{\partial n_y} dl_y - \int_{l_\varepsilon} \Phi(|x-y|) \frac{\partial u(y)}{\partial n_y} dl_y. \tag{2.50}$$

The last integrand here has a weak singularity only, which permits the estimate (see the small-argument asymptotics of the Green's function (2.16))

$$\left| \int_{l_\varepsilon} \Phi(|x-y|) \frac{\partial u(y)}{\partial n_y} dl_y \right| \leq \max_{l_\varepsilon} \left| \frac{\partial u(y)}{\partial n_y} \right| \frac{1}{2\pi} \ln \varepsilon \, 2\pi \varepsilon \to 0, \quad \varepsilon \to +0, \tag{2.51}$$

since the function $u(y)$ is analytic outside l.

The first integral on the right-hand side of Eq. (2.50) with $\varepsilon \to +0$ behaves as

$$\lim_{\varepsilon \to +0} \int_{l_\varepsilon} \frac{\partial \Phi(|x-y|)}{\partial n_y} dl_y = \frac{1}{2\pi} \int_0^{2\pi} \frac{\partial (\ln r)}{\partial r} 2\pi r \, d\psi = 1, \tag{2.52}$$

where we have introduced the polar coordinate system (r, ψ) with origin at x.

Finally, Eq. (2.44) yields

$$\int_l \left[u(y) \frac{\partial \Phi(|x-y|)}{\partial n_y} - \Phi(|x-y|) \frac{\partial u(y)}{\partial n_y} \right] dl_y$$

$$= -\int_{l_\varepsilon} \left[u(y) \frac{\partial \Phi(|x-y|)}{\partial n_y} - \Phi(|x-y|) \frac{\partial u(y)}{\partial n_y} \right] dl_y = -u(x). \tag{2.53}$$

If we change simultaneously the direction of the normal $\mathbf{n}_y$ and the sign of the integral in (2.53), we obtain exactly what was to be proved.

The Kirchhoff–Helmholtz integral formula (2.43) implies that an arbitrary scalar wave field, satisfying the radiation condition, can be represented outside the boundary ∂D as a combination of single and double layer potentials with some densities distributed on the boundary ∂D. Thus, if we knew the values of both u and $\partial u/\partial n$ on the boundary, then we would be able to directly determine the unknown wave function at an arbitrary point x from Eq. (2.43). Unfortunately, these two quantities are never known simultaneously. Actually, if we study a boundary value problem of the Dirichlet type, then the unknown function $u(y)$ itself is given on the boundary. If we consider a boundary value problem of the Neumann type, then the value of the normal derivative $\partial u/\partial n$ is known on the boundary. In this sense, the Kirchhoff–Helmholtz formula does not give any solution to the boundary value problem.

However, the result of the proved theorem, in conjunction with the properties of the potentials of the single and double layers, permits immediately the derivation of the basic boundary integral equation (BIE) of diffraction theory within the framework of the so-called *direct BIE method.* Let us move the point x to the boundary, $x \to y_0 \in \partial D$, from outside. Then, according to Eq. (2.43) and the boundary values of the potentials of the single and double layers from the previous section, we have

$$u(y_0) = \int_{\partial D} \left[u(y) \frac{\partial \Phi(|y_0 - y|)}{\partial n_y} - \Phi(|y_0 - y|) \frac{\partial u(y)}{\partial n_y} \right] ds_y + \frac{u(y_0)}{2}$$

$$\sim u(y_0) = 2 \int_{\partial D} \left[u(y) \frac{\partial \Phi(|y_0 - y|)}{\partial n_y} - \Phi(|y_0 - y|) \frac{\partial u(y)}{\partial n_y} \right] ds_y, \quad y_0 \in \partial D. \qquad (2.54)$$

Below we give a formulation of a problem that covers the principal subject of the present chapter.

Formulation of diffraction problem

Let a given incident monochromatic acoustic wave with angular frequency ω and (complex-valued) amplitude of acoustic pressure $p^{\mathrm{inc}}(x)$ fall on an obstacle occupying a domain D (see Fig. 2.3). When the wave encounters this obstacle, the wave structure begins to

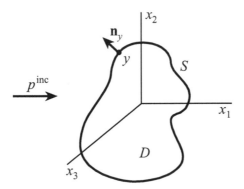

Figure 2.3. Incident acoustic wave and obstacle D with the boundary surface S

change. Any change in the incident wave due to its interaction with the obstacle is called *diffraction*. Synonyms to diffraction are *scattering* and (in the short-wave regime) *reflection*. A diffracted wave is marked in the present book with the superscript sc. Thus, due to the presence of an obstacle and the arising of a scattered wave, the structure of the full wave field, $p = p^{\mathrm{inc}} + p^{\mathrm{sc}}$, is different from that of the incident wave.

All three introduced wave pressures satisfy the Helmholtz equation (2.42), but only p^{sc} of them satisfies the radiation condition. The incident wave comes from the far-zone and hence can be assumed to arrive from infinity. Therefore, the radiation condition is broken for full pressure also, which is the sum of the pressures of the incident and the scattered waves. Hence, the derived Kirchhoff–Helmholtz integral representation (2.43) is valid only for the scattered wave,

$$p^{\mathrm{sc}}(x) = \int_{\partial D} \left[p^{\mathrm{sc}}(y) \frac{\partial \Phi(|x - y|)}{\partial n_y} - \Phi(|x - y|) \frac{\partial p^{\mathrm{sc}}(y)}{\partial n_y} \right] ds_y, \quad x \in \mathbb{R}^n \backslash D \ (n = 2, 3), \quad (2.55)$$

and so is the (BIE) boundary equation (2.54),

$$p^{\mathrm{sc}}(y_0) = 2 \int_{\partial D} \left[p^{\mathrm{sc}}(y) \frac{\partial \Phi(|y_0 - y|)}{\partial n_y} - \Phi(|y_0 - y|) \frac{\partial p^{\mathrm{sc}}(y)}{\partial n_y} \right] ds_y, \quad y_0 \in \partial D. \quad (2.56)$$

Now the two different natural types of boundary conditions are physically related to the cases of acoustically hard and acoustically soft boundary of the obstacle. In the first case (recall that $v_n = (i\rho_0\omega)^{-1}(\partial p/\partial n)$ if the wave process is harmonic in time)

$$v_n|_{\partial D} = 0 \quad \sim \quad \frac{\partial p}{\partial n}\bigg|_{\partial D} = 0 \quad \sim \quad \frac{\partial p^{sc}}{\partial n}\bigg|_{\partial D} = -\frac{\partial p^{inc}}{\partial n}\bigg|_{\partial D}, \tag{2.57}$$

and in the second case

$$p\,|_{\partial D} = 0 \quad \sim \quad p^{sc}\,|_{\partial D} = -p^{inc}|_{\partial D}. \tag{2.58}$$

In both cases we arrive at a BIE by substituting the respective boundary condition (2.57) or (2.58) into Eq. (2.56). The only inconvenience will be that in both cases the right-hand side is represented by quadrature. The following slight modification of the Kirchhoff–Helmholtz treatment allows us to avoid this inconvenience.

To this end, let us come back once again to our standard operation with Green's formula applied to a pair of functions $u_1(y) = p^{inc}(y)$ and $u_2(y) = \Phi(|x-y|)$, with both of them being analytic and satisfying the Helmholtz equation inside the domain D if x lies outside D. Then the application of Green's formula just over D proves that

$$0 = \int_D \left[p^{inc}(y)\Delta_y\Phi(|x-y|) - \Phi(|x-y|)\Delta_y p^{inc}(y) \right] dV_y$$

$$= -\int_{\partial D} \left[p^{inc}(y)\frac{\partial\Phi(|x-y|)}{\partial n_y} - \Phi(|x-y|)\frac{\partial p^{inc}(y)}{\partial n_y} \right] ds_y, \tag{2.59}$$

$$x \in \mathbb{R}^n \setminus D \quad (n = 2, 3).$$

By adding (2.59) and (2.56) together, we obtain a more advanced representation for the scattered wave field in terms of the total wave pressure:

$$p^{sc}(x) = \int_{\partial D} \left[p(y)\frac{\partial\Phi(|x-y|)}{\partial n_y} - \Phi(|x-y|)\frac{\partial p(y)}{\partial n_y} \right] ds_y, \quad x \in \mathbb{R}^n \setminus D \quad (n = 2, 3). \tag{2.60}$$

Now, using the boundary limit values of single and double layer potentials and by letting $x \to y_0 \in \partial D$, we arrive at an alternative boundary representation, instead of Eq. (2.56):

$$p^{sc}(y_0) = \int_{\partial D} \left[p(y)\frac{\partial\Phi(|y_0-y|)}{\partial n_y} - \Phi(|y_0-y|)\frac{\partial p(y)}{\partial n_y} \right] ds_y + \frac{p(y_0)}{2}, \quad y_0 \in \partial D. \tag{2.61}$$

It becomes clear from Eqs. (2.57), (2.58), (2.61) that the diffraction problem for an acoustically hard obstacle (Neumann-type boundary value problem) is reduced to a Fredholm integral equation of the second kind

$$p(y_0) - 2\int_{\partial D} \frac{\partial\Phi(|y_0-y|)}{\partial n_y} p(y)\,ds_y = 2p^{inc}(y_0), \quad y_0 \in \partial D, \tag{2.62}$$

since $p^{sc} = p - p^{inc}$. For the same reason, the acoustically soft case (Dirichlet-type boundary value problem) can be reduced to a Fredholm equation of the first kind

$$\int_{\partial D} \Phi(|y_0-y|)\frac{\partial p(y)}{\partial n_y}\,ds_y = p^{inc}(y_0), \quad y_0 \in \partial D. \tag{2.63}$$

Helpful remarks

1°. Let us briefly outline qualitative properties of equations (2.62) and (2.63). As remarked above, the first of them is a Fredholm integral equation of the second kind. As stated in Eq. (2.34), for a piecewise smooth boundary the kernel is piecewise continuous, so the integral operator in Eq. (2.62) is compact in the space $C(\partial D)$ and equation (2.62) can be studied within the framework of the Fredholm theory (Section 1.5). This implies the application of both theoretical results and well-developed numerical methods.

The integral operator (2.63) is of the first kind, and it is obvious that it possesses a weak singularity. Hence, it is again of the Fredholm type, and equation (2.63) is a Fredholm integral equation of the first kind with operator compact in $C(\partial D)$.

2°. Equation (2.62) is valid if the point y_0 belongs to the smooth part of the boundary. If y_0 hits a sharp angle of the value α, then the coefficient 2 in front of the integral operator should be changed by one expressed through α, in accordance with the described properties of the double-layer potential (see the previous section).

3°. Since only the second-kind Fredholm integral equations are described by so harmonious Fredholm theory, much effort was directed to the formulation of the Dirichlet boundary value problem in terms of the second-kind Fredholm BIE. Fortunately, this can be achieved by applying the so-called *indirect BIE method*, which is as follows. The function

$$p^{\text{sc}}(x) = \int_{\partial D} \frac{\partial \Phi(|x-y|)}{\partial n_y}\, \psi(y)\, ds_y, \qquad x \in \mathbb{R}^n \setminus D \quad (n=2,3), \tag{2.64}$$

satisfies the Helmholtz equation and the radiation condition with any continuous density ψ. If in the case of acoustically soft obstacle the point x approaches the boundary ∂D, $x \to y_0 \in \partial D$, then by using the properties of the double-layer potential and the boundary condition (2.58), we arrive at the following equation:

$$\psi(y_0) + 2\int_{\partial D} \frac{\partial \Phi(|y_0-y|)}{\partial n_y}\, \psi(y)\, ds_y = -2p^{\text{inc}}(y_0), \qquad y_0 \in \partial D, \tag{2.65}$$

which is an equation of the second kind.

2.3. Properties of Integral Operators of Diffraction Theory: General Case and Low Frequencies

Qualitative properties of the integral equations obtained are closely connected with general properties of the exterior Dirichlet and Neumann boundary value problems established by classicists (see, for example, Weyl, 1952; Kupradze, 1950; Atkinson, 1949), which we quote without proof for the sake of brevity.

THEOREM 1. *Both the Dirichlet and Neumann exterior boundary value problems for the Helmholtz equation have a unique solution in the class of functions satisfying the radiation condition.*

THEOREM 2. *The interior Dirichlet and Neumann boundary value problems for the Helmholtz equation have a unique solution for all values of the wave number k, except a countable set of positive values of k (related to eigen, or resonance frequencies).*

From this classical result we can extract some important properties of the considered integral operators.

THEOREM 3. *The second-kind integral equation (2.65) of the indirect BIE method for the Dirichlet exterior boundary value problem has a unique solution for any k, except those corresponding to eigenvalues of the interior Neumann problem.*

Proof. According to the Fredholm theory for integral equations of the second kind, equation (2.65) is surely uniquely solvable if the homogeneous equation (2.65) has only the trivial solution. So, let the parameter k be such that there is a nontrivial solution ψ of the homogeneous equation (2.65), i.e.,

$$\frac{\psi(y_0)}{2} + \int_{\partial D} \frac{\partial \Phi(|y_0 - y|)}{\partial n_y}\, \psi(y)\, ds_y = 0, \qquad y_0 \in \partial D. \tag{2.66}$$

Let us introduce the double-layer potential with this density

$$u(x) = \int_{\partial D} \frac{\partial \Phi(|x - y|)}{\partial n_y}\, \psi(y)\, ds_y. \tag{2.67}$$

Then, due to the limit boundary values of the double-layer potential, we have (the subscript "+" refers to the outer limit)

$$u_+(y_0) = \frac{\psi(y_0)}{2} + \int_{\partial D} \frac{\partial \Phi(|y_0 - y|)}{\partial n_y}\, \psi(y)\, ds_y = 0. \tag{2.68}$$

According to Theorem 1, $u(x) \equiv 0$ outside D, as a solution of the exterior Dirichlet problem with the trivial boundary condition. This immediately implies $\partial u_+(y_0)/\partial n = 0$, $y_0 \in \partial D$. Further, the continuity of the normal derivative of the double-layer potential (Theorem 4, Section 2.1) proves that $\partial u_-(y_0)/\partial n = 0$, $y_0 \in \partial D$. We can thus finally see that a nontrivial solution of the exterior Dirichlet problem generates a nontrivial (i.e., eigen) solution of the interior Neumann problem.

A similar technique can be applied to prove a similar theorem for the exterior Neumann problem.

THEOREM 4. *The second-kind integral equation (2.62) of direct BIE for the Neumann exterior boundary value problem has a unique solution for any k, except eigenvalues of the interior Dirichlet problem.*

It is somewhat unexpected that a similar result is valid for the first-kind BIE (2.63). A proof to the following theorem can be found, for example, in Colton and Kress (1983).

THEOREM 5. *The Fredholm integral equation of the first kind (2.63) with a kernel containing a weak singularity is uniquely solvable for all values of the parameter k, except those corresponding to eigenvalues of the interior Dirichlet problem.*

Low-frequency diffraction problem

From the general variational principles it follows that the countable set of eigenvalues for both types of interior boundary value problems has a limit point at infinity (see Courant and Hilbert, 1953; and our discussion in Chapter 4). Therefore, there is always a minimum eigenfrequency. Then it directly follows from Theorems 3–5 that the considered integral equations (2.62), (2.63), (2.65) have a unique solution if the frequency is less than the first eigenvalue of the corresponding interior problem. Thus, for sufficiently low frequencies we may directly treat these equations, both analytically and numerically.

Example. Low-frequency diffraction by a hard round disk.
We consider this example as a 2D diffraction problem for acoustically hard obstacle. Let a be a radius of the disk and let the incident wave be plane propagating in direction of the x_1 axis. Then we have

$$\Phi = \frac{i}{4}H_0^{(1)}(kr) \sim -\frac{1}{2\pi}\left[\ln(kr) + \gamma - \ln 2\right] + \frac{i}{4} + O\left(k^2 \ln k\right), \quad k \to 0; \quad \text{hence}$$

$$\frac{\partial\Phi}{\partial n_y} = \frac{\partial\Phi}{\partial r}\frac{\partial r}{\partial n_y} = -\frac{1}{2\pi r}\frac{\partial r}{\partial n_y} = -\frac{\cos(r,n_y)}{2\pi r}, \quad r = |y_0 - y|, \quad p^{\text{inc}}(x) = e^{ikx_1}. \tag{2.69}$$

Representation in the polar coordinate system implies

$$y = \{a\cos\theta, a\sin\theta\}, \quad y_0 = \{a\cos\psi, a\sin\psi\},$$
$$r = y - y_0 = \{a(\cos\theta - \cos\psi), a(\sin\theta - \sin\psi)\}, \quad n_y = \{\cos\theta, \sin\theta\},$$
$$\cos(r,n_y) = \frac{(r\cdot n_y)}{r} = a\frac{\cos\theta(\cos\theta-\cos\psi)+\sin\theta(\sin\theta-\sin\psi)}{r}, \tag{2.70}$$

hence

$$\frac{\partial\Phi}{\partial n_y} = -\frac{1}{2\pi a}\frac{1-\cos(\theta-\psi)}{(\cos\theta-\cos\psi)^2+(\sin\theta-\sin\psi)^2} = -\frac{1}{2\pi a}\frac{1-\cos(\theta-\psi)}{2[1-\cos(\theta-\psi)]} = -\frac{1}{4\pi a}. \tag{2.71}$$

This yields the following integral equation of low-frequency diffraction for a hard round disk, which follows from Eq. (2.62):

$$p(\psi) + \frac{1}{2\pi}\int_0^{2\pi} p(\theta)\,d\theta = 2p^{\text{inc}}(\psi) \quad (dl_y = a\,d\theta)$$
$$\sim \quad p(\psi) + \frac{P}{2\pi} = 2\,e^{ika\cos\psi}, \quad P = \int_0^{2\pi} p(\theta)\,d\theta, \tag{2.72}$$

where P is some constant.

Now, by integrating both sides of the last equation over the interval $[0, 2\pi]$, we obtain (J_0 is the Bessel function)

$$\int_0^{2\pi} p(\psi)\,d\psi + P = 2\int_0^{2\pi} e^{ika\cos\psi}\,d\psi \quad \sim \quad P = 2\pi J_0(ak), \tag{2.73}$$

where we have used the value of the tabulated integral arisen in Eq. (2.73). Now, combining Eqs. (2.72) and (2.73) together, we arrive at the following solution of the basic integral equation in the case of low frequencies:

$$p(\psi) = 2\,e^{ika\cos\psi} - J_0(ak) \approx 1 + 2ika\cos\psi, \quad k \to 0. \tag{2.74}$$

Note that the highest order of error for this solution was caused by approximation of the kernel in Eq. (2.69), so the real error of expression (2.74) that gives the solution of BIE is $O(k^2 \ln k)$.

A plane incident wave has been taken in the above example as the most standard type of incident waves used in theoretical study and in practice. Perhaps, it is so widespread in diffraction theory due to the fact that any wave can locally be treated as plane. Another reason is that all other simple types of waves (cylindrical, spherical, etc.) can be, as a rule, represented as a superposition of plane waves (see, for example, Brekhovskikh, 1980).

Arbitrary 3D obstacle: Acoustically hard boundary

Let us derive a general low-frequency form of an integral equation in the 3D case up to the linear (with respect to the small parameter k) small terms. To be more specific, we will consider a plane wave incident on an obstacle D with acoustically hard boundary, i.e., we will treat Eq. (2.62). If the direction of propagation of the plane coincides, for instance, with the axis x_3, then we have $p^{\text{inc}}(x) = \exp(iky_{0_3})$ in Eq. (2.62), or $p^{\text{inc}}(x) = 1 + iky_{0_3} + O(k^2)$ for small k.

Let us write out the first asymptotic terms of the normal derivative in the kernel of Eq. (2.62):

$$
\begin{aligned}
\frac{\partial \Phi(|y_0 - y|)}{\partial n_y} &= \frac{\partial}{\partial r}\left(\frac{e^{ikr}}{4\pi r}\right)\cos(n_y, r) = \frac{\partial}{\partial n_y}\left(\frac{1 + ikr}{4\pi r}\right)\cos(n_y, r) + O(k^2) \\
&= -\frac{1}{4\pi r^2}\cos(n_y, r) + O(k^2) = \frac{\partial \Phi_0(|y_0 - y|)}{\partial n_y} + O(k^2), \quad \text{where}
\end{aligned}
\tag{2.75}
$$

$$
\Phi_0(|y_0 - y|) = \frac{1}{4\pi r}, \quad \frac{\partial \Phi_0(|y_0 - y|)}{\partial n_y} = -\frac{1}{4\pi r^2}\cos(n_y, r), \quad r = |y_0 - y|,
$$

so, for small frequencies, with an $O(k^2)$ error, Eq. (2.62) becomes

$$
p(y_0) - 2\int_S \frac{\partial \Phi_0(|y_0 - y|)}{\partial n_y} p(y)\, ds_y = 2(1 + iky_{0_3}), \quad y_0 \in S.
\tag{2.76}
$$

It is very interesting that within the framework of this approximation the kernel is independent of the parameter k. Moreover, it coincides with the limit $(k \to 0)$ low-frequency value. Hence, the first two asymptotic terms of a low-frequency solution to the considered integral equation may be obtained by solving Eq. (2.76) just for the two respective terms of the right-hand side, so that $p(y) = p_0(y) + ikp_1(y) + O(k^2)$.

Let us prove that for arbitrary smooth shape of the obstacle boundary surface S the leading asymptotic term, which is determined as a solution of the equation

$$
p_0(y_0) - 2\int_S \frac{\partial \Phi_0(|y_0 - y|)}{\partial n_y} p_0(y)\, ds_y = 2, \quad y_0 \in S,
\tag{2.77}
$$

is a certain constant $p_0(y) \equiv p_0$. In order to prove this statement, let us evaluate how the integral operator in Eq. (2.77) acts on a constant. To this end, let us take, following our standard technique, the two functions $u_1(y) \equiv p_0$ and $u_2(y) = \Phi_0(|x - y|)$, with both of them being analytic inside D when x is located outside D. Then Green's formula, applied over a domain D, yields (cf. Eq. (2.28))

$$
\begin{aligned}
0 &= \int_D \left[u_1(y)\Delta_y u_2(y) - u_2(y)\Delta_y u_1(y)\right] dy_1\, dy_2 \\
&= \int_S \frac{\partial \Phi_0(|x - y|)}{\partial n_y} p_0\, ds_y, \quad x \in \mathbb{R}^3 \setminus D.
\end{aligned}
\tag{2.78}
$$

Now, by applying the outer limit, $x \to y_0 \in S$, to both sides of Eq. (13), we have

$$
\frac{p_0}{2} + \int_S \frac{\partial \Phi_0(|x - y|)}{\partial n_y} p_0\, ds_y = 0 \quad \sim \quad \int_S \frac{\partial \Phi_0(|x - y|)}{\partial n_y} p_0\, ds_y = -\frac{p_0}{2},
\tag{2.79}
$$

so that Eq. (2.77) becomes

$$
2\, p_0 = 2 \quad \sim \quad p_0(y) \equiv p_0 = 1.
\tag{2.80}
$$

The obtained result is quite natural from the intuitive (heuristic) point of view. Indeed, the low-frequency limit is equivalent to the case of a very small obstacle with a fixed frequency. But if the obstacle is infinitely small, then the incident wave does not feel it, and so the total pressure remains without any perturbation, i.e., $p_0|_S = p^{inc}|_{k=0} = 1$.

Unfortunately, the next asymptotic term, related to the second term on the right-hand side of Eq. (2.76), cannot be constructed for arbitrary shape, but it admits explicit treatment for some canonical shapes. The case of a spherical obstacle is considered in the next section.

Helpful remarks

1°. Theorems 3–5 state that the considered integral equations are uniquely solvable for all k excluding a countable set of eigenvalues of the corresponding interior boundary value problem. The results stated in the final part of Section 1.5 guarantee that, at least for second-kind equations, you may easily construct a stable numerical solution for arbitrary boundary of noncanonical shape in this case. A great deal of papers were devoted to proposing a theory of boundary integral equations that would be correctly solvable for all k, since exterior problems are always uniquely solvable, in contrast to their integral equations (see, for example, Jones, 1974; Kleinman and Roach, 1982). In a formal theory these results are perhaps very important; however they are less important from a practical aspect. If you create your own computer code in any algorithmic language, without regard for the value of the parameter k, then your algorithm written for the exterior problem must formally crash for the described set of eigenfrequencies, However, in practice this never happens so. It is not so easy to compel the algorithm to crash with a particular value of k. In practice, you will not feel that the algorithm behaves distinctively from regular cases until you set an "irregular" value of k with an accuracy of 10^{-6}–10^{-7}. So, if you take any k with three or four significant digits, you may use your code for arbitrary k without any problem.

2°. Analogous observations take place when you operate with a piecewise smooth boundary surface (3D case) or contour (2D case). All derived equations of the second kind, as they are stated, are valid whenever y_0 belongs to the smooth portion of the boundary. If y_0 is at a corner or cusp point, the factor in front of the unknown function outside the integral must be changed, and you should take this fact into account when writing your computer code. But we advise that you forget forever about this feature of the problem and choose instead such a grid that its nodes do not hit any sharp edge. Then you will not face any trouble in practice. The precision of your computations will be even higher than if you take a different coefficient only for one ("sharp") node.

2.4. Full Low-Frequency Solution for Spherical Obstacle

Acoustically hard obstacle

The second asymptotic term of the solution to Eq. (2.76) ought to be found from the following equation:

$$p_1(y_0) - 2 \int_S \frac{\partial \Phi_0(|y_0 - y|)}{\partial n_y} p_1(y) \, ds_y = 2y_{0_3}, \qquad y_0 \in S. \qquad (2.81)$$

Let us introduce a spherical coordinate system for the points y and y_0 on the surface of a sphere of radius a:

$$y = a(\sin\theta\cos\varphi,\ \sin\theta\sin\varphi,\ \cos\theta), \quad y_0 = a(\sin\theta_0\cos\varphi_0,\ \sin\theta_0\sin\varphi_0,\ \cos\theta_0),$$

$$r^2 = |y_0 - y|^2 = 2a^2\left[1 - \cos\theta\cos\theta_0 - \sin\theta\sin\theta_0\cos(\varphi - \varphi_0)\right],$$

$$\cos(n_y, r) = \frac{a}{r}\left[1 - \cos\theta\cos\theta_0 - \sin\theta\sin\theta_0\cos(\varphi - \varphi_0)\right], \qquad (2.82)$$

$$\frac{\partial\Phi_0(|y_0 - y|)}{\partial n_y} = -\frac{\cos(n_y, r)}{4\pi r^2} = -\frac{1}{8\pi\, a\, r}.$$

The problem in hand is axially symmetric. So we seek a solution that does not depend upon the angle φ. Then equation (2.81) in the spherical coordinate system becomes

$$p_1(\theta_0) + \frac{1}{4\pi\sqrt{2}}\int_0^\pi K_0(\theta_0, \theta)\, p_1(\theta)\, d\theta = 2a\cos\theta_0, \qquad 0 \le \theta_0 \le \pi,$$

$$K_0(\theta_0, \theta) = \sin\theta\int_0^{2\pi}\frac{d\varphi}{\sqrt{1 - \cos\theta\cos\theta_0 - \sin\theta\sin\theta_0\cos\varphi}} \qquad (2.83)$$

$$= \frac{2\sqrt{2}\,\sin\theta}{|\sin[(\theta + \theta_0)/2]|}\,\mathbf{K}\left(\frac{\sqrt{\sin\theta\sin\theta_0}}{|\sin[(\theta + \theta_0)/2]|}\right),$$

where $\mathbf{K}$ is the full elliptic integral of the first kind (Gradshteyn and Ryzhik, 1994).

One can hardly believe that this equation admits exact analytical solution. Many mathematicians spend much effort to prove that an exact solution of this equation can be found in the form

$$p_1(\theta) = B\, a\cos\theta = By_3, \qquad (2.84)$$

where B is some constant. The discussed question demonstrates an excellent example of what is discussed in the Preface: it happens so frequently that heuristic ideas help to achieve a breakthrough in some problems of pure or applied mathematics that could not be resolved directly by strict formal methods. It was Lord Rayleigh who first discovered that a small sphere behaves like a dipole when irradiated by a plane wave (a good presentation can be found, for instance, in Morse and Feshbach, 1953; Hönl et al., 1961).

So, let us prove that the solution of Eq. (2.83), or, what is the same, of Eq. (2.81), has the form (2.84). For this purpose, we apply (over a domain D) once again our standard approach using Green's formula, with two functions $u_2(y) = y_3$ and $u_2(y) = \Phi_0(|x - y|)$ (the point x is fixed outside D). Then Green's formula gives

$$0 = \int_D \left[u_1(y)\Delta_y u_2(y) - u_2(y)\Delta_y u_1(y)\right] dy_1\, dy_2$$

$$= \int_S \left[y_3\frac{\partial\Phi_0(|x - y|)}{\partial n_y} - \Phi_0(|x - y|)\frac{\partial y_3}{\partial n_y}\right] ds_y, \qquad x \in \mathbb{R}^3 \setminus D. \qquad (2.85)$$

Now, by letting $x \to y_0 \in S$ in Eq. (2.85) and by using the boundary properties of the single and double layer potentials, we obtain

$$\frac{y_{0_3}}{2} + \int_S \left[y_3\frac{\partial\Phi_0(|y_0 - y|)}{\partial n_y} - \Phi_0(|y_0 - y|)\frac{\partial y_3}{\partial n_y}\right] ds_y = 0. \qquad (2.86)$$

It is obvious that on a spherical surface of radius a, the following relations hold (cf.

Eq. (2.82)):

$$y_3 = a\cos\theta, \qquad \frac{\partial y_3}{\partial n_y} = \cos\theta = \frac{y_3}{a},$$

$$\frac{\partial\Phi_0(|y_0 - y|)}{\partial n_y} = -\frac{\cos(n_y, r)}{4\pi r^2} = -\frac{1}{8\pi\, a\, r}, \tag{2.87}$$

$$\Phi_0(|y_0 - y|) = \frac{1}{4\pi\, r} = -2a\,\frac{\partial\Phi_0(|y_0 - y|)}{\partial n_y},$$

and consequently the integral in (2.86) becomes

$$-\frac{y_{0_3}}{2} = \int_S \left[y_3 \frac{\partial\Phi_0(|y_0 - y|)}{\partial n_y} - \Phi_0(|y_0 - y|)\frac{\partial y_3}{\partial n_y} \right] ds_y = 3\int_S y_3 \frac{\partial\Phi_0(|y_0 - y|)}{\partial n_y}\, ds_y. \tag{2.88}$$

Hence, the integral operator in Eq. (2.81), in the case of a small spherical obstacle, acts on the function y_3 as follows:

$$\int_S \frac{\partial\Phi_0(|y_0 - y|)}{\partial n_y}\, y_3\, ds_y = -\frac{y_{0_3}}{6}. \tag{2.89}$$

It now becomes clear that an exact solution to Eq. (2.81), and also to that written in an equivalent form (2.83), can be obtained by substituting the potentially appropriate structure (2.84) to Eq. (2.81), which together with Eq. (2.89) leads to the relation

$$\frac{4}{3}Bay_3 = 2ay_3 \quad \sim \quad B = \frac{3}{2} \quad \sim \quad p_1 = \frac{3}{2}y_3 = \frac{3}{2}a\cos\theta, \tag{2.90}$$

which proves the hypothesis about the structure of the function $p_1(\theta)$ written in the form (2.84).

Finally, taking into account that $p_0(\theta) \equiv 1$ and also what was written in the paragraph right after Eq. (2.76), we arrive at the full two-term low-frequency asymptotics in the case of a spherical obstacle in the following form:

$$p(\theta) = 1 + \frac{3}{2}ika\cos\theta. \tag{2.91}$$

Acoustically soft obstacle

In this case the problem may be reduced to Eq. (2.63). The low-frequency expansion for the kernel is here represented as

$$\Phi(|y_0 - y|) = \frac{e^{ikr}}{4\pi r} = \frac{1}{4\pi r} + \frac{ik}{4\pi} + O(k^2) = \Phi_0(|y_0 - y|) + \frac{ik}{4\pi} + O(k^2) \quad (r = |y_0 - y|), \tag{2.92}$$

so, with an $O(k^2)$ error, equation (2.63) is equivalent to

$$\int_S \left[\Phi_0(|y_0 - y|) + \frac{ik}{4\pi} \right] g(y)\, ds_y = 1 + iky_{0_3}, \qquad y_0 \in S, \quad g(y) = \frac{\partial p(y)}{\partial n_y}. \tag{2.93}$$

The second term in the kernel generates a certain constant; hence the solution of Eq. (2.93) admits the following decomposition:

$$\int_S \Phi_0(|y_0 - y|)\, g(y)\, ds_y = 1 + iky_{0_3} - \frac{ikI}{4\pi}, \qquad y_0 \in S,$$

$$I = \int_S g(y)\, ds_y = \int_S g_0(y)\, ds_y + O(k), \qquad g(y) = g_0(y) + ikg_1(y), \tag{2.94}$$

where the functions g_0 and g_1 are independent of the wave number k and give, respectively, the leading (zeroth) and the first asymptotic terms of the solution. The first of them is a solution of Eq. (2.94) with the unit right-hand side, and the second with the right-hand side in the form $y_{0_3} - I/(4\pi)$.

Let us show that a constant is still a solution to the first equation, and this is also true in the case of acoustically soft obstacle. To this end, let us study how operator (2.94) acts on the constant $g_0(y) \equiv g_0$. The answer to this question can be given immediately in the case of spherical shape (see Eq. (2.87) and Eq. (2.79)):

$$\int_S \Phi_0(|y_0 - y|) \, g_0 \, ds_y = -2a \int_S \frac{\partial \Phi_0(|y_0 - y|)}{\partial n_y} \, g_0 \, ds_y = (-2a)\left(-\frac{g_0}{2}\right) = ag_0, \qquad (2.95)$$

hence $g_0(y) \equiv g_0 = 1/a$.

Further, if all terms of the order of $O(k^2)$, including that arising for ikI in Eq. (2.94), are neglected, the first-term function $g_1(y)$ is found from the equation

$$\int_S \Phi_0(|y_0 - y|) \, g_1(y) \, ds_y = -\frac{I}{4\pi} + y_{0_3}, \quad \text{where}$$

$$I = \int_S g_0 \, ds_y = 4\pi a^2 \, g_0 = 4\pi a, \quad y_0 \in S. \qquad (2.96)$$

Of course, a solution of this equation can be constructed as the sum $g_1(y) = h_1(y) + h_2(y)$, where the first of these new functions is responsible for the constant right-hand side in (2.96): $h_1(y) \equiv -I/(4\pi a) = -1$, and the second one is to be defined from the equation

$$\int_S \Phi_0(|y_0 - y|) \, h_2(y) \, ds_y = y_{0_3}, \qquad y_0 \in S. \qquad (2.97)$$

Since $\Phi_0(|y_0 - y|) = -2a \, \partial \Phi_0(|y_0 - y|)/\partial n_y$ for a spherical obstacle (see Eq. (2.87)) and due to relation (2.89), we conclude that $h_2(y) = 3y_3/a$.

At last, by combining together the expressions for the functions g_0, h_1, h_2, we arrive at the following low-frequency asymptotic solution:

$$\frac{\partial p(y)}{\partial n_y} = g(y) = \frac{1}{a} + ik\left(-1 + \frac{3\,y_3}{a}\right) = \frac{1 + ika(3\cos\theta - 1)}{a}. \qquad (2.98)$$

Helpful remarks

$1°$. Interestingly, the method discussed here allows us to construct a solution to the very complex integral equation (2.83) for two specific right-hand sides, which is not so easy to obtain by any alternative method. Very often general principles of mathematical physics (like Green's theorem) allow one to derive explicit-form solutions of some complex differential and integral equations.

$2°$. Some other canonical shapes admit explicit-form solutions of the diffraction problem in the low-frequency range. Among others, we mention here diffraction by a half-plane, by ellipses, by a wedge with infinite faces, etc. A good survey of the known analytical solutions can be found in Bowman et al. (1987).

2.5. Application: Scattering Diagram for Obstacles of Canonical Shape

The basic integral representation of the solution to the diffraction problem given by the Kirchhoff–Helmholtz formula (2.60) permits the derivation of the so-called *scattering diagram*, or *scattering pattern*. The latter is defined as a function that represents the dependence of the real scattered amplitude upon the angle of observation.

Let us consider again, as in formula (2.46), the far-field representation for the Green's function, now in the 3D case and for large $|x|$:

$$r = |x - y| = \sqrt{(x_1 - y_1)^2 + (x_2 - y_2)^2 + (x_3 - y_3)^2}$$
$$= \sqrt{|x|^2 - 2(x, y) + |y|^2} = |x| - \frac{(x, y)}{|x|} + O\left(\frac{1}{|x|}\right), \quad |x| \to \infty. \tag{2.99}$$

Then in a far zone we have, in the asymptotic sense,

$$\Phi(|x - y|) = \frac{e^{ikr}}{4\pi r} \approx \frac{e^{ik|x|}}{4\pi|x|} e^{-ik(x,y)/|x|},$$
$$\frac{\partial\Phi(|x - y|)}{\partial n_y} \approx \frac{ik}{4\pi r} e^{ikr} \cos(n_y, y - x) \approx -\frac{e^{ik|x|}}{4\pi|x|} e^{-ik(x,y)/|x|} ik \cos(n_y, x), \tag{2.100}$$

so that the scattered amplitude given by Eq. (2.60) takes in the far zone the following form:

$$p^{sc}(x) \sim -\frac{e^{ik|x|}}{4\pi|x|} \int_S \left[p(y) ik \cos(n_y, x) + \frac{\partial p(y)}{\partial n_y} \right] e^{-ik(x,y)/|x|} ds_y, \quad |x| \to \infty. \tag{2.101}$$

Recall that for acoustically hard obstacle the second term in the square brackets vanishes, and so does the first term for acoustically soft body. It should also be noted that expression (2.101) confirms that any scattered wave field in a far zone behaves like a spherical wave.

An analogous formula can be derived in the 2D case:

$$p^{sc}(x) \sim -\frac{(1 + i) e^{ik|x|}}{4\sqrt{\pi k|x|}} \int_l \left[p(y) ik \cos(n_y, x) + \frac{\partial p(y)}{\partial n_y} \right] e^{-ik(x,y)/|x|} dl_y, \quad |x| \to \infty. \tag{2.102}$$

Both expressions are valid for arbitrary boundary shape and arbitrary value of the frequency parameter (wave number) k. However they admit explicit analytical treatment only for canonical shapes and low frequencies.

Low-frequency scattering from a hard round disk

Recall that in this case $\partial p(y)/\partial n_y$ in Eq. (2.102) vanishes and $p(y)$ is given in the polar coordinate system by Eq. (2.74). If the far point x is represented in the same polar system as $x = (x_1 = R\cos\alpha, \ x_2 = R\sin\alpha)$, $R = |x|$, then by using the evident relations

$$\cos(n_y, x) = \frac{(n_y, x)}{|x|} = \frac{R(\cos\psi\cos\alpha + \sin\psi\sin\alpha)}{R} = \cos(\psi - \alpha),$$
$$\frac{(x, y)}{|x|} = \frac{Ra(\cos\alpha\cos\psi + \sin\alpha)\sin\psi}{R} = a\cos(\psi - \alpha), \quad y = (a\cos\psi, \ a\sin\psi) \tag{2.103}$$

and by omitting an inessential factor, finding the real-valued scattered amplitude can be reduced to the calculation of the following integral:

$$
\begin{aligned}
|p^{sc}(\alpha)| &\sim \frac{k}{\sqrt{k|x|}} \left| \int_l p(y) \cos(n_y, x)\, e^{-ik(x,y)/|x|}\, dl_y \right| \\
&= \frac{ak}{\sqrt{k|x|}} \left| \int_0^{2\pi} (1 + 2iak \cos\psi) \cos(\psi - \alpha)\, e^{-iak\cos(\psi-\alpha)}\, d\psi \right| \\
&\sim \frac{(ak)^2}{\sqrt{k|x|}} \left| 1 - 2\cos\alpha \right|,
\end{aligned}
\tag{2.104}
$$

where only the leading asymptotic term (with respect to k) is given here. The last line in Eq. (2.104) shows that as the frequency decreases the scattered amplitude decreases like $O(k^{3/2})$. The dependence upon the observation angle α is demonstrated as a diagram in Fig. 2.4.

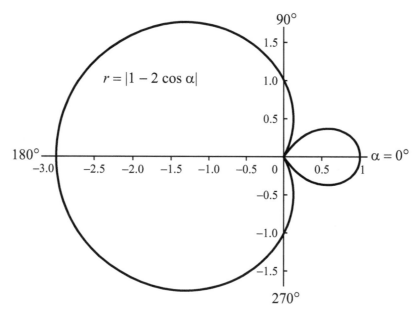

Figure 2.4. Scattering diagram for hard disk as a function of polar angle α

Low-frequency diagram for acoustically hard sphere

Here we return again to the spherical coordinate system (see Eq. (2.82)) and substitute appropriate expressions into formula (2.101). Then, by setting the angles of observation (α, β) for the far-field point x so that

$$
x = (R \sin\alpha \cos\beta,\ R \sin\alpha \sin\beta,\ R \sin\alpha), \qquad R = |x|,
\tag{2.105}
$$

taking into account the evident relations

$$
p(y) = p(\theta) = 1 + \frac{3}{2} ika \cos\theta, \qquad \frac{\partial p(y)}{\partial n_y} = 0,
$$

$$
\cos(n_y, x) = \frac{(n_y, x)}{|x|} = \frac{(y, x)}{|y||x|} = \sin\alpha \sin\theta \cos(\varphi - \beta) + \cos\alpha \cos\theta,
\tag{2.106}
$$

$$
\frac{(x, y)}{|x|} = a\left[\sin\alpha \sin\theta \cos(\varphi - \beta) + \cos\alpha \cos\theta\right],
$$

and omitting some inessential factor, we arrive at the leading asymptotic term in the form

$$
\begin{aligned}
|p^{\mathrm{sc}}(\alpha)| &\sim \frac{k}{|x|} \left| \int_S p(y)\,\cos(n_y, x)\,e^{-ik(x,y)/|x|}\,ds_y \right| \\
&= \frac{ak}{|x|} \left| \int_0^{2\pi} \int_0^{\pi} \left(1 + \frac{3}{2}iak\cos\theta \right) [\sin\alpha\sin\theta\cos(\varphi - \beta) + \cos\alpha\cos\theta] \right. \\
&\quad \left. \times e^{-iak[\sin\alpha\sin\theta\cos(\varphi - \beta) + \cos\alpha\cos\theta]}\sin\theta\,d\varphi\,d\theta \right|.
\end{aligned}
\tag{2.107}
$$

It is easily proved that the last integral does not depend on the polar angle β, and calculating the integral with respect to φ, we obtain the following dependence on the polar angle α:

$$
|p^{\mathrm{sc}}(\alpha)| \sim \frac{(ak)^2}{|x|} \left| 1 - \frac{3}{2}\cos\alpha \right|
\tag{2.108}
$$

which is a small quantity of the order of $O(k^2)$ as the frequency decreases.

It should be noted that the diagrams in Figs. 2.4, 2.5 are selected so that $\alpha = 0$ refers to direct scattering and $\alpha = 180°$ to back scattering.

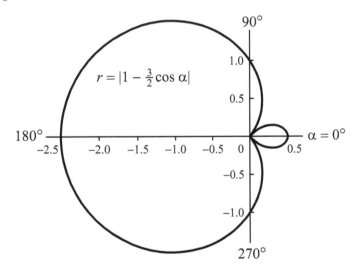

Figure 2.5. Scattering diagram for hard sphere as a function of polar angle α

Scattering by acoustically soft sphere

In this case we should take into account that $p(y) = 0$, $\partial p(y)/\partial n_y = [1 + ika(3\cos\theta - 1)]/a$ (see Eq. (2.98)) in Eq. (2.101). So we arrive at the following expression for the leading asymptotic term:

$$
\begin{aligned}
|p^{\mathrm{sc}}(\alpha)| &\sim \frac{1}{|x|} \left| \int_S \frac{\partial p(y)}{\partial n_y} e^{-ik(x,y)/|x|}\,ds_y \right| \\
&= \frac{1}{|x|} \left| \int_0^{2\pi} \int_0^{\pi} [1 + ika(3\cos\theta - 1)]\,e^{-iak(\sin\alpha\sin\theta\cos\varphi + \cos\alpha\cos\theta)}\sin\theta\,d\varphi\,d\theta \right| \\
&\sim \frac{a}{|x|}|1 - ika| + O(k^2) = \frac{a}{|x|} + O(k^2).
\end{aligned}
\tag{2.109}
$$

Helpful remarks

$1°$. It is very important for applications that for acoustically hard obstacles there are always some angles of scattering where (in the asymptotic approximation) the object does not radiate any energy. This means that while staying in a far zone at these angles of observation, one cannot detect the presence of the object. For the 2D problem these angles are $\alpha = \pm \arccos(1/2) = \pm 60°$, and in the 3D case $\alpha = \pm \arccos(2/3) \approx \pm 48°$; see Figs. 2.4 and 2.5. By contrast, a small soft sphere radiates equal energy in all directions.

$2°$. A low-frequency amplitude of scattering for an acoustically hard obstacle seems to be credible, since in both the 2D and 3D cases it vanishes as the frequency decreases. However, in the case of an acoustically soft obstacle the scattered amplitude does not vanish for infinitely low k, which looks paradoxical. But the clue to this paradox is rather simple. The constructed scattered amplitude is obtained under two assumptions: 1) small k and 2) large distance $|x|$, which may under certain conditions contradict each other, so the result obtained is not uniform with respect to these two asymptotic parameters. Actually, in a far zone the only dimensionless parameter is $k|x|$, where the first factor is small and the second one is large, so we cannot *a priori* predict whether their product is small or large. Mathematically, this is equivalent to the alternative notation of Eq. (2.109) in the form

$$|p^{\text{sc}}(\alpha)| \sim \frac{ak}{|kx|} + O(k^2), \tag{2.110}$$

and it becomes now clear that the scattered amplitude decreases with decreasing dimensionless parameter ak. Physically, this paradox is connected with the fact that a smaller frequency implies a longer wave. As regards the far-field approximation, it should be remembered that this implies also that the distance $|x|$ is much greater than the (long) wavelength.

In the final part of the present discussion we would like to notice once again that many examples with applications to scattering by various canonical shapes can be found in Bowman et al. (1987); and in Felsen and Marcuvitz (1973).

2.6. Asymptotic Character of the Kirchhoff Physical Diffraction Theory

As we already mentioned, the BIE method in the generic case of arbitrary obstacle's shape can be solved by one or another numerical method. In regular cases (for instance, Fredholm equations of the second kind) a standard direct numerical treatment (see Section 1.5) is quite applicable. The application of numerical methods in irregular problems will be discussed in Chapter 9. We saw above that canonical geometries admit analytical analysis at low frequencies. Another extreme case where the problem can be studied analytically refers to high-frequency (or, short-wavelength) diffraction by convex obstacles. The foundations of the theory were created by Kirchhoff, and his theory is called the *physical diffraction theory*. We first give heuristic foundations of the theory and then prove its asymptotic character.

There are a number of different formulations of the Kirchhoff theory (see Hönl et al., 1961). The key points can be understood from Fig. 2.6.

If there is a plane acoustic wave $p^{\text{inc}}(x) = e^{ik(m \cdot x)}$ incident to a convex obstacle, then the boundary of the obstacle is naturally divided into "light" (l^+) and "shadow" (l^-) zones, which for simplicity are shown for a 2D problem in Fig. 2.6. The light zone l^+ is defined as one containing the boundary points with $\mathbf{n} \cdot \mathbf{m} < 0$ and the shadow zone l^- contains the boundary points with $\mathbf{n} \cdot \mathbf{m} > 0$, where $\mathbf{n}$ is the outward normal to the obstacle contour and

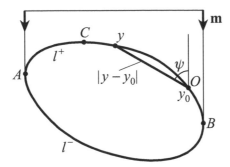

Figure 2.6. Incidence of a high-frequency acoustic wave onto a convex obstacle

m is the unit vector defining the incident wave direction. The structure of the solution is absolutely different in the light and in the shadow. Asymptotically, as $k \to \infty$, the total pressure $p(x) = p^{\text{inc}}(x) + p^{\text{sc}}(x)$ in the shadow zone is zero: $p(x) = 0$, $x \in l^-$. The solution on l^+ is more complex. The Kirchhoff idea was founded on the heuristic assumption that at high frequencies the surface becomes locally plane in the neighborhood of any point $x_0 \in l^+$, and so $p(x_0)$ can be asymptotically obtained from the solution to the problem on the reflection of a plane incident wave $p^{\text{inc}}(x)$ from a plane tangent to the boundary contour at the point x_0. To be more specific, let us assume that the contour l is acoustically hard: $(\partial p / \partial n)|_l = 0$, where $p = p^{\text{inc}} + p^{\text{sc}}$ is the total acoustic wave field. Then the solution to the problem on the reflection from an infinite plane reflector is easily constructed (see, for example, Brekhovskikh, 1980). It can be verified that the total wave field,

$$p_\infty(x) = p^{\text{inc}}(x) + p^{\text{sc}}_\infty(x) = e^{ik}(m \cdot x) + p^{\text{sc}}_\infty(x) \text{ is } p_\infty(x) = e^{ik(m \cdot x)} + e^{ik[m - 2(m \cdot n)n] \cdot x} \quad (2.111)$$

if local Cartesian coordinate axes coincide with the (τ, n) directions (τ is the unit tangential vector and n is the unit outward normal), and the origin is placed at the point x_0.

We can see from Eq. (2.111) that

$$p_\infty(x_0) = p_\infty(0) = 2 = 2p^{\text{inc}}(0) = 2p^{\text{inc}}(x_0), \quad (2.112)$$

and consequently the Kirchhoff approximation implies

$$p(x_0) \approx 2p^{\text{inc}}(x_0), \qquad k \to \infty. \quad (2.113)$$

The Kirchhoff–Helmholtz integral representation (2.43) with the Kirchhoff approximation (2.113) now contains both functions $p|_l$ and $(\partial p / \partial n)|_l$ to be known, and the diffraction problem is now reduced to the calculation of some quadratures.

For a long time there were numerous heated debates whether this theory is asymptotic at $k \to \infty$ or not. Recently it was proved that the leading asymptotic term of the exact solution coincides with Kirchhoff's prediction (see, for example, Taylor, 1981), but the proofs required too abstract mathematical instruments. We give here a simple and graphical proof. It is based upon the basic boundary integral equation (see Section 2.2). The only restriction is the established fact that the BIEs studied in this chapter are uniquely solvable for all $k > 0$, except a countable set of values $\{k_n\}$ ($k_n \to \infty$, $n \to \infty$) corresponding to the eigenvalues of the corresponding interior boundary value problem. So, strictly speaking, the asymptotic property of Kirchhoff's theory stated by the following theorem is valid only for regular high-frequency values of k. A more refined analysis should be carried out in order to prove that Kirchhoff's solution is asymptotically valid also for irregular values of k ($k \to \infty$), i.e., Kirchhoff's approximation represents the leading asymptotic term of the solution for all large k. Assuming that $p(x_0) = o(1)$, $k \to \infty$, $x_0 \in l^-$, we prove the following

THEOREM. *Let the boundary contour l be smooth, convex, and acoustically hard, and $x_0 \in l^+$. Then*

$$p(x_0) = 2p^{\text{inc}}(x_0) + o(1), \quad k \to \infty \quad (k \neq k_n). \tag{2.114}$$

Proof. If $k \neq k_n$, then the boundary integral equation in the case of the Neumann boundary condition $(\partial p/\partial n) = 0$, $x \to l$, and with $p(x_0) \equiv 0(1)$, $k \to \infty$, is asymptotically equivalent to

$$p(y_0) - 2 \int_{l^+} p(y) \frac{\partial \Phi(|y_0 - y|)}{\partial n_y} \, dl_y = 2p^{\text{inc}}(y_0) + o(1), \qquad y_0 \to l^+, \tag{2.115}$$

being uniquely solvable. Here $\Phi(r) = (i/4)H_0^{(1)}(kr)$, $r = |y - y_0|$, is the Green's function for full 2D space. Moreover, for the considered k, the integral operator of the left-hand side, $I - G$, has a bonded continuous operator $(I - G)^{-1}$ in the Banach space $C(l)$.

Let $p(y) = 2p^{\text{inc}}(y) + \varphi(y)$, $y \in l^+$, where $\varphi(y)$ is a new unknown function. Then equation (2.115) takes the form

$$\varphi(y_0) - 2 \int_{l^+} \varphi(y) \frac{\partial \Phi(|y_0 - y|)}{\partial n_y} \, dl_y = 4 \int_{l^+} p^{\text{inc}} \frac{\partial \Phi(|y_0 - y|)}{\partial n_y} \, dl_y + o(1), \quad y_0 \in l. \tag{2.116}$$

Let us prove that

$$\lim_{k \to \infty} \int_{l^+} p^{\text{inc}}(y) \frac{\partial \Phi(|y_0 - y|)}{\partial n_y} \, dl_y = O(1), \quad y_0 \in l^+ = AB, \quad k \neq k_n. \tag{2.117}$$

Indeed,

$$\int_{l^+} p^{\text{inc}}(y) \frac{\partial \Phi(|y_0 - y|)}{\partial n_y} \, dl_y = \int_{l^+ \setminus l_\varepsilon^\circ} p^{\text{inc}}(y) \frac{\partial \Phi(|y_0 - y|)}{\partial n_y} \, dl_y + \int_{l_\varepsilon^\circ} p^{\text{inc}}(y) \frac{\partial \Phi(|y_0 - y|)}{\partial n_y} \, dl_y, \tag{2.118}$$

where l_ε° is a small ε-neighborhood of the point x_0. Since the integrand is continuous (see Section 2.1), the second integral can be made arbitrarily small.

Further,

$$\Phi(|y_0 - y|) = \frac{i}{4} H_0^{(1)}(k|y_0 - y|) = \frac{i}{4} H_0^{(1)}(kr)$$

$$\sim \quad \frac{\partial \Phi}{\partial n_y} = -\frac{ik}{4} H_1^{(1)}(kr) \frac{\partial r}{\partial n_y} \tag{2.119}$$

$$= -\frac{ik}{4} \sqrt{\frac{2}{\pi kr}} \frac{e^{i(kr - \pi/4)}}{\sqrt{kr}} \cos(r, n_y) \left[1 + \left(\frac{1}{kr} \right) \right], \quad r = |y_0 - y|,$$

and this representation can be used when estimating the first integral in (2.118). Hence,

$$I = \int_{l^+} p^{\text{inc}}(y) \frac{\partial \Phi(|y_0 - y|)}{\partial n_y} \, dl_y$$

$$= e^{-3\pi i/4} \frac{\sqrt{k}}{2\sqrt{2\pi}} \int_{l^+ \setminus l_\varepsilon^\circ} e^{ik(m \cdot y + |y - y_0|)} \cos(y - y_0, n_y) \frac{dl_y}{\sqrt{|y_0 - y|}} + O\left(\frac{1}{\sqrt{k}} \right), \tag{2.120}$$

since the integrand is integrable and so the contribution of the l_ε°-integral is small.

To estimate (2.120) we have to study the behavior of the phase function (see the stationary phase method in Section 1.4)

$$S(y) = (m \cdot y) + |y - y_0|, \tag{2.121}$$

since any stationary point makes a contribution of the order of $O(1/\sqrt{k})$, which renders I nonsmall. Otherwise, the l^+-integral is at least $O(1/k)$, so I is small.

Let us divide the integral (2.120) into three parts (see Fig. 2.6): $\int_{l^+} = \int_0^C + \int_C^A + \int_0^B$, where the point $O = y_0$ is the origin of the chosen Cartesian coordinate system, A and B are points that separate the light zone l^+ from the shadow zone l^-, and C is the summit of the contour in direction towards the incident wave. It is obvious that the phase $S(y) = (m \cdot y) + |y|$ monotonically increases over the arcs CA and OB, since both terms increase. Therefore, there is no stationary point. By contrast, when $y \in OC$, the first term decreases and the second term increases so that, strictly speaking, there is a chance that a stationary point y^* may arise on the arc OC. However, the phase S here is

$$S(y) = (m \cdot y) + |y| = -|y| \cos \psi + |y| = |y|(1 - \cos \psi), \qquad (2.122)$$

where ψ is the angle between y and the direction towards the incident wave: $\psi = \widehat{y, -\mathbf{m}}$. It is clear that $\psi(y)$ is a monotonically increasing function if $y \in OC$, and $0 \le \psi \le \pi$. Since $\cos \psi$ over this variation range of ψ is a decreasing function of ψ, so $\cos \psi$ decreases monotonically, but $(1 - \cos \psi)$ increases monotonically. Since $S(y)$ is the product of two positive and monotonically increasing functions, $|y|$ and $[1 - \cos \psi(y)]$, it also increases monotonically, and hence there is no stationary point on the arc OC. This proves that the integral I in Eq. (2.120) admits the estimate $I = O(1/\sqrt{k})$, $k \to \infty$, uniformly over $y_0 \in l^+$, i.e., equation (2.116) may be rewritten in the form

$$\varphi(y_0) - 2 \int_{l^+} \varphi(y) \frac{\partial \Phi(|y_0 - y|)}{\partial n_y} \, dl_y = f(y_0), \qquad y_0 \in l^+, \qquad (2.123)$$

$$f(y_0) = 4I(y_0) + o(1) = O\left(\frac{1}{\sqrt{k}}\right) + o(1) = o(1). \qquad (2.124)$$

The final part of the proof is based on the invertibility of the integral operator on the left-hand side of Eq. (2.123) and on the fact that the right-hand side is asymptotically small, uniformly over $y_0 \in l^+$. Obtaining precise asymptotic estimates is not so simple here and a detailed residue analysis can be found in (Taylor, 1981).

Example. High-frequency scattering pattern of a round disk.

Let a plane incident acoustic wave $p^{\text{inc}} = e^{ikx_1}$ be incident upon an acoustically round disk (2D problem) of radius a. As follows from Eq. (2.104), if we consider the far-field scattering with the observation point $x = (R \cos \alpha, R \sin \alpha)$, we obtain

$$|p^{\text{sc}}| \sim \sqrt{\frac{k}{|x|}} \left| \int_l p(y) \cos(n_y, x) e^{-ik(x \cdot y)/|x|} dl_y \right|. \qquad (2.125)$$

If we substitute into this integral the boundary value of the total acoustic pressure predicted by the Kirchhoff theory, then we notice that

$$p(y) = p(\psi) = 2p^{\text{inc}} = 2e^{ika \cos \psi}, \qquad \frac{\pi}{2} < \psi < \frac{3}{2}\pi,$$

$$p(\psi) = 0, \quad |\psi| < \frac{\pi}{2}, \qquad y = (a \cos \psi, \, a \sin \psi). \qquad (2.126)$$

Let us also recall (see Section 2.5) that $\cos(n_y, x) = \cos(\psi - \alpha)$, $(x \cdot y)/|x| = a\cos(\psi - \alpha)$, so Eq. (2.125) is equivalent to

$$|p^{\text{sc}}(\alpha)| \sim \sqrt{\frac{k}{|x|}} \left| 2a \int_{\pi/2}^{3\pi/2} \cos(\psi - \alpha) \, e^{ika \cos \psi} \, e^{-ika \cos(\psi - \alpha)} \, d\psi \right|$$

$$\sim a\sqrt{\frac{k}{|x|}} \left| \int_{\pi/2}^{3\pi/2} \cos(\psi - \alpha) \, e^{-2ika \sin(\alpha/2) \sin(\psi - \alpha/2)} \, d\psi \right|. \qquad (2.127)$$

We calculate the integral (2.127) by the stationary phase method. If $ka \gg 1$, then the phase $S(\psi) = -2a\sin(\alpha/2)\sin(\psi - \alpha/2)$ has the only stationary point $\psi = (\pi + \alpha)/2$ on the interval $\pi/2 < \psi < 3\pi/2$ for any $0 < \alpha \leq \pi$. Moreover, $S(\psi^*) = -2a\sin(\alpha/2)$, $\cos(\psi^* - \alpha) = \sin(\alpha/2)$, $S''(\psi^*) = 2a\sin(\alpha/2)$, so the stationary phase method (Section 1.4) leads to the following asymptotic expression for $|p^{sc}(\alpha)|$:

$$|p^{sc}(\alpha)| \sim a\sqrt{\frac{k}{|x|}}\,A(\alpha), \quad (0 < \alpha \leq \pi), \quad A(\alpha) \sim \frac{|\sin(\alpha/2)|}{ak}, \quad k \to \infty. \qquad (2.128)$$

Due to the evident evenness, this formula can be symmetrically extended to the interval $-\pi \leq \alpha < 0$.

The only range where the representation (2.128) is not valid is related to the case when the argument of the exponential function in (2.127) is not large for large ka. This takes place when $ka\sin(\alpha/2) = O(1)$, i.e., $\alpha = O(1/ka)$, $k \to \infty$. However, this special case admits a different estimate of the integral (2.127):

$$|p^{sc}(\alpha)| \sim a\sqrt{\frac{k}{|x|}}\left| \int_{\pi/2}^{3\pi/2} \cos\psi\, e^{-2ika\sin(\alpha/2)\sin\psi}\,d\psi \right|$$

$$= a\sqrt{\frac{k}{|x|}}\left| \int_{\pi/2}^{3\pi/2} e^{-2ika\sin(\alpha/2)\sin\psi}\,d(\sin\psi) \right| \qquad (2.129)$$

$$= a\sqrt{\frac{k}{|x|}}\,\frac{\sin[2ka\sin(\alpha/2)]}{2ka\sin(\alpha/2)}.$$

Hence it follows that

$$A(\alpha) = \frac{\sin[2ka\sin(\alpha/2)]}{2ka\sin(\alpha/2)}. \qquad (2.130)$$

As you can see, $A(\alpha) = 1$ for $\alpha = 0$, then outside of a small neighborhood of this value $\alpha = 0$ the scattering pattern abates and becomes a rapidly oscillating function with an amplitude of the order of $O(1/k)$. This sharp peak in the direction of the incident wave is called the *shadow formed leaf*.

Helpful remarks

$1°$. Evidently, the basic ideas of Kirchhoff's theory can be simply extended to a nonplane incident wave, since any wave at high frequency can be locally considered to be plane.

$2°$. You may ask two very important and interesting questions: 1) Why does the proved theorem require convexity of the boundary? 2) Why the same arguments cannot be applied for $x_0 \in l^-$? In fact, both the condition of convexity and that of belonging of the point x_0 to the right side of the boundary are essential. Indeed, we can insure that the angle $\psi(y)$ monotonically increases along the arc x_0C only for convex shapes. On the other hand, if you undertake the same analysis as in the theorem, for the case $x_0 \in l^-$, you may see that $\psi(y)$ is monotonically decreasing, and hence $-\cos\psi$ ($0 < \psi < 180°$) is surely monotonically decreasing. Therefore, the phase $S(y) = |y| - |y|\cos\psi$ is the sum of a monotonically increasing ($|y|$) and monotonically decreasing functions. There is a good chance for a stationary point, where $S' = 0$, to appear. This idea is clearly seen by the example of a circle of radius a. Here $S(y) = S(\theta) = |y|(1 - \sin(\theta/2) = 2a\sin(\theta/2)\,[1 - \sin(\theta/2)]$, whose stationary point is $\theta^* = 60°$, as follows from the equation $S'(\theta^*) = 0$.

Chapter 3

Wave Fields in a Layer of Constant Thickness

3.1. Wave Operator in Acoustic Layer: Mode Expansion, Homogeneous and Inhomogeneous Waves

The present chapter will be devoted to the analytic properties of wave fields in an acoustic layer of constant thickness h. Since, for the sake of simplicity, we restrict our consideration to a 2D problem, we slightly change notations and choose a Cartesian coordinate system (x, y) so that its horizontal axis coincides with the lower boundary of the layer $y = 0$, and the upper boundary surface is $y = h$. The considered geometry represents a problems where the Fourier transform (Section 1.1) brilliantly demonstrates its high power.

Let us construct a Green's function in the considered acoustic layer. As noted in the last paragraph of Section 2.1, it suffices to solve the nonhomogeneous Helmholtz equation with a point source represented by Dirac's delta function on the right-hand side and placed at (x_0, y_0):

$$\Delta\Phi(x, y) + k^2\Phi(x, y) = \delta(x - x_0)\,\delta(y - y_0). \tag{3.1}$$

If we apply the Fourier transform with respect to the variable x, then by using Eq. (1.3) and the properties of the delta function (1.62), we arrive at the ordinary differential equation

$$\frac{d^2}{dy^2}\,\widetilde{\Phi}(s, y) - (s^2 - k^2)\,\widetilde{\Phi}(s, y) = e^{isx_0}\,\delta(y - y_0), \tag{3.2}$$

where s is a parameter of the Fourier transform and the tilde denotes the Fourier image of the sought Green's function. The solution of this equation may be expressed as the sum of the general solution to the homogeneous equation (3.2) and a particular solution of the full equation:

$$\widetilde{\Phi}(s, y) = A(s)\,e^{\gamma y} + B(s)\,e^{-\gamma y} + \widetilde{\Phi}_{\text{part}}(s, y), \qquad \gamma = \gamma(s) = \sqrt{s^2 - k^2}, \tag{3.3}$$

where $A(s)$ and $B(s)$ are two arbitrary constants, which typically arise when solving ordinary differential equations (here, of the second order).

To find $\widetilde{\Phi}_{\text{part}}$, we notice that a particular solution may be chosen as one for the whole $\mathbb{R}^2$ space, since it may not satisfy any boundary conditions on the faces of the layer. So, let us apply again the Fourier transform to Eq. (3.2), this time with respect to y. Then we obtain

$$\widetilde{\Phi}_{\text{part}}^{*}(s, \alpha) = \frac{e^{i(sx_0 + \alpha y_0)}}{\alpha^2 + \gamma^2}, \tag{3.4}$$

where the superscript $*$ denotes a Fourier image taken with respect to the variable y. Now, the application of the inverse transformation to Eq. (3.4) (see formula (1.20)) gives

$$\widehat{\Phi}_{\text{part}}(s, y) = \frac{e^{isx_0}}{2\pi}\int_{-\infty}^{\infty}\frac{e^{i\alpha(y_0 - y)}}{\alpha^2 + \gamma^2}\,d\alpha = \frac{e^{isx_0}\,e^{-|y_0 - y|\gamma}}{2\gamma} \tag{3.5}$$

(the modulus $|y_0 - y|$ can replace the difference since the denominator is even in α). Hence, the general solution (3.3) reads

$$\widetilde{\Phi}(s, y) = A(s)\, e^{\gamma y} + B(s)\, e^{-\gamma y} + \frac{e^{isx_0}\, e^{-|y_0 - y|\gamma}}{2\gamma}. \tag{3.6}$$

Now, the functions A and B should be determined from the boundary conditions on the faces of the layer $y = 0$ and $y = h$. These may be of various type, and below we quote the most important cases with solutions. A solution is given by the inverse Fourier transform and has the form

$$\Phi(x, y) = \frac{1}{2\pi} \int_{-\infty}^{\infty} L(s, y)\, e^{-is(x - x_0)}\, ds. \tag{3.7}$$

1) The acoustic layer is enclosed between two rigid plates. Then both boundary surfaces are acoustically hard, therefore the boundary conditions and a relevant function $L(s, y)$ are

$$\left.\frac{\partial p}{\partial y}\right|_{y=0} = 0, \quad \left.\frac{\partial p}{\partial y}\right|_{y=h} = 0, \quad L_1(s, y) = \frac{\cosh[\gamma(h - |y - y_0|)] + \cosh[\gamma(h - y - y_0)]}{2\gamma \sinh(\gamma h)}. \tag{3.8}$$

2) Both boundary faces are free of applied pressure. Then

$$p\big|_{y=0} = 0, \quad p\big|_{y=h} = 0, \quad L_2(s, y) = \frac{\cosh[\gamma(h - |y - y_0|)] - \cosh[\gamma(h - y - y_0)]}{2\gamma \sinh(\gamma h)}. \tag{3.9}$$

3) One boundary of the layer is in contact with an acoustically hard plate and the other one is free of applied load. In this case

$$\left.\frac{\partial p}{\partial y}\right|_{y=0} = 0, \quad p\big|_{y=h} = 0, \quad L_3(s, y) = \frac{\sinh[\gamma(h - |y - y_0|)] + \sinh[\gamma(h - y - y_0)]}{2\gamma \cosh(\gamma h)}. \tag{3.10}$$

What can be concluded from these expressions? As can be seen, a branching function $\gamma(s)$ is present in these structures (cf. Section 1.1). However, the most important conclusion is that all these functions $L(s, y)$ are meromorphic with respect to s. This means that they are analytic, since their Laurent series about the branching points $s = \pm k$ contain only even powers of γ (cf. Eq. (1.18)), except no more than a countable set of poles.

Let us note that the poles of the quoted functions (3.8)–(3.10) can be expressed explicitly. They coincide with zeros of the respective hyperbolic sine or cosine in the denominator and are given by

$$\gamma \sinh(\gamma h) = 0 \quad \sim \quad s = \pm s_m^{(1,2)},$$

$$s_m^{(1,2)} = \sqrt{k^2 - \left(\frac{\pi m}{h}\right)^2} = i\sqrt{\left(\frac{\pi m}{h}\right)^2 - k^2}, \quad m = 0, 1, 2, \ldots, \tag{3.11}$$

in the first two cases, and

$$\cosh(\gamma h) = 0 \quad \sim \quad s = \pm s_m^{(3)},$$

$$s_m^{(3)} = \sqrt{k^2 - \left[\frac{\pi(m + 1/2)}{h}\right]^2} = i\sqrt{\left[\frac{\pi(m + 1/2)}{h}\right]^2 - k^2}, \qquad m = 0, 1, 2, \ldots, \tag{3.12}$$

in the third case.

It is obvious that for relatively low frequencies (small values of the wave number k) there are no real poles, and hence all singularities lie outside of the integration contour.

However, as the frequency increases, more and more real poles hit the real axis $\text{Im}(s) = 0$. A natural question arises: In which sense should integrals of the type (3.7) be correctly treated if the integration contour crosses poles of the meromorphic integrand?

It will be shown in the next section that the integration path should bend a little around the positive poles from below and the negative ones from above. Then we can apply integration with residues at simple poles s_n situated in the upper complex half-plane (see the generalized Jordan lemma and the final example in Section 1.1). In order to have the possibility of arranging deformation of the integration contour in the upper half-plane (which is more often applied in the available literature) instead of the lower one, we note that all the functions $L(s, y)$ in (3.8)–(3.10) are even with respect to s and hence we can write the positive argument $is|x - x_0|$ of the exponential function in Eq. (3.7) instead of the negative one $-is|x - x_0|$:

$$\frac{1}{2\pi} \int_{-\infty}^{\infty} L(s, y) e^{is|x-x_0|} ds = \frac{2\pi i}{2\pi} \sum_{m=0}^{\infty} \text{Res}\left[L(s, y) e^{is|x-x_0|}, is_m\right]. \qquad (3.13)$$

This yields

$$\Phi^{(1,2)}(x, y) = \sum_{m=0}^{\infty} \frac{\cos[\pi m |y - y_0|/h] \pm \cos[\pi m(y + y_0)/h]}{2\sqrt{(\pi m)^2 - (kh)^2}} (-1)^m e^{is_m^{(1,2)}|x-x_0|},$$

$$\Phi^{(3)}(x, y) = \sum_{m=0}^{\infty} \frac{\sin[\pi(m + 1/2)|y - y_0|/h] + \sin[\pi(m + 1/2)(y + y_0)/h]}{2\sqrt{\pi(m + 1/2)^2 - (kh)^2}} \qquad (3.14)$$

$$\times (-1)^m e^{is_m^{(3)}|x-x_0|},$$

which is called a *mode expansion*. A structure that possesses mode solutions is called a *waveguide*. We can thus conclude that a layer of constant thickness represents a typical waveguide.

It is clear from the expansions (3.14) and Eqs. (3.11), (3.12) that if the frequency is less than a certain positive quantity, then only waves exponentially decaying with the distance $|x - x_0|$ can exist in the layer (except that with $m = 0$ in symmetric cases 1 and 2; see Eq. (3.11)). Such decaying waves are called *inhomogeneous mode waves*. Otherwise, if the wave number is large enough, there is a finite number of waves that can propagate far along the length of the layer. Such nondecaying waves are called *homogeneous mode waves*. They appear to be some plane waves for any fixed $0 < y < h$ if we write them in their complete form with the time-dependent factor,

$$\Phi(x, y, t) = A(y) e^{i(s_m|x-x_0|-\omega t)}, \qquad (3.15)$$

where $s_m = s_m^{(1,2)}$ or $s_m = s_m^{(3)}$, with some velocity of propagation, which for the mth mode is equal to $v_g = \omega/s_m$, or

$$v_g^{(1,2)} = \frac{\omega}{s_m^{(1,2)}} = c\left[\sqrt{1 - \left(\frac{\pi m c}{\omega h}\right)^2}\right]^{-1}, \quad v_g^{(3)} = \frac{\omega}{s_m^{(3)}} = c\left[\sqrt{1 - \left(\frac{\pi(m + 1/2) c}{\omega h}\right)^2}\right]^{-1}. \qquad (3.16)$$

This quantity is called the *group velocity* of the mth mode. Obviously, the group velocity is always greater than the wave speed in the medium: $v_g \geq c$. Only in symmetric cases, with $m = 0$, the group velocity is equal to c: $v_g^{(1,2)} = c$ ($m = 0$).

It should also be noted that all constructed mode solutions satisfy the radiation condition, since for $x \to +\infty$ they behave like a wave radiated to the right and for $x \to -\infty$ like that radiated to the left:

$$\Phi(x, y, t) = B(y) e^{i(s_m x - \omega t)}, \quad x \to +\infty,$$

$$\Phi(x, y, t) = D(y) e^{-i(s_m x + \omega t)}, \quad x \to -\infty, \qquad (3.17)$$

which is directly seen by considering the respective wave front set (cf. Section 1.9).

Helpful remarks

$1°$. It is conventionally recognized that wave problems in closed structures, as in a layer (possibly with some discontinuities), can generate only meromorphic symbolic functions $L(s, y)$, and those in open structures (like a half-plane, possibly containing some discontinuities) always lead only to certain functions with branching points. We do not know whether this statement has been rigorously proved anywhere, but we know no example that refutes this statement.

$2°$. The powerful methods of the theory of complex-valued analytic functions (see Section 1.1) can be understood well by the example of the problem studied in the present section. An explicit-form solution was first obtained as a Fourier integral, whose numerical implementation is too hard, since the integrand is a rapidly oscillating function with some singularities (poles) on the real axis. Besides, the interval of integration is infinite. By applying the residue theory combined with the Jordan lemma, we reduce this integral to an equivalent form represented by exponentially convergent series, which makes the problem of calculation of the solution much easier.

3.2. Principles of Selection of Unique Solution in Unbounded Domain

It is quite natural that, when solving the wave (or Helmholtz) equation in any unbounded domain, it is necessary to set some boundary condition at infinity in order to select a unique solution. Previously, we already treated the radiation condition in Section 2.2 when studying the exterior diffraction problem in the full unbounded space $\mathbb{R}^n$ ($n = 2, 3$), but it has not yet been formulated as a strict mathematical condition.

Sommerfeld's radiation condition

It is conventionally accepted in the literature that a mathematically correct form of the radiation condition, where the time-dependent factor is taken in the form $\exp(-i\omega t)$, is given by the following pair of asymptotic identities:

$$p(x) = O\left(\frac{1}{R}\right), \quad \frac{\partial p}{\partial R} - ikp = o\left(\frac{1}{R}\right), \quad R = |x| \to \infty \quad (x \in \mathbb{R}^3), \tag{3.18}$$

in the 3D case, and

$$p(x) = O\left(\frac{1}{\sqrt{R}}\right), \quad \frac{\partial p}{\partial R} - ikp = o\left(\frac{1}{\sqrt{R}}\right) \quad R = |x| \to \infty \quad (x \in \mathbb{R}^2), \tag{3.19}$$

in the 2D case.

Indeed, if these two conditions hold, then, by using Green's formula as in Section 2.2, we easily prove that Eq. (2.49) is valid, and consequently the Kirchhoff–Helmholtz formula (2.43) gives a unique solution to the exterior diffraction problem. This conclusion is based on the evident estimates valid if both functions $u(y)$ and $\Phi(|x-y|)$ satisfy conditions (3.19), since in the considered 2D case

$$u(y)\frac{\partial \Phi(|x-y|)}{\partial n_y} - \Phi(|x-y|)\frac{\partial u(y)}{\partial n_y} = u(y)\frac{\partial \Phi}{\partial R_y} - \Phi\frac{\partial u}{\partial R_y}$$

$$= u(y)\left[ik\Phi + o\left(\frac{1}{\sqrt{R}}\right)\right] - \Phi\left[iku + o\left(\frac{1}{\sqrt{R}}\right)\right] = o\left(\frac{1}{R}\right), \tag{3.20}$$

and hence

$$\int_{l_R} \left[u(y) \frac{\partial \Phi(|x-y|)}{\partial n_y} - \Phi(|x-y|) \frac{\partial u(y)}{\partial n_y} \right] dl_y \sim 2\pi R o\left(\frac{1}{R}\right) \to 0 \text{ as } R \to \infty. \quad (3.21)$$

Unfortunately, it cannot be guaranteed that conditions (3.18), (3.19) are universal, in the sense that they provide selection of a unique solution for arbitrary unbounded domain different from the whole space. In particular, it is not clear how one may treat integral representations for a mode solution in a layer (3.7)–(3.10) when the integrand has simple poles on the real axis.

An inexperienced researcher may try to treat solution (3.7) in the layer, in the case where there is a finite number of simple poles on the real axis, as a singular Cauchy-type principal value (see Section 1.7). In doing so, the explicit expression (3.7) can be again calculated by adding an (upper half-plane) semi-circle of a large radius R, in order to make the integration contour closed. Then by applying the generalized Jordan lemma (Section 1.1) and the boundary values of Cauchy-type singular integrals of analytic integrand (see Eq. (1.155)), we can conclude that the function $\Phi(x, y)$ in Eq. (3.7) is

$$\Phi(x,y) = \frac{1}{2\pi} \int_{-\infty}^{\infty} L(s,y) e^{is|x-x_0|} ds = i \sum \text{Res}[L(s,y) e^{is|x-x_0|}, is_m]$$

$$+ \frac{i}{2} \sum_{m \le N^*} \text{Res}[L(s,y) e^{is|x-x_0|}, s_m] + \frac{i}{2} \sum_{m \le N^*} \text{Res}[L(s,y) e^{is|x-x_0|}, -s_m], \quad (3.22)$$

where the first sum is taken over all imaginary poles in the upper complex half-plane, the second sum is related to N^* positive poles, and the third one, to N^* negative poles.

It becomes now evident that the first sum contains inhomogeneous mode waves decaying with distance, the second sum contains homogeneous waves satisfying Sommerfeld's radiation condition, and the third sum generates homogeneous waves that contradict the radiation condition. Therefore, the integral (3.7) cannot be considered as a principal Cauchy value.

A correct approach should be founded on the evident observation that the residues at positive poles are in accordance with the radiation condition, and the residues at negative poles yield mode waves arriving from infinity, thus running into a contradiction to Sommerfeld's principle. This leads to the clear understanding that, when arranging a large closed semi-circle in the upper complex half-plane, all negative poles must be outside this circle, and the positive ones may lie inside the circle. The natural symmetry in the disposition of poles (even in the integrand!) allows us to come to the following conclusion: in order to satisfy Sommerfeld's radiation condition, the integration contour should slightly bend around the positive poles from below and the negative ones from above.

In order to establish a more physically justified principle for selecting the unique solution, some authors proposed alternative conditions at infinity, and it is very interesting to investigate, to what extent they conform to the radiation condition.

Principle of extremely low absorption (Ignatowsky's principle)

It naturally arises in electromagnetic theory (see Section 1.9). If in Eq. (1.219) the parameter σ is a small positive constant ($0 < \sigma \ll 1$), then the wave number $\tilde{k} = k + i\delta$, $k = \omega\sqrt{\mu\varepsilon}/c$, $\delta = \sqrt{\omega\sigma}/(2c)$, $0 < \delta \ll 1$, is a complex-valued quantity with positive imaginary part. This causes a finite number of simple poles s_n, $n = 1, 2, \ldots, N^*$, in Eqs. (3.11), (3.12), that were real-valued for real k to become complex-valued and slightly shifted off the real axis. Moreover, positive poles shift a little up, and negative ones shift a little down, so that all mode waves (3.13) decay with distance due to absorption. The principle of extremely low absorption (first proposed by Ignatowsky, see Tikhonov and Samarsky, 1977) states that a correct solution of the problem in an ideal (i.e., nonabsorbent) medium is the limit of the corresponding solution in the absorbent medium as $\delta \to +0$.

In ideal hydroacoustic and elastic cases, where there is no parameter σ, this principle may be used too, by introducing an artificially added absorption, which in the final structure of the solution should be set equal to zero. For the problem considered in the previous section this principle leads to the same solution as the radiation condition. Indeed, if with a low absorption the positive poles move up and the negative ones down, then the natural integration contour of the inverse Fourier transform $(-\infty, \infty)$ passes below the positive poles and above the negative ones. With the absorption tending to zero, this results in the same relative disposition as in Sommerfeld's principle.

Energy radiation condition (Mandelshtam's principle)

This principle asserts that the energy flux taken for the period of oscillations and flowing out to the outside through an appropriate cross-section of the domain must be positive. Let us calculate the energy flux through a vertical cross-section of the layer $x = x^*$. To this end, we first derive a general formula for energy power W averaged over the period of oscillations and considered on a small area with normal $\mathbf{n}$:

$$W = \frac{\omega}{2\pi} \int_0^{2\pi/\omega} p v_n \, dt. \tag{3.23}$$

Then we have

$$
\begin{aligned}
p(x, y, t) &= \mathrm{Re}[e^{-i\omega t} \tilde{p}(x, y)] = \tilde{p}^{\,\mathrm{re}} \cos \omega t + \tilde{p}^{\,\mathrm{im}} \sin \omega t, \\
\frac{\partial v_n(x, y, t)}{\partial t} &= -\frac{1}{\rho_0} \, \mathrm{grad}\, p, \ \sim \ \frac{\partial v_n}{\partial t} = -\frac{1}{\rho_0} \left(\frac{\partial \tilde{p}^{\,\mathrm{re}}}{\partial n} \cos \omega t + \frac{\partial \tilde{p}^{\,\mathrm{im}}}{\partial n} \sin \omega t \right), \\
\sim \ v_n &= \frac{1}{\rho_0 \omega} \left(-\frac{\partial \tilde{p}^{\,\mathrm{re}}}{\partial n} \sin \omega t + \frac{\partial \tilde{p}^{\,\mathrm{im}}}{\partial n} \cos \omega t \right),
\end{aligned}
\tag{3.24}
$$

so the substitution of relation (3.24) into Eq. (3.23) and subsequent integration determines the averaged energy in the form

$$W = \frac{1}{2\rho_0 \omega} \left(\tilde{p}^{\,\mathrm{re}} \frac{\partial \tilde{p}^{\,\mathrm{im}}}{\partial n} - \tilde{p}^{\,\mathrm{im}} \frac{\partial \tilde{p}^{\,\mathrm{re}}}{\partial n} \right). \tag{3.25}$$

On the other hand, if we operate in the harmonic regime with the standard representation expressed in terms of complex amplitudes for appropriate functions, namely

$$p(x, y, t) = e^{-i\omega t} \tilde{p}(x, y), \quad \mathbf{v}(x, y, t) = e^{-i\omega t} \tilde{\mathbf{v}}(x, y), \quad \tilde{v}_n = \frac{1}{i\omega\rho_0} \frac{\partial \tilde{p}}{\partial n}, \tag{3.26}$$

then (an asterisk designates here the complex conjugate)

$$
\begin{aligned}
\mathrm{Re}(\tilde{p}^* \tilde{v}_n) &= \mathrm{Re} \left[(\tilde{p}^{\,\mathrm{re}} - i\tilde{p}^{\,\mathrm{im}}) \frac{1}{i\omega\rho_0} \left(\frac{\partial \tilde{p}^{\,\mathrm{re}}}{\partial n} + i \frac{\partial \tilde{p}^{\,\mathrm{im}}}{\partial n} \right) \right] \\
&= \frac{1}{\omega\rho_0} \left(\tilde{p}^{\,\mathrm{re}} \frac{\partial \tilde{p}^{\,\mathrm{im}}}{\partial n} - \tilde{p}^{\,\mathrm{im}} \frac{\partial \tilde{p}^{\,\mathrm{re}}}{\partial n} \right);
\end{aligned}
\tag{3.27}
$$

hence

$$W = \frac{1}{2} \operatorname{Re}(\tilde{p}^* \tilde{v}_n) = \frac{1}{2} \operatorname{Re}(\tilde{p}\, \tilde{v}_n^*). \tag{3.28}$$

Formula (3.28) allows us to calculate the energy flux through the cross-section $x = x^* > x_0$ over the period, taken as the contribution of the mth homogeneous mode. To be more specific, we consider cases 1 and 2 only (tildes are omitted below):

$$p = \frac{\cos[\pi m |y - y_0|/h] \pm \cos[\pi m(y + y_0)/h]}{-2i\sqrt{(kh)^2 - (\pi m)^2}}\, e^{i s_m (x - x_0)},$$

$$v_n = \frac{1}{i\omega\rho_0} \frac{\partial p}{\partial x} = \frac{\cos[\pi m |y - y_0|/h] \pm \cos[\pi m(y + y_0)/h]}{-2i\omega\rho_0\sqrt{(kh)^2 - (\pi m)^2}}\, s_m\, e^{i s_m (x - x_0)}, \tag{3.29}$$

$$s_m = \sqrt{k^2 - \left(\frac{\pi m}{h}\right)^2} > 0 \quad (m = 1, \ldots, N^*),$$

and consequently

$$W = \frac{1}{2} \operatorname{Re} \int_0^h p^* v_n \, dy = \frac{B s_m}{8\omega\rho_0[(kh)^2 - (\pi n)^2]} > 0, \tag{3.30}$$

since

$$B = \int_0^h \left\{ \cos[\pi |y - y_0|/h] \pm \cos[\pi n(y + y_0)/h] \right\}^2 dy > 0. \tag{3.31}$$

The energy flux is thus found to be positive. The case of $x = x^* < x_0$ can be proved likewise.

Therefore, the constructed solution satisfies Mandelshtam's energy radiation principle. In this sense both Sommerfeld's and Mandelshtam's radiation conditions give equivalent formulations of correct boundary conditions at infinity.

Principle of amplitude for extremely large time (Tikhonov–Samarsky principle)

This principle asserts that any harmonic in a time problem where $p(x, y, t) = e^{-i\omega t} \tilde{p}(x, y)$ and

$$\Delta\tilde{p} + k^2\tilde{p} = \delta(x - x_0)\,\delta(y - y_0), \qquad k = \frac{\omega}{c}, \tag{3.32}$$

which is a partial differential equation of the elliptic type and so does not require any initial condition, may be posed in a transient regime,

$$\Delta p - \frac{1}{c^2} \frac{\partial^2 p}{\partial t^2} = \delta(x - x_0)\,\delta(y - y_0)\, e^{-i\omega t}, \tag{3.33}$$

as a nonstationary problem with the trivial initial conditions

$$p(x, y, 0) = \frac{\partial p(x, y, 0)}{\partial t} = 0. \tag{3.34}$$

This new initial Cauchy problem has a unique solution under certain conditions. Then the asymptotics of this transient solution $p(x, y, t)$ as $t \to \infty$ gives a correct solution to the monochromatic problem in hand. Some papers were devoted to the clarification of the relationship between this principle and Sommerfeld's radiation condition (see Tikhonov and Samarsky, 1977). However, it is noteworthy that this principle is hardly applicable in practice, since solving the arising transient problem is a more difficult task compared with the one under consideration, because the dimension of the new problem increases by one due to the introduction of a new variable, t.

Helpful remarks

$1°$. If you take a look at the described principles you may come to the following conclusions.

Sommerfeld's and Mandelshtam's principles give a correct condition at infinity in the case of full acoustic space or acoustic layer of constant thickness. It is not quite clear how they can be applied in more general cases.

Principles of Ignatowsky and Tikhnov–Samarsky are more general; they give a concrete algorithm how to construct the desired solution. However, Tikhonov–Samarsky principle is hardly realizable.

So, we advise that you operate always with the extremely low absorption principle.

$2°$. Nevertheless, there arises a natural question if there is known any problem where the described principles may lead to different results.

In order to clarify this question, let us rewrite, in the case of scalar acoustics, the derived formula for energy flux (3.28) starting from governing equations. Let us consider an $x > x_0$. Then it is obvious that the constructed solution with any mode number m gives a nontrivial solution

$$p(x, y) = p_m(y) \, e^{is\,(x-x_0)}, \qquad s = s_m > 0, \tag{3.35}$$

to the homogeneous value problem for equation (3.1) since the right-hand side there disappears if $x > x_0$. Reformulated for the function $p_m(y)$, it gives a nontrivial solution to the homogeneous boundary value problem for the ordinary differential equation

$$p_m''(y) + \left(\frac{\omega^2}{c^2} - s^2 \right) p_m(y) = 0, \tag{3.36}$$

with a certain type of homogeneous boundary conditions:

$$1) \ p_m'(0) = p_m'(h) = 0; \quad 2) \ p_m(0) = p_m(h) = 0; \quad 3) \ p_m'(0) = 0, \ p_m(h) = 0. \tag{3.37}$$

Let us differentiate Eq. (3.35) with respect to s, taking into account that $s = s_m(\omega) \sim \omega = \omega(s)$ and $p_m(y) = p_m(y, s)$:

$$g''(y) + \left(\frac{\omega^2}{c^2} - s^2 \right) g(y) = 2 \left(s - \frac{\omega}{c^2} \frac{d\omega}{ds} \right) p_m(y), \tag{3.38}$$

where we have denoted $g(y) = dp_m/ds$.

The differentiation of the boundary conditions (3.37), 1)–3), gives

$$1) \ g'(0) = g'(h) = 0; \quad 2) \ g(0) = g(h) = 0; \quad 3) \ g'(0) = 0, \ g(h) = 0. \tag{3.39}$$

Now, let us take the scalar product of the conjugate of Eq. (3.36) with the function $g(y)$ over the interval $0 \le y \le h$ and the analogous scalar product of Eq. (3.38) with the function $p_m^*(y)$ and calculate the difference of the two relations obtained:

$$\int_0^h \left[g''(y)\, p_m^*(y) - p_m^*{''}(y)\, g(y) \right] = 2 \left(s - \frac{\omega}{c^2} \frac{d\omega}{ds} \right) \int_0^h p_m(y)\, p_m^*(y) \, dy. \tag{3.40}$$

Here we consider the mode with positive $s = s_m$ (so that $s^* = s$).

It is easily seen that for any pair of boundary conditions 1)–3) in Eqs. (3.39), integration by parts in the first integral in Eq. (3.40) makes it trivial. This immediately leads to the relation

$$s \int_0^h p_m(y)\, p_m^*(y) \, dy = \frac{\omega}{c^2} \frac{d\omega}{ds} \int_0^h |p_m(y)|^2 \, dy, \tag{3.41}$$

where, taking into account that

$$v_x(x, y) = \frac{1}{i\omega\rho_0} \frac{\partial p}{\partial x}; \qquad v_x(x, y) = v_m(y) e^{is(x-x_0)}, \qquad \sim \qquad v_m(y) = \frac{s}{\omega\rho_0} p_m(y), \quad (3.42)$$

the identity (3.41) finally gives

$$\omega\rho_0 \int_0^h v_m(y) p_m^*(y)\, dy = \frac{\omega}{c^2} \frac{d\omega}{ds} \int_0^h |p_m(y)|^2\, dy$$

$$\sim \quad W = \frac{1}{2} \operatorname{Re} \int_0^h v_m p_m^*\, dy = \frac{1}{\rho c^2} \frac{d\omega}{ds} \int_0^h |p_m(y)|^2\, dy. \tag{3.43}$$

It becomes clear now that any mode with real group velocity (i.e., a nondecaying wave mode) provides a positive energy flux if and only if $d\omega/ds > 0 \sim ds/d\omega > 0$ for any positive $s(\omega)$.

DEFINITION. *The function $s = s(\omega)$ that determines the dependence of the group velocity upon frequency is called a* dispersive function. *Its graph in the Cartesian coordinates is called a* dispersive curve.

In the considered problem for the scalar acoustic layer (see Eqs. (3.11), (3.12)), we have

$$\frac{ds}{d\omega} = \frac{\omega}{cs} > 0 \quad \text{if} \quad s > 0. \tag{3.44}$$

Graphically, dispersive lines are shown in Fig. 3.1, and it is obvious that they represent monotonically increasing functions, so that $ds/d\omega > 0$ for $s > 0$. This proves once again that all modes propagating to the right with positive group velocities and so satisfying Sommerfeld's radiation condition provide also a positive energy flux, i.e., they also satisfy the Mandelshtam's energy radiation condition. Simultaneously, this proves that the integration contour in Eq. (3.7) must bend around the positive poles from below, so that the positive poles lie above the contour to give mode waves satisfying both radiation principles. Due to the evident symmetry, a similar analysis of modes propagating to the left (i.e., with $s < 0$) justifies that the integration contour in Eq. (3.7) should bend around the negative poles from above.

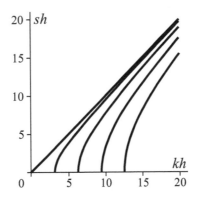

Figure 3.1. Dispersive curves of the scalar acoustic problem

3°. From this consideration it is clear that the integration contour, being symmetric about the origin, must pass through the origin. Therefore, all above results remain valid if no pole hits the origin. Otherwise, we deal with a so-called *resonance wave process*; a detailed analysis of various resonance cases, both in acoustic and elastic theory, is presented in

(Vorovich and Babeshko, 1979). Resonance frequencies for the three considered scalar problems can be found explicitly:

$$
\begin{aligned}
s_m^{(1,2)} &= 0, \quad \omega_m = \pi m \frac{c}{h}, \qquad m = 1, 2, \dots, \\
s_m^{(3)} &= 0, \quad \omega_m = \pi(m + 1/2)\frac{c}{h}, \qquad m = 0, 1, 2, \dots
\end{aligned}
\tag{3.45}
$$

Coming back to the question formulated in the beginning of paragraph 2°, in accordance with what is expressed by formula (3.43), we may expect that such a "nontypical" regime with negative energy flux can arise in the case where $ds/d\omega < 0$ for positive s. Figure 3.1 shows that for the considered geometry of a scalar acoustic layer of constant thickness this is never realized. We do not know any scalar acoustic or vector electromagnetic problem with such a feature. However, Mindlin (1955) first demonstrated that Sommerfeld's radiation condition and the energy radiation condition may contradict each other in a problem for an elastic layer. The next section deals with some aspects of wave propagation in an elastic layer.

3.3. Waves in Elastic Layer

As follows from equations of Sections 1.9, the 2D (in-plane) dynamic elastic problem is governed by the relations

$$
\begin{aligned}
u_x &= \frac{\partial \varphi}{\partial x} + \frac{\partial \psi}{\partial y}, \qquad u_y = \frac{\partial \varphi}{\partial y} - \frac{\partial \psi}{\partial x}, \\
\Delta\varphi &+ k_p^2 \varphi = 0, \qquad \Delta\psi + k_s^2 \psi = 0, \qquad k_p = \frac{\omega}{c_p}, \qquad k_s = \frac{\omega}{c_s}, \\
\frac{\sigma_{xy}}{\rho c_s^2} &= 2\frac{\partial^2\varphi}{\partial x \partial y} + \frac{\partial^2\psi}{\partial y^2} - \frac{\partial^2\psi}{\partial x^2}, \qquad \frac{\sigma_{yy}}{\rho c_s^2} = 2\left(\frac{\partial^2\varphi}{\partial x^2} + \frac{\partial^2\psi}{\partial x \partial y}\right) + k_s^2 \varphi, \\
\frac{\sigma_{xx}}{\rho c_s^2} &= 2\left(\frac{\partial^2\varphi}{\partial y^2} + \frac{\partial^2\psi}{\partial x \partial y}\right) + k_s^2 \varphi.
\end{aligned}
\tag{3.46}
$$

Here c_p, c_s are the longitudinal and transverse wave speeds; $\{u_x(x, y),\, u_y(x, y)\}$ are the components of the displacement vector; φ and ψ are the Lamé potentials; σ_{xx}, σ_{yy}, σ_{xy} are the components of the stress tensor involved in the 2D problem; and ρ is the mass density.

The number of various combinations of boundary conditions here is greater than that for the scalar acoustic layer. For simplicity, we will consider only one concrete boundary value problem:

$$
\begin{aligned}
y = 0: &\quad \sigma_{xy} = 0, \quad u_y = 0, \\
y = h: &\quad \sigma_{xy} = 0, \quad u_y = \delta(x - x_0),
\end{aligned}
\tag{3.47}
$$

whose physical meaning is that the layer rests (along the line $y = 0$) upon a rigid foundation and the upper surface of the layer is subjected to harmonic vibrations caused by an oscillating point indentor.

The application of the Fourier transform with respect to the x-variable reduces the Helmholtz equations in (3.46) to ordinary differential equations with constant coefficient, whose solutions in Fourier images are given by

$$
\begin{aligned}
\Phi &= A(s)\cosh(\gamma y) + B(s)\sinh(\gamma y), \qquad \gamma = \sqrt{s^2 - k_p^2}, \\
\Psi &= C(s)\cosh(qy) + D(s)\sinh(qy), \qquad q = \sqrt{s^2 - k_s^2},
\end{aligned}
\tag{3.48}
$$

The boundary conditions (3.47) for the Fourier transforms have the form (a prime denotes a derivative with respect to y)

$$
\begin{aligned}
-2is\Phi' + \Psi'' + s^2\Psi &= 0, & y &= 0, \\
\Phi' + is\Psi &= 0, & y &= 0, \\
-2is\Phi' + \Psi' + s^2\Psi &= 0, & y &= h, \\
(k_s^2 - 2s^2)\,\Phi - 2is\,\Psi' &= \frac{e^{ix_0 s}}{\rho c_s^2}, & y &= h.
\end{aligned}
\tag{3.49}
$$

System (3.49) with Φ and Ψ defined by (3.48) leads to a 4×4 linear algebraic system for the coefficients A, B, C, D:

$$
\begin{cases}
-2is\gamma B + (2s^2 - k_s^2)\,C = 0, \\
\gamma B + isC = 0, \\
-2\,is\gamma A \sinh(\gamma h) + (2s^2 - k_s^2)\,D \sinh(qh) = 0, \\
-(2s^2 - k_s^2)\,A\,\sinh(\gamma h) - 2isq \sinh(qh) = \dfrac{e^{ix_0 s}}{\rho c_s^2},
\end{cases}
\tag{3.50}
$$

whose solution is

$$
B = C = 0, \quad A = -\frac{e^{ix_0 s}}{\rho c_s^2}\,\frac{(2s^2 - k_s^2)\sinh(qh)}{\Delta(s)}, \quad D = -\frac{e^{ix_0 s}}{\rho c_s^2}\,\frac{2is\gamma \sinh(\gamma h)}{\Delta(s)},
\tag{3.51}
$$

$$
\Delta(s) = (2s^2 - k_s^2)^2 \sinh(qh)\cosh(\gamma h) - 4s^2\gamma q \sinh(\gamma h)\cosh(qh).
\tag{3.52}
$$

This finally determines all elastic functions in the layer. In particular, the Lamé potentials are

$$
\begin{aligned}
\varphi(x, y) &= -\frac{1}{2\pi\rho c_s^2}\int_{-\infty}^{\infty} \frac{(2s^2 - k_s^2)\sinh(qh)\cosh(\gamma y)}{\Delta(s)}\,e^{is(x-x_0)}\,ds, \\
\psi(x, y) &= -\frac{1}{2\pi\rho c_s^2}\int_{-\infty}^{\infty} \frac{2is\gamma \sinh(\gamma h)\sinh(qy)}{\Delta(s)}\,e^{is(x-x_0)}\,ds,
\end{aligned}
\tag{3.53}
$$

where the evenness of the integrand allows us again to write $\exp[is(x - x_0)]$ rather than $\exp[-is(x - x_0)]$.

In order to clarify what is the relation between the Sommerfeld and energy radiation conditions here in the case of elastic layer, let us first derive a representation for the energy flux averaged over the period $T = 2\pi/\omega$:

$$
W = -\frac{\omega}{2\pi}\int_0^{2\pi/\omega}\left(\sigma_{xx}\frac{du_x}{dt} + \sigma_{xy}\frac{du_y}{dt}\right)dt,
\tag{3.54}
$$

where the sign "minus" is put in front of this formula because v_{xx} and p are directed oppositely in the scalar acoustics.

Now, in terms of real quantities, the components of the displacement vector are

$$
u_i(x, y, t) = \mathrm{Re}\{e^{-i\omega t}\tilde{u}_i(x, y)\} = \tilde{u}_i^{\mathrm{re}}\cos\omega t + \tilde{u}_i^{\mathrm{im}}\sin\omega t \quad (i = 1, 2),
$$

$$
\sigma_{xy} = \mu\left(\frac{\partial u_x}{\partial y} + \frac{\partial u_y}{\partial x}\right) = \mu\left(\frac{\partial \tilde{u}_x^{\mathrm{re}}}{\partial y} + \frac{\partial \tilde{u}_y^{\mathrm{re}}}{\partial x}\right)\cos\omega t + \mu\left(\frac{\partial \tilde{u}_x^{\mathrm{im}}}{\partial y} + \frac{\partial \tilde{u}_y^{\mathrm{im}}}{\partial x}\right)\sin\omega t
\tag{3.55}
$$

$$
= \tilde{\sigma}_{xy}^{\mathrm{re}}\cos\omega t + \tilde{\sigma}_{xy}^{\mathrm{im}}\sin\omega t, \quad \frac{du_y}{dt} = \omega\left(-\tilde{u}_y^{\mathrm{re}}\sin\omega t + \tilde{u}_y^{\mathrm{im}}\cos\omega t\right),
$$

and consequently

$$\frac{\omega}{2\pi}\int_0^{2\pi/\omega}\sigma_{xy}\frac{du_y}{dt}\,dt = \frac{\omega}{2}\mu\left[\left(\frac{\partial\tilde{u}_x^{re}}{\partial y}+\frac{\partial\tilde{u}_y^{re}}{\partial x}\right)\tilde{u}_y^{im}-\left(\frac{\partial\tilde{u}_x^{im}}{\partial y}+\frac{\partial\tilde{u}_y^{im}}{\partial x}\right)\tilde{u}_y^{re}\right]$$
$$= \frac{\omega}{2}\left(\tilde{\sigma}_{xy}^{re}\tilde{u}_y^{im}-\tilde{\sigma}_{xy}^{im}\tilde{u}_y^{re}\right). \tag{3.56}$$

By analogy,

$$\frac{\omega}{2\pi}\int_0^{2\pi/\omega}\sigma_{xx}\frac{du_x}{dt}\,dt = \frac{\omega}{2}\left(\tilde{\sigma}_{xx}^{re}\tilde{u}_x^{im}-\tilde{\sigma}_{xx}^{im}\tilde{u}_x^{re}\right); \tag{3.57}$$

hence

$$W = -\frac{\omega}{2}\left[\left(\tilde{\sigma}_{xx}^{re}\tilde{u}_x^{im}-\tilde{\sigma}_{xx}^{im}\tilde{u}_x^{re}\right)+\left(\tilde{\sigma}_{xy}^{re}\tilde{u}_x^{im}-\tilde{\sigma}_{xy}^{im}\tilde{u}_x^{re}\right)\right]. \tag{3.58}$$

On the other hand, when dealing with the complex amplitudes

$$u_i(x,t)=e^{-i\omega t}\tilde{u}_i(x),\quad \sigma_{ij}(x,t)=e^{-i\omega t}\tilde{\sigma}_{ij}(x)\quad (i,j=1,2), \tag{3.59}$$

we have

$$\mathrm{Im}(\tilde{\sigma}_{xy}^*\tilde{u}_y+\tilde{\sigma}_{xx}^*\tilde{u}_x)=(\tilde{\sigma}_{xx}^{re}\tilde{u}_x^{im}-\tilde{\sigma}_{xx}^{im}\tilde{u}_x^{re})+(\tilde{\sigma}_{xy}^{re}\tilde{u}_y^{im}-\tilde{\sigma}_{xy}^{im}\tilde{u}_y^{re}); \tag{3.60}$$

hence, taking into account Eq. (3.54), we obtain

$$W = -\frac{\omega}{2}\mathrm{Im}(\tilde{\sigma}_{xx}^*\tilde{u}_x+\tilde{\sigma}_{xy}^*\tilde{u}_y)=\frac{\omega}{2}\mathrm{Im}(\tilde{\sigma}_{xx}\tilde{u}_x^*+\tilde{\sigma}_{xy}\tilde{u}_y^*). \tag{3.61}$$

This formula is very convenient for the calculation of the energy flux, to avoid transfer from complex to real-valued quantities.

Further, we will derive a different representation for W containing $d\omega/ds$, where $\omega(s)$ and $s(\omega)$ are dispersive functions, i.e., solutions of the Rayleigh–Lamb equation $\Delta(s,\omega)=0$. Let us consider any homogeneous mode obtained as the residue at a simple positive pole in Eq. (3.53), i.e., as a root of the equation $\Delta=0$ in the denominator. Then the structure of any mode in terms of the displacement vector in the case $x>x_0$ is

$$u_i(x,y)=u_{i_m}(y)\,e^{is(x-x_0)},\quad s=s_m,\quad \sigma_{ij}(x,y)=\sigma_{ij_m}e^{is(x-x_0)}, \tag{3.62}$$

where m designates the mode number.

Since for $x>x_0$ all right-hand sides in Eqs. (3.46)–(3.47) are homogeneous, the constructed mode wave (3.62) is a solution to the homogeneous boundary value problems, which is now convenient to be taken as equations of motion (see Section 1.9) expressed through components of the stress tensor, $\sigma_{ij,j}+\rho\omega^2u_i=0$ ($i=1,2$), with summation over the repeated index j. Then with the help of Eq. (3.62) we have

$$\begin{cases} is\sigma_{xx_m}+\sigma'_{xy_m}+\rho\omega^2u_{x_m}=0,\\ is\sigma_{xy_m}+\sigma'_{yy_m}+\rho\omega^2u_{y_m}=0, \end{cases} \tag{3.63}$$

with the boundary conditions

$$y=0:\ u_{y_m}=0,\quad \sigma_{xy}=0:\ u'_{x_m}+isu_{y_m}=0,$$
$$y=h,\ \sigma_{xy}=0:\ u'_{x_m}+isu_{y_m}=0,\quad \sigma_{yy}=0:\ c_p^2\,is\,u_{x_m}+(c_p^2-2c_s^2)u'_{y_m}=0 \tag{3.64}$$

(a prime denotes a derivative with respect to y).

The differentiation of system (3.63) with respect to s gives

$$\begin{cases} isg_{xx} + g'_{xy} + \rho\omega^2 g_x = -i\sigma_{xx_m} - 2\rho\omega\dfrac{d\omega}{ds} u_{x_m}, \\ isg_{xy} + g'_{yy} + \rho\omega^2 g_y = -i\sigma_{xy_m} - 2\rho\omega\dfrac{d\omega}{ds} u_{y_m}, \end{cases} \qquad (3.65)$$

where we have denoted

$$g_{xx} = \dfrac{d\sigma_{xx_m}}{ds}, \quad g_{xy} = \dfrac{d\sigma_{xy_m}}{ds}, \quad g_{yy} = \dfrac{d\sigma_{yy_m}}{ds}, \quad g_x = \dfrac{du_{x_m}}{ds}, \quad g_y = \dfrac{du_{y_m}}{ds}. \qquad (3.66)$$

Analogously, the differentiation of the boundary conditions (3.64) yields

$$y = 0: \quad g_y = 0, \quad g'_x + isg_y = -iu_{y_m},$$
$$y = h: \quad g'_x + isg_y = -iu_{y_m}, \quad c_p^2 isg_x + (c_p^2 - 2c_s^2)g'_y = -ic_p^2 u_{x_m}. \qquad (3.67)$$

Now, let us take the conjugate of system (3.63):

$$\begin{cases} -is\sigma^*_{xx_m} + \sigma^{*\prime}_{xy_m} + \rho\omega^2 u^*_{x_m} = 0, \\ -is\sigma^*_{xy_m} + \sigma^{*\prime}_{yy_m} + \rho\omega^2 u^*_{y_m} = 0, \end{cases} \qquad (3.68)$$

and combine the following scalar products:

$$(3.65)_1 \cdot u^*_{x_m} + (3.65)_2 \cdot u^*_{y_m} - (3.68)_1 \cdot g_x - (3.68)_2 \cdot g_y, \qquad (3.69)$$

where the sign of the scalar product implies the integration of an equation multiplied by an appropriate function. This procedure leads to the identity

$$is \int_0^h (g_{xx} u^*_{x_m} + \sigma^*_{xx_m} g_x)\, dy + \int_0^h (g'_{xy} u^*_{x_m} - \sigma^{*\prime}_{xy_m} g_x)\, dy$$
$$+ is \int_0^h (g_{xy} u^*_{y_m} + \sigma^*_{xy_m} g_y)\, dy + \int_0^h (g'_{yy} u^*_{y_m} - \sigma^{*\prime}_{yy_m} g_y)\, dy \qquad (3.70)$$
$$= -i \int_0^h (\sigma_{xx_m} u^*_{x_m} + \sigma^*_{xy_m} u^*_{y_m})\, dy - 2\rho\omega\dfrac{d\omega}{ds} \int_0^h (|u_{x_m}|^2 + |u_{y_m}|^2)\, dy.$$

Very scrupulous mathematical manipulations with the integrals on the left-hand side in (3.70), with the help of the boundary conditions (3.64), (3.67), show that, as in the scalar acoustic case, this left-hand side is equal to zero. Then it follows from Eqs. (3.70), (3.61) that

$$W = \dfrac{\omega}{2} \int_0^h \text{Im}(\tilde{\sigma}_{xx}\tilde{u}^*_x + \tilde{\sigma}_{xy}\tilde{u}^*_y)\, dy = \rho\omega^2 \dfrac{d\omega}{ds} \int_0^h (|\tilde{u}_{x_m}|^2 + |u_{y_m}|^2)\, dy. \qquad (3.71)$$

Formula (3.71) allows us to clarify whether there is any correlation between Sommerfeld's and Mandelshtam's principles. Let $s = s_m > 0$ be a positive root of the Rayleigh–Lamb equation $\Delta(s) = 0$ (since only squares of s are present in the structure of the function $\Delta(s)$, if $s = s_m$ is a root of the equation, then $s = -s_m$ is also a root of it). Then the point $s = s_m$ is a simple pole in all integrands in the corresponding Fourier-transform representations of the type (3.14) for all physical quantities, so that the pair of functions

$$\varphi(x, y, t) = \tilde{\varphi}(y)\, e^{i[s_m(x-x_0)-\omega t]}, \qquad \psi(x, y, t) = \tilde{\psi}(y)\, e^{i[s_m(x-x_0)-\omega t]}, \qquad (3.72)$$

determines a homogeneous mode that propagates, according to Sommerfeld's radiation condition without decay far to the right. It is now clear from Eq. (3.71) that this mode is in agreement with Mandelshtam's energy radiation condition if and only if

$$\frac{d\omega}{ds}(s = s_m) = \frac{1}{ds/d\omega} > 0. \tag{3.73}$$

In order to check whether condition (3.73) holds or not, we need to construct dispersive curves. In the scalar acoustics, as we remember, all dispersive curves are monotonically increasing graphs. For an elastic layer, an example of dispersive curves, as solutions to the Rayleigh–Lamb equation $\Delta(s, \omega) = 0$, $c^2 = c_s^2/s_p^2$, $0 < c^2 < 1$, is shown in Fig. 3.2.

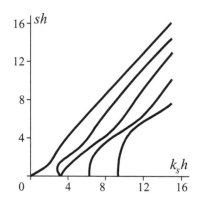

Figure 3.2. Dispersive curves of the dynamic elastic problem

It can be seen from this graph that there are cases with $ds/d\omega < 0$. As follows from the described theory, in such cases the modes propagating to the right bring energy from infinity, and hence contradict the energy radiation condition. Physically, this looks mystical, and this is indeed so, confirming that physical intuition may sometimes be in contradiction to strict mathematical analysis. Anyway, we should agree that the energy radiation condition is more fundamental than Sommerfeld's principle, and so we have to reject the modes for which

$$\frac{ds}{d\omega}(s = s_m) < 0, \quad s_m > 0. \tag{3.74}$$

This conclusion influences the choice of the integration contour in Eq. (3.7). As regards the positive half of the contour, it should bend around the poles s_m of the integrand where $ds/d\omega > 0$ from below and in the case of $ds/d\omega < 0$, from above.

It should be noted once again that, apparently, Mindlin (1955) was first to indicate a case where a wave satisfying Sommerfeld's radiation condition can bring energy from infinity. A fundamental and very impressive analysis of correlations between various radiation conditions can be found in Vorovich and Babeshko (1979); and also in Babeshko (1971).

Helpful remarks

The Rayleigh–Lamb equation $\Delta(s, \omega) = 0$, where the function $\Delta(s, \omega)$ is defined by Eq. (3.52), determines the behavior of the dispersive lines, as in Fig. 3.2. In contrast with the dispersion equation of the scalar acoustic theory, which admits simple solution (see Section 3.2), the Rayleigh–Lamb equation cannot be solved explicitly. That is why this dispersion equation is studied in the literature in detail. Many papers were devoted to asymptotic estimates for its large roots and many authors dealt with numerical analysis of the Rayleigh–Lamb equation. Some helpful results with further references can be found in Mindlin (1955), Vorovich and Babeshko (1979), and other papers. Detailed numerical implementation allowed the authors of those papers to discover cases of the cited irregular behavior of the dispersive curves. Note that the only dimensionless parameter whose variation changes the configuration of the dispersive lines is a relation between the longitudinal and the transverse wave speeds. Here in Fig. 3.2 we demonstrate these curves in the case where $c_p/c_s = 2$, which is very realistic for metals.

3.4. Generalized Riemann's Zeta Function and Summation of Some Oscillating Series

As we could see in Section 3.1, in many problems the application of Poisson's summation formula transforms some given slowly convergent series to series convergent much more rapidly, so that the new series obtained can be calculated more efficiently. Unfortunately, this technique is not applicable, as a rule, for series with oscillating terms, at least in its straightforward application.

Here we study the following series that will be encountered in our forthcoming analysis of high-frequency processes in a layer of constant thickness:

$$C_\beta(\xi) = \sum_{m=1}^{\infty} \frac{\cos m\xi}{m^\beta}, \qquad S_\beta(\xi) = \sum_{m=1}^{\infty} \frac{\sin m\xi}{m^\beta}. \tag{3.75}$$

As noted in the Section 1.3, these series converge in the classical sense if $\mathrm{Re}(\beta) > 0$. However, practical calculation for relatively small β is a difficult problem. Concrete computations show that in order to guarantee three significant digits, one should take nearly a few hundred thousand terms. Naturally, the question arises about the convergence acceleration.

To this end, we attract some results from the theory of Riemann's zeta function. Recall that ordinary and generalized Riemann's zeta functions are defined, respectively, by the following series:

$$\zeta(s) = \sum_{m=1}^{\infty} \frac{1}{m^s}, \qquad \zeta(s, v) = \sum_{m=0}^{\infty} \frac{1}{(m+v)^s}. \tag{3.76}$$

It can be shown that both functions, as functions of the argument s, are analytic in the whole complex plane except the only singular point $s = 1$, which is a simple pole for both of them (see Bateman and Erdelyi, 1953).

In what follows we will use the profound classical result given by the Hurwitz formula (see Bateman and Erdelyi, 1953)

$$
\begin{aligned}
\zeta\left(1 - \beta, \frac{\xi}{2\pi}\right) &= \frac{2\,\Gamma(\beta)}{(2\pi)^\beta} \sum_{m=1}^{\infty} \frac{\cos(m\xi - \pi\beta/2)}{m^\beta} \\
&= \frac{2\Gamma(\beta)}{(2\pi)^\beta} \left(\cos \frac{\pi\beta}{2} \sum_{m=1}^{\infty} \frac{\cos m\xi}{m^\beta} + \sin \frac{\pi\beta}{2} \sum_{m=1}^{\infty} \frac{\sin m\xi}{m^\beta} \right) \\
&= \frac{2\Gamma(\beta)}{(2\pi)^\beta} \left(\cos \frac{\pi\beta}{2} C_\beta(\xi) + \sin \frac{\pi\beta}{2} S_\beta(\xi) \right), \qquad \mathrm{Re}\,\xi > 0.
\end{aligned}
\tag{3.77}
$$

Our final goal is to extract from Eq. (3.77) an expansion in powers of ξ. To this end, we will try to expand the second relation in (3.76) into a Taylor series about the point $v = 0$, taking into account that the derivative of $\zeta(s, v)$ with respect to v at $v = 0$ is the ordinary Riemann's zeta function with some power s. Namely, we obtain from (3.76)

$$
\begin{aligned}
\zeta(s, v) &= \frac{1}{v^s} + \sum_{m=1}^{\infty} \frac{1}{(m+v)^s} = \frac{1}{v^s} + \sum_{m=0}^{\infty} (-1)^m \frac{s(s+1)\ldots(s+m-1)}{m!} \zeta(s+m)\, v^m \\
&= \frac{1}{v^s} + \sum_{m=0}^{\infty} (-1)^m \frac{\Gamma(s+m)}{\Gamma(s)\, m!} \zeta(s+m)\, v^m \qquad (0! = 1).
\end{aligned}
$$
(3.78)

Therefore, the left-hand side of (3.77) admits the alternative representation

$$
\zeta(1-\beta, \xi/2\pi) = \left(\frac{2\pi}{\xi}\right)^{1-\beta} + \sum_{m=0}^{\infty} (-1)^m \frac{\Gamma(m+1-\beta)}{\Gamma(1-\beta)\, m!} \zeta(m+1-\beta) \left(\frac{\xi}{2\pi}\right)^m.
$$
(3.79)

Since the left-hand sides of Eqs. (3.77) and (3.79) are identical, their right-hand sides should coincide with each other, too. This allows us to construct two independent power expansions in $(\xi/2\pi)$, separately for $\sum_{m=1}^{\infty} \cos(m\xi)/m^\beta$ and $\sum_{m=1}^{\infty} \sin(m\xi)/m^\beta$, since the even powers of ξ in (3.79) refer to the former and the odd powers, to the latter. The only obstacle consists in the presence of the irregular term $(2\pi/\xi)^{1-\beta}$ in (3.79). One part of this term belongs to a series of cosines and the other part to a series with sines. Correct separation of the irregular term is not simple. To do this, we have to use Poisson's formula (see, for example, Collin, 1960), which is usually useless in the treatment of oscillating series. In our case it is applied in the following way:

$$
\sum_{m=1}^{\infty} \frac{e^{im\xi}}{m^\beta} = e^{i\xi} \sum_{m=0}^{\infty} \frac{e^{im\xi}}{(m+1)^\beta} = e^{i\xi} \left[\frac{1}{2} + H(0) + 2 \sum_{m=1}^{\infty} H(2\pi m) \right],
$$
(3.80)

where

$$
H(2\pi m) = \int_0^\infty \cos(2\pi m u) \frac{e^{iu\xi}}{(u+1)^\beta}\, du, \qquad m = 0, 1, 2, \ldots.
$$
(3.81)

If m is a positive integer here ($m \geq 1$), then the last relation yields only a regular power expansion in ξ for $H(2\pi m)$. This fact can be proved by expanding $\exp(iu\xi)$ into a Taylor series, since the integral

$$
\int_0^\infty \cos(2\pi m u) \frac{u^p}{(u+1)^\beta}\, du
$$
(3.82)

has a finite generalized value for arbitrary large positive integer p (cf. Section 1.3), which implies an analytic continuation with respect to p, from the range $-1 < \text{Re}(p) < 0$, where it converges in the classical sense. The irregular term is therefore extracted from the case $m = 0$ as follows:

$$
\begin{aligned}
H(0) &= \int_0^\infty \frac{e^{iu\xi}}{(u+1)^\beta}\, du = \xi^{\beta-1} \int_0^\infty \frac{e^{iu}\, du}{(u+\xi)^\beta} \\
&\sim \xi^{\beta-1} \int_0^\infty \frac{e^{iu}\, du}{u^\beta} = \Gamma(1-\beta)\, \xi^{\beta-1} e^{\pi i(1-\beta)/2} \qquad (\xi \to +0),
\end{aligned}
$$
(3.83)

and hence

$$
\sum_{m=1}^{\infty} \frac{e^{im\xi}}{m^\beta} = \Gamma(1-\beta)\, \xi^{\beta-1}\, e^{\pi i(1-\beta)/2} + \sum_{m=0}^{\infty} a_m \xi^m \qquad (\xi \to +0),
$$
(3.84)

with some coefficients a_m of the regular power part. By separating the real and imaginary parts in the last relation, we see that

$$\sum_{m=1}^{\infty} \frac{\cos m\xi}{m^\beta} = \Gamma(1-\beta) \cos\left[\frac{\pi}{2}(1-\beta)\right] \xi^{\beta-1} + \sum_{m=0}^{\infty} b_m \xi^{2m}, \qquad (3.85)$$

$$\sum_{m=1}^{\infty} \frac{\sin m\xi}{m^\beta} = \Gamma(1-\beta) \sin\left[\frac{\pi}{2}(1-\beta)\right] \xi^{\beta-1} + \sum_{m=0}^{\infty} d_m \xi^{2m+1}. \qquad (3.86)$$

As stated above, the coefficients of the regular expansions can be defined by comparing the right-hand sides in formulas (3.77) and (3.79), which finally yields

$$C_\beta(\xi) = \sum_{m=1}^{\infty} \frac{\cos m\xi}{m^\beta} = \Gamma(1-\beta) \cos\left[\frac{\pi}{2}(1-\beta)\right] \xi^{\beta-1} + \frac{(2\pi)^\beta}{2\,\Gamma(\beta)\,\Gamma(1-\beta)\,\cos(\pi\beta/2)}$$

$$\times \sum_{m=0}^{\infty} \frac{\Gamma(2m+1-\beta)\,\zeta(2m+1-\beta)}{(2m)!} \left(\frac{\xi}{2\pi}\right)^{2m} \quad (\xi,\,\beta > 0), \qquad (3.87)$$

$$S_\beta(\xi) = \sum_{m=1}^{\infty} \frac{\sin m\xi}{m^\beta} = \Gamma(1-\beta) \sin\left[\frac{\pi}{2}(1-\beta)\right] \xi^{\beta-1} - \frac{(2\pi)^\beta}{2\,\Gamma(\beta)\,\Gamma(1-\beta)\,\sin(\pi\beta/2)}$$

$$\times \sum_{m=0}^{\infty} \frac{\Gamma(2m+2-\beta)\,\zeta(2m+2-\beta)}{(2m+1)!} \left(\frac{\xi}{2\pi}\right)^{2m+1} \quad (\xi,\,\beta > 0). \qquad (3.88)$$

In order to estimate how rapidly the terms of the last series decrease with increasing m, let us apply the Sterling formula (see Abramowitz and Stegun, 1965)

$$\Gamma(z) \sim \sqrt{\frac{2\pi}{z}}\, e^{-z} z^z \quad (z \to +\infty); \quad \text{hence} \quad \frac{\Gamma(z+\alpha)}{\Gamma(z+\gamma)} \sim z^{\alpha-\gamma} \quad (z \to +\infty). \quad (3.89)$$

This estimate shows that the terms of the series (3.87) and (3.88) decrease according to $O[m^{-\beta}(\xi/2\pi)^{2m}]$ with $m \to \infty$, because $(2m)! = \Gamma(2m+1)$ and $(2m+1)! = \Gamma(2m+2)$. It should also be noted that it suffices to evaluate them on the interval $0 < \xi \le \pi$ only, owing to the evident periodicity: $C_\beta(2\pi - \xi) = C_\beta(\xi)$, $S_\beta(2\pi - \xi) = -S_\beta(\xi)$. Therefore, the rate of convergence of the series (3.13) is the same as that of a geometric progression $\{q^m\}$ with basis $q \le 1/2$. In practice, it suffices to keep 3 or 4 terms to provide a good accuracy.

Let us quote explicit expressions for particular cases, which will appear in the next section:

$$C_{1/2}(\xi) \approx \sqrt{\frac{\pi}{2\xi}} - 1.460 + 0.0255\,\frac{\xi^2}{2!} + 0.00444\,\frac{\xi^4}{4!} + 0.00267\,\frac{\xi^6}{6!}, \qquad (3.90)$$

$$S_{1/2}(\xi) \approx \sqrt{\frac{\pi}{2\xi}} - 0.208\,\xi - 0.00852\,\frac{\xi^3}{3!} - 0.00309\,\frac{\xi^5}{5!}, \qquad (3.91)$$

$$C_{3/2}(\xi) \approx -\sqrt{2\pi\xi} + 2.612 + 0.208\,\frac{\xi^2}{2!} + 0.00852\,\frac{\xi^4}{4!} + 0.00309\,\frac{\xi^6}{6!}, \qquad (3.92)$$

$$S_{3/2}(\xi) \approx \sqrt{2\pi\xi} - 1.460\,\xi + 0.0255\,\frac{\xi^3}{3!} + 0.00444\,\frac{\xi^5}{5!}, \qquad (3.93)$$

$$C_{5/2}(\xi) \approx -\frac{2\sqrt{2\pi}}{3}\,\xi^{3/2} + 1.341 + 1.460\,\frac{\xi^2}{2!} - 0.0255\,\frac{\xi^4}{4!} - 0.00444\,\frac{\xi^6}{6!}, \qquad (3.94)$$

$$S_{5/2}(\xi) \approx -\frac{2\sqrt{2\pi}}{3}\,\xi^{3/2} + 2.612\,\xi + 0.208\,\frac{\xi^3}{3!} + 0.00852\,\frac{\xi^5}{5!}. \qquad (3.95)$$

In practice all of them are to be used for $0 < \xi \leq \pi$ and then to be continued periodically. The relative error of these formulas is less than 1%. The accuracy of these formulas can be improved by taking more terms in the expansions. Below we give a comparison of results predicted by Eqs. (3.90)–(3.95) with those obtained by direct computations by formulas (3.75) with a double-precision Fortran code and 5×10^5 terms retained in the series:

ξ	$\pi/8$	$\pi/4$	$3\pi/8$	$\pi/2$	$5\pi/8$	$3\pi/4$	$7\pi/8$
$C_{1/2}$, (3.75)	0.5418	−0.0380	−0.2874	−0.4275	−0.5136	−0.5665	−0.5954
$C_{1/2}$, (3.90)	0.5420	−0.0379	−0.2872	−0.4274	−0.5135	−0.5664	−0.5956
$S_{1/2}$, (3.75)	1.916	1.249	0.9067	0.6672	0.4743	0.3058	0.1501
$S_{1/2}$, (3.91)	1.918	1.250	0.9073	0.6675	0.4745	0.3060	0.1506
$C_{3/2}$, (3.75)	1.058	0.4552	0.0366	−0.2705	−0.4938	−0.6465	−0.7357
$C_{3/2}$, (3.92)	1.057	0.4548	0.0363	−0.2708	−0.4939	−0.6466	−0.7359
$S_{3/2}$, (3.75)	0.9975	1.077	1.007	0.8645	0.6783	0.4653	0.2363
$S_{3/2}$, (3.93)	0.9977	1.077	1.008	0.8651	0.6790	0.4659	0.2366
$C_{5/2}$, (3.75)	1.043	0.6283	0.2160	−0.1533	−0.4573	−0.6825	−0.8207
$C_{5/2}$, (3.94)	1.042	0.6278	0.2153	−0.1542	−0.4585	−0.6840	−0.8223
$S_{5/2}$, (3.75)	0.6167	0.9054	0.9976	0.9486	0.7961	0.5700	0.2966
$S_{5/2}$, (3.95)	0.6166	0.9051	0.9972	0.9481	0.7954	0.5691	0.2952

Helpful remarks

The above ideas can be applied to the calculation of sums of divergent series. As described in Section 1.3, identities (3.87) and (3.88) can be analytically continued over the parameter β in the complex plane, taking into account those β for which both the left- and the right-hand sides are finite. As an example, we quote here two remarkable divergent series

$$C_{-1/2}(\xi) = \sum_{m=1}^{\infty} m^{1/2} \cos m\xi, \qquad S_{-1/2}(\xi) = \sum_{m=1}^{\infty} m^{1/2} \sin m\xi, \qquad (3.96)$$

and compare results predicted by the direct summation and the method using the derived formulas (3.87), (3.88) with $\beta = -1/2$. These series are expressed as

$$C_{-1/2}(\xi) \approx -\frac{1}{2}\sqrt{\frac{\pi}{2}} \xi^{-3/2} - 0.208 - 0.00852 \frac{\xi^2}{2!} - 0.00309 \frac{\xi^4}{4!} - 0.00346 \frac{\xi^6}{6!}, \quad (3.97)$$

$$S_{-1/2}(\xi) \approx \frac{1}{2}\sqrt{\frac{\pi}{2}} \xi^{-3/2} - 0.0255\,\xi - 0.00444 \frac{\xi^3}{3!} - 0.00267 \frac{\xi^5}{5!}. \quad (3.98)$$

In the direct summation we have applied ideas proposed in Section 1.3. More precisely, we set $x = 0.999$ and take the upper limit of summation 5×10^4 in the series

$$\sum_{m=1}^{\infty} x^m m^{1/2} \cos m\xi, \qquad \sum_{m=1}^{\infty} x^m m^{1/2} \sin m\xi, \qquad (3.99)$$

treated as a generalized value for the series (3.96). Results of the comparison are reflected in the following table:

ξ	$\pi/8$	$\pi/4$	$3\pi/8$	$\pi/2$	$5\pi/8$	$3\pi/4$	$7\pi/8$
$C_{-1/2}$, (3.75)	−2.745	−1.109	−0.7035	−0.5372	−0.4540	−0.4094	−0.3869
$C_{-1/2}$, (3.97)	−2.755	−1.111	−0.7041	−0.5376	−0.4543	−0.4096	−0.3870
$S_{-1/2}$, (3.75)	2.5461	0.8816	0.4594	0.2755	0.1715	0.1017	0.0476
$S_{-1/2}$, (3.98)	2.5364	0.8799	0.4588	0.2752	0.1714	0.1019	0.0485

An impressive table for divergent series, isn't it?

3.5. Application: Efficient Calculation of Wave Fields in a Layer of Constant Thickness

Let us come back to the problem of efficient calculation of wave fields in a layer of constant thickness. We will deal here only with acoustic case, since the proposed ideas can be applied to elastic case, too.

To begin with, we note that for arbitrary fixed k only a finite number of terms, N, in the mode representation (3.14) have a real group velocity s_n. For all other terms when $n > N$, the quantities s_n are imaginary and the modulus $|s_n|$ grows with n. Therefore, this series (3.14) converges like a geometric progression, in the case of $x \neq x_0$. Many investigations have been devoted to the question of efficient calculation of acoustic waves in a layer of constant thickness in the case of $x = x_0$, and we give here a short survey of existing methods.

In principle, a direct calculation of the integral (3.7), where $L(s)$ is defined by any of Eqs. (3.8)–(3.10), gives an acceptable tool for calculations, since all singularities of the integrand lie on the finite interval $-k < s < k$. Outside of this interval the integrand decays exponentially (we exclude the case $y = y_0$ from the consideration, because in the case that $x \to x_0$, $y \to y_0$ simultaneously, the constructed Green's function has a logarithmic singularity). But in practice this approach is never applied for the following two reasons. Firstly, a finite number of real-valued singularities are all simple poles, and the integral can be reduced to a Cauchy-type singular integral by adding together the halves of the residues at appropriate poles. However, direct numerical treatment of singular integrals is a more complex problem compared to those with continuous integrands. Secondly, numerical calculation of any integral (even with continuous integrand) is always a more difficult problem than the calculation of the sum of a series. So, now we are going to transform the integral to some series.

On the other hand, the mode expansions (3.14) with $x = x_0$ cannot be used, as they are written, for direct numerical calculations, since the common term decreases as $1/n$. The series converges, due to the presence of an oscillating factor, but it is absolutely unsuitable for direct computations. To overcome this difficulty, some authors apply the so-called *Kummer's transformation* in this case (see, for example, Jorgenson and Mittra, 1990; Mathis and Peterson, 1996). If $x = x_0$, then the mode expansions (3.14) can be represented as a combination of the following series:

$$F^{\pm}(\eta) = \sum_{m=0}^{\infty} \frac{(-1)^m}{q_m} \left\{ \begin{array}{c} \cos(p_m^{(1,2)}\eta) \\ \sin(p_m^{(3)}\eta) \end{array} \right\}, \tag{3.100}$$

$$q_m = \sqrt{p_m^2 - k^2}, \qquad p_m^{(1,2)} = \pi m, \qquad p_m^{(3)} = \pi \left(m + \tfrac{1}{2} \right).$$

More precisely,

$$\Phi^{(1)}(x_0, y) = \frac{1}{2} \left[F^+ \left(\frac{|y - y_0|}{h} \right) + F^+ \left(\frac{y + y_0}{h} \right) \right],$$

$$\Phi^{(2)}(x_0, y) = \frac{1}{2} \left[F^+ \left(\frac{|y - y_0|}{h} \right) - F^+ \left(\frac{y + y_0}{h} \right) \right], \tag{3.101}$$

$$\Phi^{(3)}(x_0, y) = \frac{1}{2} \left[F^- \left(\frac{|y - y_0|}{h} \right) + F^- \left(\frac{y + y_0}{h} \right) \right],$$

and the problem can be reduced to the calculation of the series (3.100). This is based on the observation that

$$F^{\pm}(\eta) = \sum_{n=0}^{\infty} (-1)^n \left\{ \begin{matrix} \cos(p_n^{(1,2)}\eta) \\ \sin(p_n^{(3)}\eta) \end{matrix} \right\} \left(\frac{1}{q_n} - \frac{1}{r_n} \right) + \sum_{n=0}^{\infty} \frac{(-1)^n}{r_n} \left\{ \begin{matrix} \cos(p_n^{(1,2)}\eta) \\ \sin(p_n^{(3)}\eta) \end{matrix} \right\}, \quad (3.102)$$

$$q_n = \sqrt{p_n^2 - k^2}, \qquad r_n = \sqrt{p_n^2 + b^2}.$$

Now, for any fixed value of the parameter k the common term in the first series decreases as $1/n^3$, and the series is quite suitable for efficient calculations, but the second series converges still as slowly as in (3.14). However, this admits the application of the *Poisson summation formula* (see, for example, Titchmarsh, 1948). We already treated some series by this formula in the previous section, and give here a more complete overview. If $f(x) \in C[0, \infty)$, $f(x) \in L_1[0, \infty)$ and a is a parameter, then

$$a \sum_{n=0}^{\infty} f(an) = \frac{a}{2} f(0) + F(0) + 2 \sum_{n=1}^{\infty} F\left(\frac{2\pi n}{a} \right), \quad (3.103)$$

where $F(s)$ is the cosine Fourier transform of $f(x)$:

$$F(s) = \int_0^{\infty} \cos(sx) f(x) \, dx. \quad (3.104)$$

This formula has a finite-form analogue:

$$a \sum_{n=0}^{n} f(an) = \frac{a}{2} [f(0) + f(a)] + F(0) + 2 \sum_{n=1}^{\infty} F\left(\frac{2\pi n}{a} \right). \quad (3.105)$$

Since

$$\sum_{n=0}^{\infty} \frac{(-1)^n}{r_n} \left\{ \begin{matrix} \cos(p_n^{(1,2)}\eta) \\ \sin(p_n^{(3)}\eta) \end{matrix} \right\} = \sum_{n=0}^{\infty} \frac{1}{r_n} \left\{ \begin{matrix} \cos(p_n^{(1,2)}) \\ \sin(p_n^{(3)}) \end{matrix} \right\} \left\{ \begin{matrix} \cos(p_n^{(1,2)}\eta) \\ \sin(p_n^{(3)}\eta) \end{matrix} \right\}, \quad (3.106)$$

the transformation of the products of trigonometric functions to their sums reduces the calculation of the second series (3.102) to the summation of some other series, which we consider separately in the $(1, 2)$ and (3) problems.

The top line in Eq. (3.106) leads to a series of the form

$$S^{(1,2)} = \sum_{n=0}^{\infty} \frac{\cos(\pi n \zeta)}{\sqrt{(\pi n)^2 + b^2}}, \quad (3.107)$$

which admits immediate application of the Poisson formula (3.103)–(3.104), since here $a = \pi$, $f(x) = \cos(x\zeta)/\sqrt{x^2 + b^2}$ and

$$F(s) = \int_0^{\infty} \frac{\cos(sx) \cos(x\zeta)}{\sqrt{x^2 + b^2}} = \frac{1}{2} \int_0^{\infty} \frac{\cos[x(s + \zeta)] + \cos[x(s - \zeta)]}{\sqrt{x^2 + b^2}}$$
$$= \frac{K_0(b|s + \zeta|) + K_0(b|s - \zeta|)}{2}, \quad (3.108)$$

where K_0 is the McDonald function, which exponentially decreases with increasing argument. Therefore, in these problems we arrive at the series

$$S^{(1,2)} = \frac{1}{2b} + \frac{K_0(b|\zeta|)}{\pi} + \frac{1}{\pi} \sum_{n=1}^{\infty} [K_0(b|2n + \zeta|) + K_0(b|2n - \zeta|)], \quad (3.109)$$

which converges exponentially.

In problem (3) the bottom line in (3.106) yields a series of the form

$$
\begin{aligned}
S^{(3)} &= \sum_{n=0}^{\infty} \frac{\cos[\pi(n+1/2)\zeta]}{\sqrt{[\pi(n+1/2)]^2 + b^2}} = \frac{1}{2} \sum_{n=-\infty}^{\infty} \frac{\exp[\pi i(n+1/2)\zeta]}{\sqrt{[\pi(n+1/2)]^2 + b^2}} \\
&= \frac{\exp(\pi\zeta i/2)}{2} \sum_{n=-\infty}^{\infty} \frac{\exp(\pi n\zeta i)}{\sqrt{[\pi(n+1/2)]^2 + b^2}} \\
&= \frac{\exp(\pi\zeta i/2)}{2} \sum_{n=-\infty}^{\infty} \left[\frac{\exp(\pi n\zeta i)}{\sqrt{[\pi(n+1/2)]^2 + b^2}} - \frac{\exp(\pi n\zeta i)}{\sqrt{(\pi n)^2 + b^2}} \right] \\
&\quad + \exp(\pi\zeta i/2) \left[\sum_{n=0}^{\infty} \frac{\cos(\pi n\zeta)}{\sqrt{(\pi n)^2 + b^2}} - \frac{1}{2b} \right],
\end{aligned}
\tag{3.110}
$$

where the common term of the first series on the right-hand side (3.110) decreases again as $1/n^3$ and the last series is reduced to $S^{(1,2)}$, already calculated.

Thus, for low and moderate k, Kummer's transformation leads to quite rapid convergence. However, this transformation becomes inefficient with increasing frequency parameter k, due to a specific behavior of the difference in parentheses in Eq. (3.102). That is why we develop a different approach.

Let us note that all integrands in Eqs. (3.7)–(3.10), Section 3.1, represent some combinations of the following type:

$$
L^{\pm}(s,\zeta) = \frac{e^{-\zeta\gamma}}{2\gamma\left(1 \pm e^{-2\gamma h}\right)}, \qquad \zeta > 0.
\tag{3.111}
$$

More precisely,

$$
\begin{aligned}
L_1(s,y) &= L^-(s,|y-y_0|) + L^-(s,2h-|y-y_0|) + L^-(s,y+y_0) + L^-(s,2h-y-y_0), \\
L_2(s,y) &= L^-(s,|y-y_0|) + L^-(s,2h-|y-y_0|) - L^-(s,y+y_0) - L^-(s,2h-y-y_0), \\
L_3(s,y) &= L^+(s,|y-y_0|) - L^+(s,2h-|y-y_0|) + L^+(s,y+y_0) - L^+(s,2h-y-y_0).
\end{aligned}
\tag{3.112}
$$

Therefore, the problem is reduced to efficient calculation of the integral

$$
\begin{aligned}
G(x,\zeta) &= \frac{1}{2\pi} \int_{-\infty}^{\infty} L^{\pm}(s,\zeta)\, e^{-is(x-x_0)} \\
&= \frac{1}{4\pi} \sum_{n=0}^{\infty} (\mp 1)^n \int_{-\infty}^{\infty} \frac{e^{-\gamma(2nh+\zeta)}}{\gamma}\, e^{-is(x-x_0)}\, ds \\
&= \frac{i}{4} \sum_{n=0}^{\infty} (\mp 1)^n H_0^{(1)} \left[k\sqrt{(2nh+\zeta)^2 + (x-x_0)^2} \right] \\
&= \frac{i}{4} H_0^{(1)} \left(k\sqrt{(x-x_0)^2 + \zeta^2} \right) + G_1(x,y),
\end{aligned}
\tag{3.113}
$$

$$
G_1(x,y) = \frac{i}{4} \sum_{n=1}^{\infty} (\mp 1)^n H_0^{(1)} \left[k\sqrt{(2nh+\zeta)^2 + (x-x_0)^2} \right], \qquad 0 \le \zeta \le 2h.
$$

Recall that now we study summation for $x = x_0$, where the first N terms (N is sufficiently large) can be calculated directly as

$$
\frac{i}{4} \sum_{n=1}^{N} (\mp 1)^n H_0^{(1)}\, [k(2nh+\zeta)],
\tag{3.114}
$$

and for $n > N$ we can use the asymptotic representation of the Hankel function (see Bateman and Erdelyi, 1953)

$$H_0^{(1)}(z) \sim \sqrt{\frac{2}{\pi z}}\, e^{i(z-\pi/4)} \sum_{m=0}^{\infty} \frac{\Gamma\left(\frac{1}{2}+m\right)}{m!\,\Gamma\left(\frac{1}{2}-m\right)(-2zi)^m}, \quad z \to \infty. \tag{3.115}$$

As a result, the summation problem is reduced to the calculation of a series of the following type:

$$\sum_{n=1}^{\infty} (\mp 1)^n \frac{e^{2khni}}{(2nh+\zeta)^{m+1/2}}. \tag{3.116}$$

Let us write out a few first terms in the Taylor expansion of the denominator:

$$\frac{1}{(n+\xi)^{m+1/2}} = \frac{1}{n^{m+1/2}} - \frac{(m+1/2)\,y}{n^{m+3/2}} + \frac{(m+1/2)(m+3/2)\,y}{n^{m+5/2}}$$
$$+ \left[\frac{1}{(n+\xi)^{m+1/2}} - \frac{1}{n^{m+1/2}} + \frac{(m+1/2)\,y}{n^{m+3/2}} - \frac{(m+1/2)(m+3/2)\,y}{n^{m+5/2}} \right], \quad \xi = \frac{\zeta}{2h}. \tag{3.117}$$

The sum in square brackets here tends to zero at least as $n^{-7/2}$ with $n \to \infty$, and it suffices to keep only a few terms in the series to obtain a good accuracy, uniformly over the frequency parameter k. Thus, the problem is reduced to a slowly convergent series of the form

$$\sum_{n=1}^{\infty} \frac{e^{\delta n i}}{n^{\beta}} = C_{\beta}(\delta) + i S_{\beta}(\delta), \tag{3.118}$$

with some values of δ and β; this series was studied in the previous section.

Helpful remarks

It is interesting to note that in the representation (3.113) each term is similar to the point-source Green's function for the full unbounded space. This is in complete agreement with the heuristic idea that the problem of determining the Green's function in a layer differs from the analogous problem in unbounded space by additional boundary conditions $\partial p / \partial n$ over two straight lines. Any pair of sources displaced symmetrically with respect to any straight line automatically provides trivial normal derivative on this line. From this point of view, Eq. (3.113) can be treated as representing a sequence of required sources to arrange a symmetrical balance of the total array with respect to these boundary lines.

3.6. Waves in the Stratified Half-Plane

Here we will show that a stratified half-space possesses, under certain conditions, the properties of a layer of constant thickness established in Section 3.1.

Let us consider a 2D problem for a stratified acoustic half-plane $y > 0$, $|x| < \infty$, where the wave process is assumed to be harmonic in time with angular frequency ω. Then, the acoustic pressure p satisfies the Helmholtz equation

$$\Delta p(x,y) + \frac{\omega^2}{c^2(y)}\, p(x,y) = 0, \tag{3.119}$$

where the wave speed $c(y)$ is a function of the coordinate y varying with the depth of the half-plane.

The application of the Fourier transform with respect to the variable x reduces the partial differential equation (3.119) to the following ordinary equation of the second order with variable coefficient:

$$P''(s,y) - \tilde{\gamma}^2(s,y)\,P(s,y) = 0, \qquad \tilde{\gamma}(s,y) = \sqrt{s^2 - \frac{\omega}{c^2(y)}}. \qquad (3.120)$$

The sign of the derivative is related to the variable y.

As usual, the boundary condition may be of various type. To be more specific, we consider a point source placed on the free boundary surface $y = 0$ at a point x_0:

$$y = 0: \qquad p = \delta(x - x_0). \qquad (3.121)$$

In this scalar acoustic problem wave behavior at infinity may be accepted to satisfy Sommerfeld's radiation condition.

Generally, Eq. (3.120) cannot be solved analytically. For some specific functions $c(y)$ its solution is expressed in terms of certain special functions, and a good survey of these cases admitting exact analytical solution is given by Brekhovskikh (1980).

In the present section we construct an explicit analytical solution for high frequencies. In doing so, we assume the wave speed $c(y)$ to be a monotonic and bounded function of the argument $y > 0$, at least twice differentiable: $c(y) \in C^2\,[0, \infty)$, $m \le |c(y)| \le M$.

In ocean acoustics, a monotonic behavior of $c(y)$ is valid for Arctic zone, as well as for moderate latitudes in the winter period of the year.

Let us apply the change of variable $s = \omega\tilde{s}$ (the tildes are omitted later on), then Eq. (3.120) becomes

$$P''(s,y) - \omega^2\gamma^2(s,y)\,P(s,y) = 0, \qquad \gamma(s,y) = \sqrt{s^2 - \frac{1}{c^2(y)}}. \qquad (3.122)$$

The further considerations for the cases of monotonically decreasing and monotonically increasing functions $c(y)$ are absolutely different.

The case of monotonically decreasing wave speed c(y).
Here

$$c(0) = c_1, \quad c(\infty) = c_2, \quad c_1 > c_2, \quad c_2 \le c(y) \le c_1. \qquad (3.123)$$

Since the dependence upon the parameter s in Eq. (3.122) is even, we consider only its positive values $0 \le s < \infty$.

Case 1. Let

$$0 \le s < \frac{1}{c_1} \quad \text{or} \quad s > \frac{1}{c_2}. \qquad (3.124)$$

For such values of the parameter s, the function $\gamma(s,y)$ is nonzero. Then for high frequencies ω, Eq. (3.122) can be solved by the classical WKB method (see, for instance, Erdelyi, 1956; Brekhovskikh, 1980) to give

$$P(s,y) = \frac{1}{\gamma(s,y)}[A(s)\,e^{-\omega\xi(s,y)} + B(s)\,e^{\omega\xi(s,y)}], \qquad \xi(s,y) = \int_0^y \gamma(s,y)\,dy, \qquad (3.125)$$

where $A(s)$ and $B(s)$ are some unknown functions.

It is noteworthy that the representation (3.125) is valid both in the case $s > 1/c_2 > 1/c(y)$, where $\gamma(s,y) > 0$, and in the case $0 < s < 1/c_1 < 1/c(y)$, where $\gamma(s,y) = -iq(s,y)$ and $q(s,y) = \sqrt{[1/c(y)]^2 - s^2} > 0$. In both cases, the second term in Eq. (3.125) should be neglected, because for inhomogeneous waves ($\gamma > 0$) it gives an unbounded solution (since

$\xi(s, y) \to +\infty$ as $y \to +\infty$), and for homogeneous waves ($\gamma = -iq$, $q > 0$) this gives a solution with a far-field behavior of the type $\exp(-ib\omega y)$, $y \to +\infty$ ($b = \sqrt{(1/c_2)^2 - s^2} > 0$), which represents a wave arriving from infinity rather than a wave going to infinity. The first term in (3.125) certainly satisfies Sommerfeld's radiation condition.

Thus, in the considered range of variation of the parameter s we have $B = 0$; hence

$$P(s, y) = \frac{A(s)}{\gamma(s, y)} e^{-\omega\xi(s,y)}, \quad \text{or} \quad P(s, y) = P(s, 0) \frac{\gamma(s, 0)}{\gamma(s, y)} e^{-\omega\xi(s,y)}. \tag{3.126}$$

Case 2. Now consider the case of

$$1/c_1 < s < 1/c_2 \quad \sim \quad c_2 < 1/s < c_1. \tag{3.127}$$

Then there is always a unique solution of the equation $\gamma = 0 \sim c(y_0) = 1/s$, since $c(y)$ is monotonic and $c_2 < c(y) < c_1$. The value $y = y_0 = y_0(s)$ is a simple root of the equation $\gamma = 0$: $\gamma(s, y_0) = 0$, $\gamma'(s, y_0) \neq 0$. This point y_0 ($0 < y_0 < \infty$) is called a *turning point* of the differential equation (3.122). It should be noted that we assumed the function $c(y)$ to be monotonic just to guarantee that the root y_0 is simple.

The general solution of Eq. (3.122) in this case is known to be expressed as follows (see Erdelyi, 1956; Brekhovskikh, 1980):

$$P(s, y) = \frac{1}{\sqrt{\xi'(s, y)}} \left\{ A(s) \operatorname{Ai} [\omega^{2/3}\xi(s, y)] + B(s) \operatorname{Bi} [\omega^{2/3}\xi(s, y)] \right\},$$

$$\xi(s, y) = \left[\frac{3}{2} \int_{y_0}^{y} \gamma(s, y) \, dy \right]^{2/3}, \tag{3.128}$$

where $\operatorname{Ai}(x)$ and $\operatorname{Bi}(x)$ are the Airy functions, $A(s)$ and $B(s)$ are some unknown functions, and the derivative is applied with respect to y.

Let us study the behavior of solution (3.128) at infinity. Note that $c_2 < c(y) < c(y_0) = 1/s$ for $y > y_0$, so in this range

$$\gamma = \sqrt{s^2 - \frac{1}{c^2(y)}} = -iq, \quad q = q(s, y) = \sqrt{\frac{1}{c^2(y)} - s^2} > 0; \quad \text{hence} \quad \text{as} \quad y \to +\infty,$$

$$\xi(s, y) \sim \left[\frac{3}{2} iq_2(s)y \right]^{2/3} = -\left[\frac{3}{2} q_2(s)y \right]^{2/3}, \quad q_2(s) = \sqrt{\frac{1}{c_2^2} - s^2} > 0. \tag{3.129}$$

Then we should use the asymptotics of the Airy functions for large negative arguments (see Fedorjuk, 1977):

$$\operatorname{Ai}(-y) \sim \frac{\sin(\eta + \pi/4)}{\pi^{1/2}y^{1/4}}, \quad \operatorname{Bi}(-y) \sim \frac{\cos(\eta + \pi/4)}{\pi^{1/2}y^{1/4}}, \quad y \to +\infty, \quad \eta = \frac{2}{3} y^{3/2}. \tag{3.130}$$

In the considered case

$$\eta \sim \omega q_2 y, \quad \xi'(y) \sim -\left(\frac{2}{3} \right)^{1/3} q_2^{2/3} y^{-1/3}, \quad y \to +\infty, \tag{3.131}$$

hence, as $y \to +\infty$,

$$P(s, y) \sim \text{const} \left(q_2^{-1/3} y^{1/6} \right) \left(\omega^{-1/6} q_2^{-1/6} y^{-1/6} \right)$$
$$\times \left[A(s) \sin\left(\omega q_2 y + \tfrac{\pi}{4} \right) + B(s) \cos\left(\omega q_2 y + \tfrac{\pi}{4} \right) \right] \tag{3.132}$$
$$= \text{const} \left[A(s) \sin\left(\omega q_2 y + \tfrac{\pi}{4} \right) + B(s) \cos\left(\omega q_2 y + \tfrac{\pi}{4} \right) \right].$$

This solution must satisfy the radiation condition, i.e., must be of the form

$$P(s, y) \sim \text{const}(s)\, e^{idy}, \quad d > 0, \quad y \to +\infty. \tag{3.133}$$

This can be provided if and only if

$$A(s) = iB(s). \tag{3.134}$$

Therefore, in this case the general solution of Eq. (3.132) that satisfies the radiation condition at infinity is expressed as

$$P(s, y) = \frac{B(s)}{\sqrt{\xi'(s, y)}}\left\{i\,\text{Ai}[\omega^{2/3}\xi(s, y)] + \text{Bi}[\omega^{2/3}\xi(s, y)]\right\}; \tag{3.135}$$

hence

$$P(s, y) = P(s, 0)\sqrt{\frac{\xi'(s, 0)}{\xi'(s, y)}}\,\frac{i\,\text{Ai}[\omega^{2/3}\xi(s, y)] + \text{Bi}[\omega^{2/3}\xi(s, y)]}{i\,\text{Ai}[\omega^{2/3}\xi(s, 0)] + \text{Bi}[\omega^{2/3}\xi(s, 0)]}. \tag{3.136}$$

Let us study the structure of the solution for any fixed $0 < y < y_0$:

$$\begin{aligned}
p(x, y) = &\frac{1}{\pi}\int_{|s|\leq 1/c_1}\frac{\gamma(s, 0)}{\gamma(s, y)}e^{-\omega\xi(s,y)}\,e^{is(x-x_0)}\,ds \\
&+ \int_{1/c_1<|s|<1/c_2}\sqrt{\frac{\xi'(s, 0)}{\xi'(s, y)}}\,\frac{i\,\text{Ai}[\omega^{2/3}\xi(s, y)] + \text{Bi}[\omega^{2/3}\xi(s, y)]}{i\,\text{Ai}[\omega^{2/3}\xi(s, 0)] + \text{Bi}[\omega^{2/3}\xi(s, 0)]}e^{is(x-x_0)}\,ds \quad (3.137) \\
&+ \int_{|s|\geq 1/c_2}\frac{\gamma(s, 0)}{\gamma(s, y)}e^{-\omega\xi(s,y)}e^{is(x-x_0)}\,ds = I_1(x, y) + I_2(x, y) + I_3(x, y),
\end{aligned}$$

where we have taken into account that $p(x, y) = \delta(x - x_0)$ and hence $P(s, 0) = e^{ix_0 s}$. The integrands here contain the branching functions $\gamma(s, 0) = \sqrt{s^2 - 1/c_1^2}$ and $\gamma(s, y) = \sqrt{s^2 - 1/c^2(y)}$. The application of the principle of extremely low absorption shows that the positive branching points $s = 1/c$, $s = 1/c(y)$ shift upwards with small absorption, and the negative ones shift downwards. In the first integral I_1, we have $\xi(s, y) = -i\int_0^y q(s, y)\,dy$, $q > 0$; hence there is an oscillating phase in the argument of the exponential function. According to the asymptotic estimate of the type (1.73) with $\beta = 1/2$ and $\alpha = 1$, this integral turns out to be of the order of $O(\omega^{-1/2})$, $\omega \to \infty$, uniformly over $|x| < \infty$. In the third integral I_3, $\xi(s, y) = \int_0^y \gamma(s, y)\,dy$, $\gamma > 0$, so the phase in the argument of the exponential function is positive, and we have, as $\omega \to \infty$, a exponentially decaying kernel. Then the asymptotics (1.68) with $\beta = 1/2$ and $\alpha = 1$ gives the same estimate $O(\omega^{-1/2})$ for I_3. We thus can conclude that homogeneous waves can exist only in the case that the denominator of the second integrand itself has a pole. However, if $0 < y < y_0$, then $c(y) > c(y_0) = 1/s$ and so $\xi(s, y) = \left[\frac{3}{2}\int_{y_0}^y \gamma(s, y)\,dy\right]^{2/3} > 0$, since $\gamma = \sqrt{s^2 - [1/c(y)]^2} > 0$. It is clear from this observation that in the considered range we can apply the asymptotic expansion of the Airy functions, present in the integrand of I_2, for large positive argument:

$$\text{Ai}(y) \sim \frac{e^{-\eta}}{2\sqrt{\pi}\,y^{1/4}}, \quad \text{Bi}(y) \sim \frac{e^{\eta}}{2\sqrt{\pi}\,y^{1/4}}, \quad \eta = \frac{2}{3}y^{3/2}, \quad y \to +\infty, \tag{3.138}$$

which converts I_2 into the following expression:

$$\begin{aligned}
I_2(x, y) = &\int_{1/c_1<|s|<1/c_2}\sqrt{\frac{\xi'(s, 0)}{\xi'(s, y)}}\left[\frac{\xi(s, 0)}{\xi(s, y)}\right]^{1/4} \\
&\times \exp\left\{\omega\left[\int_{y_0}^y \gamma(s, y)\,dy - \int_{y_0}^0 \gamma(s, y)\,dy\right]\right\}e^{-is(x-x_0)}\,ds \quad (3.139) \\
= &\int_{1/c_1<|s|<1/c_2}\frac{\gamma(s, 0)}{\gamma(s, y)}e^{-\omega\xi(s,y)}e^{-is(x-x_0)}\,ds.
\end{aligned}$$

It thus asymptotically coincides with I_3, and hence is of the order of $O(\omega^{-1/2})$, $\omega \to \infty$. Mathematically, this means that there is no homogeneous mode in the structure of the solution since the integrand has no poles on the real axis $\text{Im}(s) = 0$. From heuristic point of view, this means that a gradually decreasing (with depth) stiffness cannot provide such turning points for acoustic beams, which behave like the bottom of a layer reflecting incident rays back and thus generating standing (i.e., mode) waves.

The case of monotonically increasing wave speed $c(y)$.
Here $c(0) = c_1$, $c(\infty) = c_2$, $c_1 < c_2$, and $c_1 \le c(y) \le c_2$.
The two cases for different s remain here as in the case of decreasing $c(y)$.
Case 1:

$$\{0 \le s < 1/c_2\} \cup \{s > 1/c_1\}. \tag{3.140}$$

In this range $\gamma(s, y) \ne 0$ and hence, in accordance with the WKB method,

$$P(s, y) = e^{isx_0} \frac{\gamma(s, 0)}{\gamma(s, y)} e^{-\omega \xi(s, y)}, \quad \gamma(s, y) = \sqrt{s^2 - 1/c^2(y)}. \tag{3.141}$$

Case 2:

$$1/c_2 < s < 1/c_1 \quad \sim \quad c_1 < 1/s < c_2. \tag{3.142}$$

Here, for the same reason as in the case of decreasing $c(y)$, the solution can be expressed again as

$$P(s, y) = \frac{1}{\sqrt{\xi'(s, y)}} \left\{ A(s) \, \text{Ai} \left[\omega^{2/3} \xi(s, y) \right] + B(s) \, \text{Bi} \left[\omega^{2/3} \xi(s, y) \right] \right\},$$
$$\xi(s, y) = \left[\frac{3}{2} \int_{y_0}^{y} \gamma(s, y) \, dy \right]^{2/3}. \tag{3.143}$$

With $y \to +\infty$ we have $1/s = c(y_0) < c(y) < c_2 \sim 1/c(y) < s$, so $\gamma = \sqrt{s^2 - 1/c^2(y)} > 0$ in this range. To estimate the far-field behavior of (3.143) as $y \to +\infty$, we can apply the asymptotic expansions of the Airy functions for large positive argument (see Abramowitz and Stegun, 1965):

$$\text{Ai}(y) \sim \frac{e^{-\eta}}{2\sqrt{\pi} y^{1/4}}, \qquad \text{Bi}(y) \sim \frac{e^{\eta}}{2\sqrt{\pi} y^{1/4}}, \qquad \eta = \frac{2}{3} y^{3/2}. \tag{3.144}$$

Therefore, to obtain a bounded solution in the far zone, we should set $B(s) = 0$. Then (3.143) becomes equivalent to

$$P(s, y) = \sqrt{\frac{\xi'(s, 0)}{\xi'(s, y)}} \frac{\text{Ai}[\omega^{2/3} \xi(s, y)]}{\text{Ai}[\omega^{2/3} \xi(s, 0)]} e^{isx_0}. \tag{3.145}$$

In order to discover mode properties of a stratified half-space, let us study, for example, the velocity of oscillations on the free surface: $v_y(x, 0) = (i\omega\rho)^{-1}(\partial p/\partial y)(x, 0)$. It is obvious that the leading asymptotic term for high ω is obtained by differentiating only the Airy function in the numerator:

$$P'(s, 0) = e^{isx_0} \xi'(s, 0) \, \omega^{3/2} \frac{\text{Ai}'[\omega^{2/3} \xi(s, 0)]}{\text{Ai}[\omega^{2/3} \xi(s, 0)]}. \tag{3.146}$$

Representing the inverse Fourier integral for the acoustic pressure as the sum of three integrals of the type (3.137), we can estimate that I_1 and I_3 are again of the order of $O(\omega^{-1/2})$, $\omega \to \infty$. The second integral I_2 is here

$$I_2 = \omega^{3/2} \int_{1/c_2 < |s| < 1/c_1} \xi'(s,0) \frac{\mathrm{Ai}'[\omega^{2/3}\xi(s,0)]}{\mathrm{Ai}[\omega^{2/3}\xi(s,0)]} e^{is(x-x_0)} \, ds, \qquad (3.147)$$

where $c(y) < c(y_0) = 1/s$ if $0 \le y < y_0$, and so

$$\xi(s,0) = \left[\frac{3}{2} \int_{y_0}^0 \sqrt{s^2 - \frac{1}{c^2(y)}} \, dy \right]^{2/3} = \left[\frac{3}{2} i \int_0^{y_0} \sqrt{\frac{1}{c^2(y)} - s^2} \, dy \right]^{2/3} = -\alpha(s,0),$$

$$\alpha(s,0) = \left[\frac{3}{2} \int_0^{y_0} \sqrt{\frac{1}{c^2(y)} - s^2} \, dy \right]^{2/3} > 0. \qquad (3.148)$$

Helpful remarks

$1°$. We can conclude from Eqs. (3.147), (3.148) that the argument of the Airy functions is negative, and both functions Ai and Ai' oscillate for such arguments, with a change of the sign. Mathematically, zeros s_m of the denominator, i.e., solutions of $\mathrm{Ai}[\omega^{2/3}\xi(s,0)] = 0$, determine, by the residue theorem, homogeneous mode waves $B_m e^{is_m(x-x_0)}$ that propagate along the free surface to the right ($x \to +\infty$), in accordance with Sommerfeld's radiation condition. Physically, a stratified acoustic half-plane turns out to be a layer, since the increasing stiffness of the medium makes acoustic rays reflect and turn back upwards, thus leading, with a full interference, to the formation of mode waves concentrated in an upper layer of the half-space.

$2°$. It should be noted that in the tropical seas the profile line of the wave speed with depth is nonmonotonic. Typically, in the near-surface zone, the wave speed first decreases and then, in deeper layers, becomes a monotonically increasing function. The function $c = c(y)$ has thus one minimum at a certain depth $y = y_0$. It is known that near this critical depth there can arise waveguides. Mathematically, our analysis in this case becomes more complicated, because there may appear two different turning points.

Chapter 4

Analytical Methods for Simply Connected Bounded Domains

4.1. General Spectral Properties of the Interior Problem for Laplacian

When we speak about spectral properties of any homogeneous boundary value problem for the Laplacian, this question is evidently connected with existence of eigenfrequencies and related eigenfunctions, to be more precise, any eigenfrequency ω with respective eigenfunction $p \neq 0$:

$$\Delta p + k^2 p = 0, \qquad k = \frac{\omega}{c}, \tag{4.1}$$

evidently belongs to the spectrum of the operator $-\Delta$:

$$-\Delta p = \mu p, \qquad \mu = k^2. \tag{4.2}$$

In our presentation of boundary value problems for the wave (Helmholtz) operator we started in Chapter 2 from exterior diffraction problems for a full (and so infinite) space $\mathbb{R}^n$ ($n = 2, 3$), where we could see that both the Dirichlet and Neumann exterior problems satisfying Sommerfeld's radiation condition have a unique solution. This certainly implies that there are no eigenvalues in (4.1)–(4.2); otherwise there would be a nontrivial solution to the exterior problem with the trivial boundary condition, which contradicts uniqueness.

In Chapter 3 we studied wave properties of acoustic layer of constant thickness, which could be called a *semi-infinite* domain, since it possesses some properties of the infinite domain. In particular, correct formulation requires a certain boundary condition at infinity, as in the case of full space. At the same time, in acoustic layer there can exist homogeneous waves, i.e., nontrivial solutions that satisfy a homogeneous boundary condition. This happens if the frequency is greater than a certain critical value.

Here we will study spectral properties of the same problem in the bounded domains. We will show that the spectrum is always discrete and give an asymptotic estimate for high eigenfrequencies.

The described spectral problem in the domain D is quite classical, and is studied by various methods. One of them is based on the fact that the operator $-\Delta$ is positive definite and self-adjoint in the functional space $L_2(D)$, defined on the subset of doubly differential functions $C_2(D) \subset L_2(D)$.

Symmetry of the operator $A = -\Delta$ subjected to homogeneous Dirichlet or Neumann boundary conditions is proved by using a different form of Green's formula:

$$
\int_D (Au, v) \, dy = -\int_D \Delta u(y) \, v(y) \, dy = \int_D \sum_{j=1}^{m} \frac{\partial u}{\partial y_j} \frac{\partial v}{\partial y_j} \, dy - \int_{\partial D} \frac{\partial u(y)}{\partial n} v(y) \, ds
$$
$$
= \int_D \sum_{j=1}^{m} \frac{\partial u}{\partial y_j} \frac{\partial v}{\partial y_j} \, dy, \qquad y \in \mathbb{R}^m \quad (m = 2, 3), \tag{4.3}
$$

for arbitrary functions $u, v \in C_2(D)$ satisfying homogeneous conditions of either Dirichlet or Neumann type. Now it becomes clear that the symmetry of the operator $-\Delta$ is explained by the fact that the swap of the functions u and v on the left-hand side of Eq. (4.3) does not change its right-hand side.

Positive definiteness of the operator $A = -\Delta$ also follows from Eq. (4.3) by using the well-known

Friedrichs inequality (see Mikhlin, 1964). If $u \in C_2(D) \subset L_2(D)$ and $u(y)$ satisfies either Dirichlet or Neumann homogeneous boundary conditions, then

$$\int_D \sum_{j=1}^m \left(\frac{\partial u}{\partial y_j} \right)^2 dy \geq \gamma^2 \int_{\partial D} u^2(y)\, ds, \quad y \in \mathbb{R}^m, \quad m = 2,3 \quad (\gamma^2 > 0). \tag{4.4}$$

Equations (4.3), (4.4) prove that the operator $A = -\Delta$ is symmetric (self-adjoint) and positive definite over the subspace of functions from $C_2(D)$ and satisfying either Dirichlet or Neumann homogeneous boundary conditions on ∂D. Then spectral properties of this operator are directly extracted from general theory of such operators acting in a Hilbert space.

As can be seen, here we considered the case $H = L_2(D)$, since we used the scalar product and the norm defined in this space. Unfortunately, the considered operator $A = -\Delta$ does not act on the whole $H = L_2(D)$ since it cannot be defined in the classical sense if u is not doubly differential. To overcome this difficulty, there was created a special theory of Sobolev's Hilbert functional spaces (which is well presented, for example, in Mikhlin, 1964). This new theory operates with the so-called *energetic* rather than classical solution of the considered operator equation, which permits the application of the mentioned general theory of positive definite self-adjoint operators. This theory gives also an excellent basis for constructing finite element methods.

However, we prefer to follow the line of classics-founders of modern mathematics, which is based on the potential theory and the concept of Green's function for bounded domains D.

Let us introduce a Green's function for the simply connected bounded domain D, which satisfies the given homogeneous boundary condition of Dirichlet or Neumann type.

To be more specific, we restrict the consideration by the 2D problem with the Dirichlet condition. Other cases can be studied by analogy. Then the sought Green's function $G(z, x)$, $z, x \in D \subset \mathbb{R}^2$ ought to possess the following properties:

1) To be a solution of the Laplace equation

$$\Delta_x G(z, x) = 0, \qquad x \in D \setminus D_z^\varepsilon, \tag{4.5}$$

outside of a small ε-neighborhood of the chosen point $z \in D$.

2) To satisfy the homogeneous boundary condition:

$$G(z, x) = 0, \qquad x \in \partial D = l. \tag{4.6}$$

3) When $x \to z$, the behavior of this function is absolutely the same as of the one for full space $\mathbb{R}^2$:

$$G(z, x) = -\frac{1}{2\pi} \ln |z - x| + H(z, x), \tag{4.7}$$

where $H(z, x)$ is a regular function in D with respect to both its arguments.

THEOREM 1. *If the boundary l of the bounded simply connected domain is smooth, then the Green's function (4.5)–(4.7) exists and it is unique.*

Proof. First of all, we notice that it suffices to prove that for any fixed $z \in D$ there exists a unique regular solution $H(z, x)$ of the Laplace equation

$$\Delta_x H(z, x) = 0, \qquad x \in D, \tag{4.8}$$

with the nonhomogeneous Dirichlet boundary condition

$$H(z, y) = \varphi(y) = \frac{1}{2\pi} \ln|z - y| \in C(l), \qquad y \in l. \tag{4.9}$$

By the method analogous to that applied in Sections 2.1 and 2.2, the questions of solvability of this problem can be reduced to the investigation of the BIE with a Green's function for the whole space and $k = 0$:

$$\Phi_0(|x - y|) = \Phi|_{k=0}(|x - y|) = -\frac{1}{2\pi} \ln|x - y|. \tag{4.10}$$

Namely, let us apply the indirect BIE method by introducing the potential of double layer, analogously to approach described in remark 3° for Section 2.2. Namely, the function $u(x) = H(z, x)$

$$u(x) = \int_l \frac{\partial \Phi_0(|x - y|)}{\partial n_y} \psi(y)\, dl_y, \qquad x \in D, \tag{4.11}$$

gives a solution to problem (4.8)–(4.9) if the density $\psi(y)$ satisfies the integral equation

$$\psi(y_0) - 2 \int_l \frac{\partial \Phi_0(|y_0 - y|)}{\partial n_y} \psi(y)\, dl_y = \varphi(y_0), \tag{4.12}$$

which is a consequence of the limit boundary value of the double-layer potential as $x \to y_0 \in l$. The latter is taken from inside, which involves the sign "minus" in front of the integral in Eq. (4.12), in contrast to Eq. (2.65).

Further, let us prove that $\lambda = 1$ cannot belong to spectral set of the operator K,

$$(K\psi)(y) = 2 \int_l \frac{\partial \Phi_0(|y_0 - y|)}{\partial n_y} \psi(y)\, dl_y. \tag{4.13}$$

Indeed, if there would be an eigenfunction $\mu(y)$, $y \in l$, coupled with the eigenvalue $\lambda = 1$, then we would have

$$\mu(y_0) = 2 \int_l \frac{\partial \Phi_0(|y_0 - y|)}{\partial n_y} \mu(y)\, dl_y, \qquad y_0 \in l, \qquad \mu(y) \in C(l). \tag{4.14}$$

Then the function

$$v(x) = \int_l \Phi_0(|x - y|)\, \mu(y)\, dl_y \tag{4.15}$$

would give a solution to the exterior Neumann problem for the Laplace equation $\Delta v = 0$ with the boundary condition (see Theorem 5, Section 2.1)

$$\left[\frac{\partial v(x)}{\partial n_x} \right]_{x=y_0} = -\frac{\mu(y_0)}{2} + \int_l \frac{\partial \Phi_0(|y_0 - y|)}{\partial n_{y_0}}\, dl_y = 0, \tag{4.16}$$

where we have used identity (3.14).

Finally, we have arrived at the function $v(x)$, which is a solution of the exterior Neumann problem with the trivial boundary condition. Due to the uniqueness of the solution to this problem, we have $v(x) \equiv 0$, and hence $\mu(y) \equiv 0$, which contradicts the assumption that $\mu(y)$ is an eigenfunction.

If $\lambda = 1$ is not an eigenvalue of operator (4.13), this means that the Fredholm integral equation of the second kind (4.12) is uniquely solvable. Then Eq. (4.11) determines the structure of the regular part $H(z, y)$ of the Green's function, and so Eq. (4.7) gives the Green's function itself. The theorem is proved.

THEOREM 2. *The constructed Green's function for Laplacian in the domain D is symmetric: $G(z, x) = G(x, z)$.*

Proof. Let us apply Green's formula for the pair of functions $u_1(y) = G(z, y)$ and $u_2(y) = G(x, y)$ in the domain $\widetilde{D}$ formed by D with small ε-neighborhoods of the points z and x removed. Since both functions satisfy the Laplace equation, we have

$$
\begin{aligned}
0 &= \int_{\widetilde{D}} \left[G(z, y)\Delta G_y(x, y) - G(x, y)\Delta_y G(z, y) \right] dl_y \\
&= \int_{l+l_1+l_2} \left[G(z, y)\frac{\partial G(x, y)}{\partial n_y} - G(x, y)\frac{\partial G(z, y)}{\partial n_y} \right] dl_y,
\end{aligned}
\tag{4.17}
$$

l_1 and l_2 are small circles of radius ε around points z and x, respectively. The integral along l is equal to zero, since both functions satisfy homogeneous boundary conditions. Calculation of the contribution from integration over Γ_1 and Γ_2, in the same way as in deriving the Kirchhoff–Helmholtz formula (Section 2.2), leads to the identity $G(z, x) = G(x, z)$.

THEOREM 3. *If $G(x, y)$ is a Green's function for a homogeneous Dirichlet problem in the domain D, then the homogeneous boundary value problem for the Poisson equation*

$$
\Delta p(x) = -f(x), \qquad x \in D, \qquad p|_l = 0,
\tag{4.18}
$$

has the following solution written in exact explicit form:

$$
p(x) = \int_D G(x, y)\, f(y)\, dy.
\tag{4.19}
$$

Proof. Let us apply Green's formula to the pair of functions $u_1(y)=p(y)$, $u_2(y)=G(x, y)$ in the domain $\widetilde{D}$ formed as D with a small ε-neighborhood about the point x deleted:

$$
\int_{\widetilde{D}} G(x, y)f(y)\, dy = \int_{\widetilde{D}} (p\,\Delta_y G - G\,\Delta p)\, dy = \int_{l+l_\varepsilon} \left(p\frac{\partial G}{\partial n_y} - G\frac{\partial p}{\partial n_y} \right) dl_y.
\tag{4.20}
$$

Integral over the contour l is zero since both functions satisfy homogeneous boundary conditions. Further, the second term in (4.20) gives no contribution over l_ε when $\varepsilon \to 0$ since p and $\partial p/\partial n$ are regular inside D and G has only a weak (i.e., integrable) singularity. The first term's contribution is calculated in the same way as in derivation of the Kirchhoff–Helmholtz integral formula, which results finally in Eq. (4.19). The theorem is proved.

It follows from this theorem that the spectral problem for Laplacian (4.2) can be reduced to the problem of characteristic values for a self-adjoint integral operator, whose kernel is the Green's function:

$$
p = \mu\, Gp \quad \sim \quad p(x) = \mu \int_D G(x, y)\, p(y)\, dy, \qquad x \in D.
\tag{4.21}
$$

What is the difference between the self-adjoint problem for a positive definite operator (4.2) and the self-adjoint problem (4.21)? By other words, why do we prefer to study Eq. (4.21), instead of (4.2)? The answer is quite evident. The most thorough results for operator equations are obtained in Hilbert space, and the natural functional Hilbert space is $L_2(D)$. Among two operators (4.2), (4.21) the latter acts in $L_2(D)$, but the former can be applied only to doubly-differential functions, and so it does not act in $L_2(D)$.

Now we can apply, in our spectral analysis, the well-developed Fredholm theory (see Section 1.5) to the integral equation (4.21) of the second kind with symmetric kernel

$G(x, y)$. Further results are based on a combination of the Fredholm theory with some properties of positive definite operators. This is possible since (4.2) $\sim$ (4.21).

First of all, it follows from the fourth Fredholm theorem that the frequency spectrum of the interior acoustic problem for any domain D is discrete, and the number of eigenfrequencies ω_m is countable. Further, the only accumulation point for these spectral values may be at infinity, i.e., $\omega_m \to \infty$, $m \to \infty \sim k_m \to \infty$, $m \to \infty$. What can be said additionally about these eigenvalues k_m? The following properties of eigenvalues follow from classical results of general spectral theory for operator equations (4.2) and (4.21).

THEOREM 4. *All eigenvalues of a self-adjoint operator G in the complex Hilbert space H are real-valued.*

Proof. Let $G : H \to H$ and let there exist an eigenvalue λ and $u \neq 0$ such that $Gu = \lambda u$. Then $\lambda(u, u) = (Gu, u) = (u, Gu) = (u, \lambda u) = \bar{\lambda}(u, u)$, so $\lambda = \bar{\lambda}$, which implies that λ is real-valued.

Corollary. All eigenvalues of the homogeneous interior acoustic problem are real-valued, which directly follows from the symmetry of the operator G in equation (4.21), $\mu = 1/\lambda$, acting in $L_2(D)$.

THEOREM 5. *All eigenvalues of a positive definite operator G, defined on a subset of any Hilbert space H, are positive.*

Proof. Let $G : V \to H$, $V \subset H$, and $(Gu, u) \geq \gamma^2(u, u)$. If there is a real number λ and $u \neq 0$ such that $Gu = \lambda u$, then under the conditions of the theorem:

$$(Gu, u) = \lambda(u, u) \quad \sim \quad \lambda = (Gu, u)/(u, u) \geq \gamma^2 > 0. \tag{4.22}$$

Corollary. All eigenvalues of the homogeneous interior acoustic problem are positive, which is a consequence of the positive definiteness of the operator equation (4.2) considered on the sublet $C_2(D) \subset L_2(D) = H$.

THEOREM 6. *Eigenfunctions related to different eigenvalues of any symmetric operator are orthogonal to each other.*

Proof. If we have for Eq. (4.21) $Gu_i = \lambda_i u_i$, $Gu_j = \lambda_j u_j$, $u_i \neq 0$, $u_j \neq 0$, then $\lambda_i(u_i, u_j) = (Gu_i, u_j) = (u_i, Gu_j) = \lambda_j(u_i, u_j)$. Hence, if $\lambda_i \neq \lambda_j$, then $(u_i, u_j) = 0$.

We will assume throughout this section that the set of eigenfunctions u_i, $i = 1, 2, \ldots$, is normalized: $(u_i, u_j) = \delta_{ij}$, where δ_{ij} is the Kronecker delta.

Now we cite the classical Hilbert–Schmidt theorem, which is presented in the literature in various forms. For the forthcoming applications we formulate it in the following version, which is indeed one of numerous corollaries from this theorem.

THEOREM 7. *(Hilbert–Schmidt Theorem). If a symmetric kernel of the integral equation (4.21) satisfies the condition*

$$\int_D |G(x, y)|^2 \, dy \leq A, \tag{4.23}$$

with a positive constant A, uniformly with respect to $x \in D$, then the kernel can be expanded to a series in its eigenfunctions

$$G(x, y) = \sum_{m=1}^{\infty} \frac{\varphi_m(x) \overline{\varphi_m(y)}}{\lambda_m}, \tag{4.24}$$

which converges uniformly over $(x, y) \in D$.

The proof is omitted. This is based on some constructive techniques connected with approximation of the kernel by finite-dimensional structures like some degenerate kernels.

Corollary 1. The sum of the series $\sum_{m=1}^{\infty} \left(1/\lambda_m^2\right)$ is finite.

Indeed, the double scalar product in relation (4.24) gives

$$\sum_{m=1}^{\infty} \frac{1}{\lambda_m^2} = \int_D \int_D |G(x,y)|^2 \, dx \, dy \leq A\, S(D).$$ (4.25)

Corollary 2. If, in the conditions of Theorem 7, $G(x,y)$ is continuous, then the sum of the series

$$\sum_{m=1}^{\infty} \frac{1}{\lambda_m} = \int_D G(x,x) \, dx < \infty$$ (4.26)

is finite.

This result is obtained by direct integration of Eq. (4.24).

Helpful remarks

$1°$. The principal goal of the present chapter is to study the interior acoustic problem described by the Helmholtz equation

$$\Delta p(x) + k^2 p(x) = 0, \qquad x \in D,$$ (4.27)

with some boundary conditions of Dirichlet or Neumann type, set on the boundary ∂D.

This problem can be reduced to a BIE, by the same technique as in Chapter 2. In regular cases this equation admits the application of direct numerical methods, like one described in Section 1.5. Therefore, investigation of eigenfrequencies, which make the problem irregular, is very important in this theory.

$2°$. It follows from the above analysis that every λ_i may be connected only with a finite number of eigenfunctions linearly independent. Otherwise this λ_i would be an accumulation point for the set $\{\lambda_m\}$. The number of linearly independent eigenfunctions coupled with the eigenvalue λ_i is called the *multiplicity* of λ_i.

$3°$. What do we already know about distribution of $\{\lambda_m\}$? They all are positive, form a discrete set, with the only accumulation point at infinity: $\lambda_m \to \infty, m \to \infty$. Multiplicity of any λ_m is finite. Therefore, we can arrange them in the increasing order: $0 < \lambda_1 \leq \lambda_2 \leq \lambda_3 \leq \cdots$ counting each eigenvalue as many times as its multiplicity.

$4°$. Convergence of the series (4.25) formed by inverse squares of characteristic numbers shows that the asymptotic behavior of λ_m is: $\lambda_m^{-1} = o(1/\sqrt{m})$, $m \to \infty$, but this is a too rough estimate. If the Green's function in equation (4.21) would be continuous then relation (4.26) would lead to the estimate: $\lambda_m^{-1} = o(1/m)$, $m \to \infty$. However, the Green's function $G(x,y)$ as a kernel of the integral operator in (21) has a logarithmic singularity $\sim \ln|x-y|$, when $|x-y| \to 0$, so the second estimate is not valid. Heuristically, the logarithmic singularity is extremely weak and we might expect that this estimate is very close, in the sense of increasing rate of λ_m, to the exact value. This heuristic idea appears quite reasonable, since below we will see that a correct estimate is $\lambda_m^{-1} = O(1/m)$.

$5°$. Strictly speaking, the number of eigenvalues and respective eigenfunctions may appear finite, but it can be proved that this happens only if the kernel of the symmetric integral operator is degenerate (see Courant and Hilbert, 1953). The Green's function for a bounded simply connected domain is not degenerate.

6°. Usually, a system of orthonormal functions $\{u_i\}$, $i = 1, 2, \ldots$, creates a basis in Hilbert space H if this system is complete in H. Theorem 6 asserts that eigenfunctions represent an orthonormal system, but there is no result on completeness. This question is not so simple, and in some cases it can be thoroughly proved that eigenfunctions form a complete subset (compare with the theory of Fourier series with trigonometric basis). However, generally, this requires rather refined analysis, which is beyond the scope of the present book.

4.2. Explicit Formulas for Eigenfrequencies of Round Disc

For a round disc of radius a, the Helmholtz equation $\Delta p + k^2 p = 0$ can be expressed in the polar coordinate system in the following form:

$$\frac{\partial^2 p}{\partial r^2} + \frac{1}{r}\frac{\partial p}{\partial r} + \frac{1}{r^2}\frac{\partial^2 p}{\partial \theta^2} + k^2 p = 0, \qquad 0 \le r \le a, \qquad 0 \le \theta < 2\pi. \tag{4.28}$$

One can seek a solution of the last equation as a Fourier expansion over $0 \le \theta < 2\pi$:

$$p = \sum_{n=-\infty}^{\infty} p_n(r)\, e^{in\theta}. \tag{4.29}$$

Then, for each $p_n(r)$, due to linear independence of trigonometric functions, one arrives at the ordinary differential equation

$$r^2 p_n''(r) + r p_n'(r) + (r^2 k^2 - n^2) p_n(r) = 0, \tag{4.30}$$

whose only solution regular inside the disk is the Bessel function of the first kind and of order m (see Abramowitz and Stegun, 1965):

$$p_n(r) = J_n(kr). \tag{4.31}$$

Therefore, a complete system of linearly independent solutions of Eq. (4.28) is $\{J_n(kr) \times \exp(in\theta)\}$, $|n| < \infty$. However, since $J_n(x) = (-1)^n J_n(x)$, this can be chosen as $\{J_n(kr) \times \exp(\pm in\theta)\}$, $n = 0, 1, 2, \ldots$, and a set of eigenvalues for the Dirichlet and Neumann boundary value problems could be found, respectively, from the following transcendental equations:

$$J_n(ka) = 0, \qquad J_n'(ka) = 0. \tag{4.32}$$

Our main purpose is to evaluate large eigenvalues from Eq. (4.32). To this end, we may use an asymptotic representation for the Bessel functions. The only trouble is related to the well known feature that there are absolutely different asymptotics of Bessel functions for the cases of large argument and large order.

This fact is very important because the subscript n in Eq. (4.32) may be arbitrary large. A natural consequence from this feature is that we have to apply a uniform asymptotic expansion valid for arbitrary values of n and k.

Let us calculate the number of eigenvalues $\lambda_m = k_m^2$ not exceeding n: $N(n) = \sum_{k_m \le n} 1$. Our presentation follows to Kuznetsov and Fedosov (1965), and to Babich and Buldyrev (1989).

It is clear that

$$N(n) = N_0(n) + 2\sum_{m=1}^{[n]} N_m(n), \tag{4.33}$$

since the only one eigenfunction is linked with the Bessel function of the zeroth order, and for other orders there are two different eigenfunctions corresponding to opposite signs of the exponential function depending on the angular polar coordinate. Here $N_m(n)$ designates the number of zeros of the Bessel function $J_m(x)$ from the interval $m < x \leq n$: $N_m(n) = \sum_j 1$, $J_m(x_j) = 0$, $m < x_j \leq n$. Here $[n]$ denotes integer part of n. Besides, we can claim that Eq. (4.33) is valid because it is quite a classical result that all positive zeros of the Bessel function are greater than its order: $J_i(x_j) = 0$ and hence $x_j > i$ (see Abramowitz and Stegun, 1965).

Let us introduce the two integer quantities ν_0, ν_1: $0 < \nu_0(n) < \nu_1(n) < [n]$, both being positive and less than $[n]$, so that $\nu_0(n) = [n^{1/3}]$, $\nu_1(n) = [n - n^{4/9}]$, $n \to \infty$.

Then

$$\sum_{m=1}^{[n]} N_m(n) = \sum_{m=1}^{\nu_0(n)} N_m(n) + \sum_{m=\nu_0+1}^{\nu_1(n)} N_m(n) + \sum_{m=\nu_1+1}^{[n]} N_m(n) = S_1 + S_2 + S_3, \qquad (4.34)$$

and we will estimate each of the three sums independently.

In order to estimate the last two sums S_2 and S_3, we may apply the following asymptotics of the Bessel function $J_m(x)$ (see Abramowitz and Stegun, 1965):

$$J_m(x) = \sqrt{\frac{2}{\pi\sqrt{x^2 - m^2}}} \left\{ \cos\left[\eta(x, m) - \frac{\pi}{4}\right] + O\left(\frac{1}{\eta}\right) + O\left(\frac{1}{m}\right) \right\}, \qquad (4.35)$$

valid uniformly at least on the interval $\nu_0(x) \leq m \leq \nu_1(x)$, where like above, $\nu_0(x) = x^{1/3}$, $\nu_1(x) = x - x^{4/9}$. Here, in Eq. (4.35)

$$\eta(x, m) = \sqrt{x^2 - m^2} - m \arccos\frac{m}{x}. \qquad (4.36)$$

Let us note that $\partial\eta(x, m)/\partial x = \sqrt{x^2 - m^2}/x \geq 0$ if $x \geq m$, so the function $\eta(x, m)$ is a monotonically increasing function of the first argument x. This allows us to conclude that the interval $m \leq x \leq n$ corresponds, with a one-to-one mapping, to the interval $0 \leq \eta(x, m) \leq \eta(n, m)$. Therefore, it follows from Eqs. (4.30), (4.31) that the number of zeros x_j of the function $J_m(x)$: $J_m(x_j) = 0$ on the interval $m \leq x_j \leq n$ coincides with the number of zeros of the function $\cos(\eta - \pi/4)$ on the interval $0 \leq \eta \leq \eta(n, m)$. Since the last quantity is evidently equal to $\eta(n, m)/\pi + 1/4$, we can conclude that asymptotically

$$N_m(n) = \frac{\eta(n, m)}{\pi} + \frac{1}{4}, \qquad \nu_0(n) \leq m \leq \nu_1(n). \qquad (4.37)$$

Now the first sum S_1 in Eq. (4.34) can be estimated in the following way:

$$S_1 = \sum_{m=1}^{\nu_0} N_m(n) = \sum_{m=1}^{\nu_0} [N_{\nu_0}(n) + N_m(n) - N_{\nu_0}(n)] = \nu_0 N_{\nu_0}(n) + \sum_{m=1}^{\nu_0} [N_m(n) - N_{\nu_0}(n)]. \quad (4.38)$$

The common term under the sum sign here admits the estimate:

$$N_m(n) - N_{\nu_0}(n) \leq N_0(n) - N_{\nu_0}(n). \qquad (4.39)$$

Further, we need to use a classical result about the number of positive zeros of the Bessel function $J_0(x)$ not exceeding the large quantity n. This is based on the asymptotic representation of this Bessel function for large argument (see Abramowitz and Stegun, 1965)

$$J_0(x) = \sqrt{\frac{2}{\pi x}} \left[\cos\left(x - \frac{\pi}{4}\right) + O\left(\frac{1}{x}\right) \right]; \qquad (4.40)$$

hence asymptotically the zeros of $J_0(x)$ we are seeking now coincide with zeros of cosine in Eq. (4.40):

$$N_0(n) = \frac{n}{\pi} + o(1), \qquad n \to \infty. \tag{4.41}$$

In order to finally complete the estimate (4.39), we note that

$$\eta(\nu_0, n) = \sqrt{n^2 - [n^{1/3}]^2} - ([n^{1/3}] - 1) \arccos \frac{[n^{1/3}] - 1}{n} = n + O(n^{1/3}), \tag{4.42}$$

so finally, taking into account Eqs. (4.37)–(4.39), (4.41), (4.42), we obtain

$$S_1 = \nu_0 N_{\nu_0}(n) + \nu_0 \left\{ \left[\frac{n}{\pi} + O(n^{1/3}) \right] - \left[\frac{n}{\pi} + o(1) \right] \right\} = \frac{n}{\pi} \nu_0 + O(n^{2/3}), \qquad n \to \infty. \tag{4.43}$$

In order to give an estimate of the sum S_3 in Eq. (4.29), let us notice that

$$S_3 = \sum_{m=\nu_1+1}^{[n]} N_m(n) \le \sum_{m=\nu_1+1}^{[n]} N_{\nu_1+1}(n) = \{[n] - (\nu_1 + 1)\}$$

$$\times \left\{ \frac{\eta(n, \nu_1 + 1)}{\pi} + \frac{1}{4} + o(1) \right\} = \{n^{4/9} + O(1)\} \left\{ \frac{\eta(n, \nu_1 + 1)}{\pi} + O(1) \right\}. \tag{4.44}$$

Let us calculate the value of the expression in the second braces here. We have

$$\eta(n, \nu_1 + 1) = \sqrt{n^2 - (n - n^{4/9})^2} - (n - n^{4/9}) \arccos \frac{n - n^{4/9}}{n} + O(1)$$

$$= \sqrt{2n^{13/9} - n^{8/9}} - (n - n^{4/9}) \arccos(1 - n^{-5/9}) + O(1). \tag{4.45}$$

Then we use the following Taylor series of arccos:

$$\arccos(1 - x) = \sqrt{2x} \left\{ 1 + \frac{x}{12} + O(x^2) \right\}, \qquad x \to +0. \tag{4.46}$$

Therefore, the leading asymptotic term in Eq. (4.45) is

$$\eta(n, \nu_1 + 1) = \sqrt{2} n^{13/18} \left(1 - \frac{n^{-5/9}}{2} \right)^{1/2} - (n - n^{4/9}) \sqrt{2} n^{-5/18} \left(1 + \frac{n^{-5/9}}{12} \right)$$

$$= \left\{ \sqrt{2} n^{13/18} - \frac{\sqrt{2}}{4} n^{1/6} \right\} - \left\{ \sqrt{2} n^{13/18} - \sqrt{2} n^{1/6} + \frac{\sqrt{2}}{12} n^{1/6} \right\} = \frac{2}{3} \sqrt{2} n^{1/6}, \tag{4.47}$$

so

$$S_3 = O(n^{4/9} \times n^{1/6}) = O(n^{11/18}), \qquad n \to \infty. \tag{4.48}$$

Further, the second sum S_2 in (4.29) is estimated on the basis of some refined results related to Van der Corput's theorem from the number theory (see Sierpinski, 1988). It is proved in Kuznetsov and Fedosov (1965) that

$$S_2 = \int_{\nu_0(n)}^{\nu_1(n)} \left\{ \frac{\eta(n, \nu)}{\pi} - \frac{1}{4} \right\} d\nu + O(n^{2/3}), \qquad n \to \infty. \tag{4.49}$$

Now, by collecting together the estimates (4.41), (4.45), (4.48), (4.49), we arrive at the following asymptotic relation:

$$N(n) = N_0(n) + 2(S_1 + S_2 + S_3)$$

$$= \frac{n}{\pi} + 2 \left\{ \frac{n}{\pi} \nu_0 + \int_{\nu_0(n)+1/2}^{\nu_1(n)+1/2} \left(\frac{\eta(n, \nu)}{\pi} - \frac{1}{4} \right) d\nu \right\} + O\left(n^{2/3}\right). \tag{4.50}$$

The integral here can be calculated explicitly as a tabulated integral (see Gradshteyn and Ryzhik, 1994):

$$\int \eta(n,\nu)\,d\nu = \frac{3}{4}\nu\sqrt{n^2-\nu^2} + \frac{\pi}{4}n^2 - \left(\frac{\nu^2}{2}+\frac{n^2}{4}\right)\arccos\frac{\nu}{n}, \qquad (4.51)$$

so

$$\int_{\nu_0(n)+1/2}^{\nu_1(n)+1/2}\eta(n,\nu)\,d\nu = \frac{\pi}{8}n^2 - n^{4/3} - \frac{n}{2} + O\left(n^{2/3}\right), \qquad n\to\infty, \qquad (4.52)$$

where we have used expansion (4.53) and also

$$\arccos(x) = \frac{\pi}{2} - x + O\left(x^3\right), \qquad x\to 0. \qquad (4.53)$$

Substitution of Eq. (4.53) into (4.50) yields finally the following asymptotic result:

$$N(n)=\frac{n}{\pi}+2\left\{\frac{n^{4/3}}{\pi}+\left(\frac{n^2}{8}-\frac{n^{4/3}}{\pi}-\frac{n}{2\pi}-\frac{1}{4}n\right)\right\}+O\left(n^{2/3}\right)=\frac{n^2}{4}-\frac{n}{2}+O\left(n^{2/3}\right), \quad (4.54)$$

which for the disk of radius $r=1$ is equivalent to

$$N(n) = \frac{S}{4\pi}n^2 - \frac{L}{4\pi}n + O\left(n^{2/3}\right), \qquad (4.55)$$

since $S=\pi r^2=\pi$, $L=2\pi r=2\pi$. Here S is the area of the disk and L is its perimeter.

Helpful remarks

$1°$. It can be shown by analogy to the above consideration that the case of acoustically hard boundary leads to quite similar result:

$$N(n) = \frac{S}{4\pi}n^2 + \frac{L}{4\pi}n + O\left(n^{2/3}\right). \qquad (4.56)$$

$2°$. Weyl (1912) was first to suppose that formulas (4.55) and (4.56) for the Dirichlet and Neumann boundary value problems are universal for all bounded acoustic domains. However, generically this statement is not yet proved. For some canonical shapes this is strictly proved, but existing general theories can justify only the leading asymptotic terms in these formulas. In the next section we give a short survey of this fundamental theory.

4.3. Some Variational Principles for Eigenvalues

As we could see in the previous section asymptotics of large eigenvalues in the case of canonical domains obeys some certain regularity, and it is very interesting to clarify if the asymptotic behavior remains regular also in general case. A good instrument to study this question is based on variational principles for eigenvalues. We expound this theory here following the classical ideas of Courant and Hilbert (1953), and Mikhlin (1964).

Recall that the Laplacian $A=-\Delta$ is a symmetric positive definite operator, which acts from a subset H_0 of the Hilbert space $H=L_2(D)$, i.e., $A=-\Delta: H=C_2(D)\subset H\to H$.

Recall also that we proved that A is positive definite only over the subset of $C_2(D)$ whose elements satisfy homogeneous (Dirichlet or Neumann) boundary conditions (see Eq. (4.4)). So, throughout this section we will say that any function $u\in H_A$ if it is from $C_2(D)$ and satisfies the given homogeneous boundary conditions.

DEFINITION. *The fraction $R(u) = (Au, u)/(u, u)$ is called the* Rayleigh *ratio.*

THEOREM 1. *The first (i.e., minimum among others) eigenvalue is*

$$\lambda_1 = \inf_{u \in H_A} R(u) = R(u_0) \quad \text{if } u_0 \in H_A. \tag{4.57}$$

Proof. Let $\eta \in H$ be an arbitrary element from the domain of definition of the operator A, and α be an arbitrary real number. Then $u_0 + \alpha\eta \in H_0$, and

$$R(u_0 + \alpha\eta) = \frac{(A(u_0 + \alpha\eta), (u_0 + \alpha\eta))}{(u_0 + \alpha\eta, u_0 + \alpha\eta)} = \frac{\alpha^2(A\eta, \eta) + 2\alpha(Au_0, \eta) + (Au_0, u_0)}{\alpha^2(\eta, \eta) + 2\alpha(u_0, \eta) + (u_0, u_0)}. \tag{4.58}$$

Since the minimum value of $R(u)$ is achieved with the element $u = u_0$, so $R'_\alpha(\alpha = 0) = 0$, that yields

$$(Au_0, \eta)(u_0, u_0) - (Au_0, u_0)(u_0, \eta) = 0 \quad \sim \quad (Au_0, \eta) - \lambda_1(u_0, \eta) = 0$$
$$\sim \quad (Au_0 - \lambda_1 u_0, \eta) = 0, \quad \forall \eta \in H_A. \tag{4.59}$$

We thus have found that the element $Au_0 - \lambda_1 u_0$ is orthogonal to arbitrary element η of the subset H_A which is everywhere dense in H. This is possible only if $Au_0 - \lambda_1 u_0 = 0 \sim Au_0 = \lambda_1 u_0$, i.e., λ_1 is an eigenvalue.

Let us prove that λ_1 is the minimum eigenvalue. Indeed, if there is any other λ_m such that $Au_m = \lambda_m u_m$, $u_m \in H_A$, then

$$\lambda_m = \frac{(Au_m, u_m)}{(u_m, u_m)} \geq \inf_{u \in H_A} \frac{(Au, u)}{(u, u)} = \lambda_1. \tag{4.60}$$

It should be noted that the stated variational property of the lowest eigenvalue can be applied only to elements with a unit norm:

$$\lambda_1 = \inf_{u \in H_A} (Au, u), \quad \|u\| = 1. \tag{4.61}$$

THEOREM 2. *All consequent higher eigenvalues can be constructed within the framework of the iterative variational scheme:*

$$\lambda_m = \inf_{u \in H_A} R(u) = \inf_{u \in H_A} \frac{(Au, u)}{(u, u)} = (Au_m, u_m),$$
$$(u, u_n) = 0, \quad n = 1, 2, \ldots, m - 1, \quad \|u_n\| = 1. \tag{4.62}$$

Proof. Let $\xi \in H_0$ be an arbitrary element from the domain of definition of the operator A, and we consider the element

$$\eta = \xi - \sum_{n=1}^{m-1} (\xi, u_n) u_n. \tag{4.63}$$

It is clear that $(\eta, u_n) = 0$, $n = 1, \ldots, m - 1$, since

$$(\eta, u_p) = (\xi, u_p) - \sum_{n=p} (\xi, u_n)(u_n, u_p) = (\xi, u_p) - (\xi, u_p) = 0, \quad 1 \leq p \leq m - 1 \tag{4.64}$$

(recall that the eigenfunctions u_n related to different eigenvalues are all mutually orthogonal to each other). It is also clear that for arbitrary real number α the element $u_m + \alpha\eta$ is such that

$$(u_m + \alpha\eta, u_n) = 0, \quad n = 1, \ldots, m - 1. \tag{4.65}$$

Now the function $u_m + \alpha\eta$ satisfies all conditions of orthogonality (4.62) of the present theorem, so minimum of the function

$$R(u_m + \alpha\eta) = \frac{\big(A(u_m + \alpha\eta), (u_m + \alpha\eta)\big)}{(u_m + \alpha\eta, \, u_m + \alpha\eta)} \tag{4.66}$$

is achieved when $\alpha = 0$. This requires $R'_\alpha(\alpha = 0) = 0 \sim (Au_m - \lambda_m u_m, \eta) = 0$ (compare with the proof of Theorem 1). Let us prove that the same relation is valid for functions ξ instead of η. Indeed,

$$(Au_m - \lambda_m u_m, \xi) = (Au_m - \lambda_m u_m, \eta) + \sum_{n=1}^{m-1}(\xi, u_n)(Au_m - \lambda_m u_m, u_n)$$

$$= 0 + \sum_{n=1}^{m-1}(\xi, u_n)[(Au_m, u_n) - \lambda_m(u_m, u_n)] \tag{4.67}$$

$$= \sum_{n=1}^{m-1}(\xi, u_n)(u_m, Au_n) = \sum_{n=1}^{m-1}(\xi, u_n)(u_m, \lambda u_n) = 0,$$

(recall that A is self-adjoint). Now, by analogy to Theorem 1, if $(Au_m - \lambda_m u_m)$ is orthogonal to arbitrary element of a dense subset, this implies that this element is null, i.e., $Au_m = \lambda_m u_m$. We have thus come to the conclusion that λ_m, u_m are eigenvalues and eigenfunction, respectively.

Let us prove that λ_m is the next eigenvalue after λ_{m-1}. If there is any other eigenvalue $\widetilde\lambda > \lambda_m$ such that $Av = \widetilde\lambda v$, then

$$\widetilde\lambda = \frac{(Av, v)}{(v, v)}, \quad \text{and} \quad (v, u_n) = 0, \quad n = 1, \ldots, m-1. \tag{4.68}$$

However, if λ_m is a minimum of the Rayleigh ratio under the same conditions of orthogonality, this means that $\widetilde\lambda = R(v) \geq R(u_m) = \lambda_m$. This proves the theorem.

It is rather inconvenient to construct the consequent eigenvalue λ_m in terms of orthogonality to all previous λ_n $(n = 1, \ldots, m-1)$. It is more convenient to arrange this process so that we could find the current λ_m without knowing other values λ_n, if necessary. Such an opportunity is given by the following theorem.

THEOREM 3. *(a maxi-minimum principle).*

$$\lambda_m = \max_{\{v_n\}} d\left(\{v_n\}_{n=1}^{m-1}\right), \quad \forall v_n \in H, \quad \text{where} \quad d\left(\{v_n\}_{n=1}^{m-1}\right) = \min_{u \in H_A} \frac{(Au, u)}{(u, u)}, \tag{4.69}$$

$$(u, v_n) = 0, \quad n = 1, \ldots, m-1.$$

Proof. First of all, we note that for $v_n = u_n$, $n = 1, \ldots, m-1$, according to Theorem 2, we have $\lambda_m = \min_{u \in H_A}(Au, u)/(u, u) = d\left(\{u_n\}_{n=1}^{m-1}\right)$, where u_n, $n = 1, \ldots, m-1$, are previous eigenfunctions related to all smaller eigenvalues $\lambda_n \leq \lambda_m$, $n = 1, \ldots, m-1$. This proves that $\max_{\{v_n\}} d\left(\{v_n\}_{n=1}^{m-1}\right)$ is no less than λ_m. If we prove that for any other array of elements $\{v_n\}_{n=1}^{m-1}$ with the orthogonality conditions (4.69), $d\left(\{v_n\}_{n=1}^{m-1}\right) \leq \lambda_m$, this will be the final point of the proof.

In order to justify this hypothesis, let us construct a special function $\widetilde u$ in the form

$$\widetilde u = \sum_{j=1}^{m} c_j u_j, \qquad \text{such that} \quad R(\widetilde u) \leq \lambda_m. \tag{4.70}$$

To this end, we subject this function to the orthogonality conditions with the chosen array $\{v_n\}_{n=1}^{m-1}$, $(\widetilde{u}, v_n) = 0$, $n = 1, \ldots, m-1$, which leads to the linear algebraic system

$$\sum_{j=1}^{m} c_j(u_j, v_n) = 0, \qquad n = 1, \ldots, m-1, \tag{4.71}$$

with $(m-1)$ equations for m unknowns c_j, $j = 1, \ldots, m$. It is obvious that the rank of the matrix $\{a_{nj}\} = \{u_j, v_n\}$ is not greater than $m-1$, which is strictly less than the number of unknowns m. General results of linear algebra show that in this case the system surely possesses a nontrivial solution (see, for instance, Kurosh, 1972), which we can make normalized: $\sum_{j=1}^{m} c_j^2 = 1$. Since

$$(\widetilde{u}, \widetilde{u}) = \sum_{i,j=1}^{m} c_i c_j(u_i, u_j) = \sum_{i=j} c_i c_j = \sum_{i=1}^{m} c_i^2 = 1 \tag{4.72}$$

(recall that $\{u_i\}$, $i = 1, 2, \ldots$, form an orthonormalized system), we have

$$R(\widetilde{u}) = (A\widetilde{u}, \widetilde{u}) = \sum_{i,j=1}^{m} c_i c_j(Au_i, u_j) = \sum_{i,j=1}^{m} c_i c_j \lambda_i(u_i, u_j) = \sum_{i=1}^{m} \lambda_i c_i^2 \leq \lambda_m \sum_{i=1}^{m} c_i^2 = \lambda_m. \tag{4.73}$$

Hence, we have found the element $\widetilde{u} \neq 0$, for which $(A\widetilde{u}, \widetilde{u})/(\widetilde{u}, \widetilde{u}) \leq \lambda_m$ and which is orthogonal to all v_n, $n = 1, \ldots, m-1$. Therefore, the minimum of the Rayleigh ratio $R(u) = (Au, u)/(u, u)$ over the class of functions orthogonal to all v_n, $n = 1, \ldots, m-1$, which is $d\left(\{v_n\}_{n=1}^{m-1}\right)$, cannot be wider than λ_m. The theorem is proved.

DEFINITION. *If $A : H_A \subset H \to H$ and $B : H_B \subset H \to H$, and both operators are positive definite: $(Au, u) \geq \gamma_A^2(u, u)$, $(Bu, u) \geq \gamma_B^2(u, u)$, $\forall u \in H_A, H_B$. We say that $A \geq B$ if $(Au, u) \geq (Bu, u)$ for $\forall u \in H_A, H_B$.*

THEOREM 4. *If $A \geq B$, then $\lambda_m^A \geq \lambda_m^B$, $\forall m = 1, 2, \ldots$.*

Proof. If, by definition, $(Au, u) \geq (Bu, u)$, for $\forall u \in H_A, H_B$, then

$$d_A\left(\{v_n\}_{n=1}^{m-1}\right) = \min_u \frac{(Au, u)}{(u, u)} \geq \min_u \frac{(Bu, u)}{(u, u)} = d_B\left(\{v_n\}_{n=1}^{m-1}\right), \tag{4.74}$$

$$(u, v_n) = 0, \quad n = 1, \ldots, m-1.$$

It follows from this inequality that

$$\lambda_m^A = \max_{\{v_n\}} d_A\left(\{v_n\}_{n=1}^{m-1}\right) \geq \max_{\{v_n\}} d_B\left(\{v_n\}_{n=1}^{m-1}\right) = \lambda_m^B, \qquad \forall v_n \in H, \tag{4.75}$$

as was to be proved.

As the most important corollary from this theorem, we prove the following result.

THEOREM 5. *Let D_A be a subdomain of a domain D_B: $D_A \subset D_B$. Then for homogeneous boundary value problem with the condition $u|_{\partial D} = 0$: $\lambda_m^B \leq \lambda_m^A$ $\forall m = 1, 2, \ldots$.*

Proof. We prove the theorem in the 2D case of the Dirichlet problem. The 3D case can be proved by analogy.

It is clear that $H_A \subset H_B$. Indeed, if $u \in H_A$, then $u|_{\partial A} = 0$ and as follows from Eq. (4.3)

$$(Au, u) = \int_{D_A} \left[\left(\frac{\partial u}{\partial x}\right)^2 + \left(\frac{\partial u}{\partial y}\right)^2\right] dx\, dy$$

$$= \int_{D_B} \left[\left(\frac{\partial u}{\partial x}\right)^2 + \left(\frac{\partial u}{\partial y}\right)^2\right] dx\, dy = (Bu, u) \tag{4.76}$$

if the function $u(x, y)$ is extended by the value $u(x, y) \equiv 0$ to $D_B \setminus D_A$. We thus have proved that $(Au, u) = (Bu, u)$ for all $u \in H_A$. Now, since $H_A \subset H_B$ this implies

$$d_B \left(\{v_n\}_{n=1}^{m-1} \right) = \inf_{u \in H_B} \frac{(Bu, u)}{(u, u)} \le \inf_{u \in H_A} \frac{(Au, u)}{(u, u)} = d_A \left(\{v_n\}_{n=1}^{m-1} \right), \tag{4.77}$$

and so

$$\lambda_m^B = \max_{\{v_n\}} d_B \left(\{v_n\}_{n=1}^{m-1} \right) \le \max_{\{v_n\}} d_A \left(\{v_n\}_{n=1}^{m-1} \right) \le \lambda_m^A, \tag{4.78}$$

which was to be proved.

For applications, this theorem states that eigenfrequencies of a homogeneous interior acoustic problem for some domain D are not higher than those for any of its subdomain, with the same type of homogeneous boundary condition.

A more advanced conclusion from Theorem 4 is given by the following statement.

THEOREM 6. *Let D be a domain consisting of a finite number of subdomains D_i, $i = 1, \ldots, I$, so that $D = \bigcup_{i=1}^{I} D_i$, $\bigcap_{i=1}^{I} D_i = 0$. Let $\{\lambda_m^*\}$ denote an ordered sequence of eigenvalues collected all together from those of a Dirichlet homogeneous boundary value problem for all domains D_i. If $\{\lambda_m\}$ are eigenvalues of the corresponding homogeneous problem for the domain D then $\lambda_m \le \lambda_m^*$.*

Proof. Let us compare the two problems:

1) The given eigenvalue problem in the domain D with the given homogeneous boundary condition over the boundary ∂D;

2) The same problem with the additional constrained conditions $u = 0$ (or $\partial u / \partial n = 0$) posed over boundary lines of all subdomains. By analogy with the proof of the previous theorem, we can claim that $H_2 \subset H_1$ since any doubly differential function satisfying all constraints of problem 2) satisfies also the constraints (which are the given boundary conditions on the boundary ∂D) of problem 1). Therefore, over the class of functions $u \in H_2 : (A_1 u, u) = (A_2 u, u)$ and since $H_2 \subset H_1$, we have

$$d_1 \left(\{v_n\}_{n=1}^{m-1} \right) = \inf_{u \in H_1} \frac{(A_1 u, u)}{(u, u)} \le \inf_{u \in H_2} \frac{(A_2 u, u)}{(u, u)} = d_2 \left(\{v_n\}_{n=1}^{m-1} \right). \tag{4.79}$$

The proof is completed by the observation that Eq. (4.79) involves

$$\lambda_m = \lambda_m^{(1)} = \max_{v_n} d_1 \left(\{v_n\}_{n=1}^{m-1} \right) \le \max_{\{v_n\}} d_2 \left(\{v_n\}_{n=1}^{m-1} \right) = \lambda_m^{(2)} = \lambda_m^*, \tag{4.80}$$

because the set $\{\lambda_m^{(2)}\}$ is formed by the union of all eigenvalues for all particular problems for each subdomain D_i. This proves the theorem.

Corollary. Let, under the conditions of Theorem 5, $N(\lambda)$ denote the number of eigenvalues (each counted taking into account multiplicity) not exceeding $\lambda > 0$, for the given Dirichlet homogeneous boundary value problem. Let $N^*(\lambda)$ denote the total number of analogous quantities for each subdomain D_i: $N^*(\lambda) = \sum_{i=1}^{I} N_i(\lambda)$. Then $N(\lambda) \ge N^*(\lambda)$.

In the case of the Neumann boundary condition $\partial u / \partial n |_{\partial D} = 0$ the result analogous to what is proved in Theorem 6 is given by

THEOREM 7. *Let $\{\lambda_m\}$ denote the ordered set of increasing eigenvalues, each taken according to its multiplicity, and $\{\lambda_m^*\}$ denote the ordered full set of eigenvalues for all subdomains D_i, $i = 1, \ldots, N$: $D = \bigcup_{i=1}^{I} D_i$, $\bigcap_{i=1}^{I} D_i = 0$. Then $\lambda_m \ge \lambda_m^* \ \forall m = 1, 2, \ldots$.*

Proof. Let us consider again the two problems described in the conditions of the theorem (problems 1 and 2, respectively). Recall that admissible class of functions in H_1

consists of those with piecewise continuous first-order derivatives and $(\partial u/\partial n)/\partial D = 0$. This automatically means that the functions from H_1 are continuous. Let us widen this class refusing continuity of u. It is obvious that the functional $R(u)$ on this widened class of functions looks like the one on the functions with the homogeneous boundary condition for each D_i: $(\partial u/\partial n)|_{\partial D_i} = 0$. We thus may claim that the two mentioned problems cover the sets H_1 and H_2 with $H_1 \subset H_2$, hence

$$d_2\left(\{v_n\}_{n=1}^{m-1}\right) = \inf_{u \in H_2} R_2(u) \le \inf_{u \in H_1} R_1(u) = d_1\left(\{v_n\}_{n=1}^{m-1}\right), \tag{4.81}$$

and so, by analogy to previous theorem, $\lambda_m^* \ge \lambda_m$.

Corollary. Under the conditions of this theorem $N(\lambda) \le N^*(\lambda) = \sum_{i=1}^{I} N_i(\lambda)$.

The next theorem compares eigenfrequencies of the Dirichlet and Neumann homogeneous problems.

THEOREM 8. *Let $\{\lambda_m^D\}$ designate the ordered set of eigenvalues for the Dirichlet problem, and $\{\lambda_m^N\}$ the analogous eigenvalues for the Neumann problem. Then $\lambda_m^N \le \lambda_m^D$, $m = 1, 2, \ldots$.*

Proof. Compare two maximin problems:

1) Maximin problem on the class of functions with piecewise continuous first-order derivatives with the constraint $u|_{\partial D} = 0$;

2) the same maximin problem on the same class of functions but without the condition on the boundary.

The former class of maximin problems yields all eigenvalues of the Dirichlet homogeneous problem $\{\lambda_m^D\}$. The second class is wider. Besides, it consists of functions with piecewise continuous first-order derivatives, for which the functional $R(u)$ looks as in the case where they would be subjected to the Neumann boundary condition over ∂D. Then it can be proved by sequential arguments, as in the previous theorems, that the eigenvalues on a wider class do not exceed those for a more restricted class: $\lambda_m^N \le \lambda_m^D$.

Helpful remarks

It should be noted that the quantity $N(\lambda)$ for both boundary value problems is in inverse relation to λ_m. In particular, this implies that $N_D(\lambda) \le N_N(\lambda)$. Explicit estimates of eigenvalues in the problem of round disk, formulas (4.55) and (4.56), confirm this conclusion.

4.4. Weyl–Carleman Theory of Asymptotic Distribution of Large Eigenvalues

Let D be a 2D simply connected bounded domain of the area S with a boundary of finite length l. The most direct way to describe the Weyl–Carleman theory is to consequently move from simple shapes to more and more complex ones.

Let us start with a rectangular domain of arbitrary size $a \times b$: $0 \le x \le a$, $0 \le y \le b$. The eigenfunctions of the Dirichlet and the Neumann homogeneous boundary value problem are, respectively,

$$\begin{aligned}
u_{nm}^D(x, y) &= \sin\frac{\pi n x}{a} \sin\frac{\pi m y}{b} \quad (n, m = 1, 2, \ldots), \\
u_{nm}^N(x, y) &= \cos\frac{\pi n x}{a} \cos\frac{\pi m y}{b} \quad (n, m = 0, 1, 2, \ldots).
\end{aligned} \tag{4.82}$$

Due to Weierstrass theorem, the trigonometric series (4.82) form a complete orthogonal system in D. Besides, they satisfy the corresponding homogeneous boundary conditions.

This proves that there are no other eigenfunctions, because any new eigenfunction must be orthogonal to all $u_{nm}(x, y)$ (Theorem 6, Section 4.1), which is impossible.

Eigenvalues of both problems are given explicitly by the expression

$$\lambda_{nm} = \pi^2 \left(\frac{n^2}{a^2} + \frac{m^2}{b^2} \right), \quad (n, m = 0, 1, 2, \ldots). \tag{4.83}$$

Let us calculate $N(\lambda)$, the number of eigenvalues λ_{nm}, which are less or equal to λ. These are defined by the number of nonnegative integral solutions of the inequality

$$\pi^2 \left(\frac{n^2}{a^2} + \frac{m^2}{b^2} \right) \leq \lambda \sim \frac{\pi^2}{\lambda} \frac{n^2}{a^2} + \frac{\pi^2}{\lambda} \frac{m^2}{b^2} \leq 1. \tag{4.84}$$

These solutions are distributed on the lattice of the integer-valued Cartesian coordinate system (n, m), $n, m = 0, 1, 2, \ldots$. They are related to the nodes lying inside an ellipse with semi-axes $\widetilde{a} = a\sqrt{\lambda}/\pi$ and $\widetilde{b} = b\sqrt{\lambda}/\pi$.

It can be easily shown that

$$\lim_{\lambda \to \infty} \frac{N(\lambda)}{(\pi \widetilde{a} \widetilde{b}/4)} = 1, \quad \text{so} \quad N(\lambda) = \frac{ab\lambda}{4\pi} = \frac{S}{4\pi}\lambda + o(\lambda), \tag{4.85}$$

i.e., asymptotically, the number of modes inside the quarter-ellipse is equal to the area of the latter, $\pi \widetilde{a} \widetilde{b}/4$. Here S is the area of the rectangle: $S = ab$. An estimate of the remainder shows that (see Courant and Hilbert, 1953)

$$N(\lambda) = \frac{S}{4\pi}\lambda + O(\sqrt{\lambda}), \quad \lambda \to \infty. \tag{4.86}$$

Note that this estimate is valid for both types of boundary condition. As mentioned above (see remarks for Section 4.2), Weyl (1912) conjectured that the second asymptotic term can be represented explicitly in the form

$$N_{\mathrm{D,N}}(\lambda) = \frac{S}{4\pi}\lambda \mp \frac{l}{4\pi}\sqrt{\lambda} + o(\sqrt{\lambda}), \quad \lambda \to \infty, \tag{4.87}$$

but the proof for general case is not yet known (for a survey, see Birman and Solomyak, 1979).

In the considered case of rectangular domain D with dimensions $a \times b$, the asymptotic relation can be justified as follows. The proof is based on some results of number theory connected with Landau and Van der Corput theorems (see, for example, Sierpinski, 1988; Kuznetsov, 1966; Makai, 1970), which give a more precise relation between the square of the ellipse mentioned above and the number of integer lattice nodes situated in the interior of the lattice. For our purposes this theory results in the following conclusion. The area of the elliptic sector $\pi \widetilde{a} \widetilde{b}/4$ is equal to the number of lattice points in the interior of the sector plus half a number of lattice points on the linear part of the boundary of this sector, with the error of the order of $o(\sqrt{\lambda})$:

$$\frac{\pi \widetilde{a} \widetilde{b}}{4} = N_{\mathrm{D}}(\lambda) + \frac{\widetilde{a} + \widetilde{b}}{2} + o(\sqrt{\lambda}), \tag{4.88}$$

since the first solution in (4.82) implies $n, m \geq 1$, i.e., it operates with lattice points lying in the interior of the ellipse only. On the contrary, $(\widetilde{a} + \widetilde{b})$ points with $n = 0$ and $m = 0$ on the linear boundary should be added in the case of the Neumann problem: $N_N(\lambda) = N_D(\lambda) + (\tilde{a} + \tilde{b})$. Finally, we arrive at the representation (recall that $\tilde{a} = a\sqrt{\lambda}/\pi, \tilde{b} = b\sqrt{\lambda}/\pi$)

$$
\begin{aligned}
N_D(\lambda) &= \frac{\pi \tilde{a}\tilde{b}}{4} - \frac{\tilde{a} + \tilde{b}}{2} + o(\sqrt{\lambda}) = \frac{ab}{4\pi}\lambda - \frac{(a+b)}{2\pi}\sqrt{\lambda} + o(\sqrt{\lambda}) \\
&= \frac{S}{4\pi}\lambda - \frac{l}{4\pi}\sqrt{\lambda} + o(\sqrt{\lambda}), \quad \lambda \to \infty,
\end{aligned}
\tag{4.89}
$$

$$
\begin{aligned}
N_N(\lambda) &= N_D(\lambda) + \frac{a+b}{\pi}\sqrt{\lambda} = \frac{S}{4\pi}\lambda + \frac{a+b}{2\pi}\sqrt{\lambda} + o(\lambda) \\
&= \frac{S}{4\pi}\lambda + \frac{l}{4\pi}\sqrt{\lambda} + o(\sqrt{\lambda}), \quad \lambda \to \infty,
\end{aligned}
\tag{4.90}
$$

which confirms the generic formula (4.87) for the case of rectangular domain. Do not forget that the second term in the asymptotic representation (4.87) is not yet strictly proved for arbitrary domain. However, if we compare (4.87) with respective results obtained for round circle, Eqs. (4.55), (4.56), derived for the eigenfrequencies (which are square of the eigenvalues), we can conclude that formulas for round disk also conform with the Weyl assumption (4.87).

Domains consisting of a finite number of subdomains.

Unfortunately, in the generic case with a domain of arbitrary shape, the existing strict results confirm only the leading asymptotic term in (4.87). Let the domain D be the union of a finite number of squares D_i with side a, $i = 1, \ldots, I$. Then variational principles stated in the previous section give an appropriate estimate

$$
\sum_{i=1}^{I} N_i^D(\lambda) \le N_D(\lambda) \le N_N(\lambda) \le \sum_{i=1}^{I} N_i^N(\lambda), \quad \lambda \to \infty.
\tag{4.91}
$$

Note that $N_i^D = (S_i/4\pi)\lambda + O(\sqrt{\lambda})$, $N_i^N = (S_i/4\pi)\lambda + O(\sqrt{\lambda})$, and $S = \sum_{i=1}^{I} S_i$, hence

$$
\frac{S}{4\pi}\lambda + O(\sqrt{\lambda}) \le N_D(\lambda) \le N_N(\lambda) \le \frac{S}{4\pi}\lambda + O(\sqrt{\lambda}), \quad \lambda \to \infty.
\tag{4.92}
$$

Domains of arbitrary shape.

Our treatment here will be based upon approximation of arbitrary bounded simply connected domain by shapes which represent a union of small areas. The following lemma plays a key role in this construction.

LEMMA. *Let in the plane (x, y) there be a bounded 2D domain D of the area S and the length of perimeter l. If there is a square lattice on the plane with the small step h, then the number of nodes in interior of S is*

$$
n_h(D) = \frac{S}{h^2} + O(h^{-1}), \quad h \to 0.
\tag{4.93}
$$

Proof. The proof is based on the Jarnik proposition (see, for example, Kuznetsov, 1966) which proves that in the natural 2D Cartesian coordinate system, which is provided with an orthogonal lattice of a unit step, the mentioned number of interior nodes $n(D)$ satisfies the inequality $|S - n(D)| < l$. Now, under the conditions of the lemma, let us make the change of variables $\tilde{x} = x/h$, $\tilde{y} = y/h$. Then in new coordinate system $(\tilde{x}, \tilde{y})$ we have: $\tilde{S} = S/h^2$, $\tilde{l} = l/h$, $\tilde{S} - \tilde{l} < n_h(D) < \tilde{S} + \tilde{l}$, or $S/h^2 - l/h < n_h(D) < S/h^2 + l/h$, which was to be proved. This lemma directly leads to the basic result of the quoted theory.

THEOREM 1. *The asymptotic behavior of large eigenvalues, in both the Dirichlet and the Neumann homogeneous boundary value problem for the Helmholtz operator, is given by*

$$N(\lambda) = \frac{S}{4\pi}\lambda + O(\sqrt{\lambda}), \qquad \lambda \to \infty, \tag{4.94}$$

where S is the area of the considered 2D domain of arbitrary shape.

Proof. Let us furnish the (x, y) plane with orthogonal lattice of the step $h = 1/\sqrt{\lambda}$, and consider the domain formed by the union of all elementary squares of this lattice that lie inside the given domain D. It can be seen that the number of squares in interior of D, $n_{sq}(D)$ is of the same order as the number of interior nodes $n_h(D)$: $n_h(D) = n_{sq}(D) + O(1/h) \sim n_{sq}(D) = n_h(D) + O(\sqrt{\lambda}), \lambda \to \infty$, so the total area of inner squares is $S_{sq}^D = n_h(D)h^2 = S + O(h), h \to 0$ or $S_{sq}(D) = S + O(1/\sqrt{\lambda}), \lambda \to \infty$.

Further, the application of the above variational principles leads to

$$N_{sq}^D(\lambda) \leq N_D(\lambda) \leq N_N(\lambda) \leq N_{sq}^N(\lambda), \tag{4.95}$$

where $N_{sq}^D(\lambda)$, $N_{sq}^N(\lambda)$ express the number of eigenvalues not exceeding λ in the Dirichlet and the Neumann problem in the domain created by inner squares. Now,

$$N_{sq}^D(\lambda) = \frac{S_{sq}(D)}{4\pi}\lambda + O(\sqrt{\lambda}) \tag{4.96}$$

(see Eq. (4.92)), that yields

$$N_{sq}^D(\lambda) = \frac{S}{4\pi}\lambda + O(\sqrt{\lambda}). \tag{4.97}$$

The estimate for $N_{sq}^N(\lambda)$ is achieved by analogy. Collecting together the asymptotic estimates (4.95)–(4.97) we finally arrive at the basic result

$$\frac{S}{4\pi}\lambda + O(\sqrt{\lambda}) \leq N_D(\lambda) \leq N_N(\lambda) \leq \frac{S}{4\pi}\lambda + O(\sqrt{\lambda}). \tag{4.98}$$

Helpful remarks

$1°$. Recall that in the considered theory $\lambda = \omega^2/c^2$ (ω is the frequency and c is the wave speed). So, if we estimate the number of eigenfrequencies ω_n not exceeding the given frequency ω this is determined as

$$N(\omega) = \frac{S\omega^2}{4\pi c^2} + O(\omega), \qquad \omega \to \infty. \tag{4.99}$$

We thus can see that the number of eigenfrequencies grows as ω^2. Therefore, in computer simulation the problem of operation with *singular* values ω_n, insignificant for low and moderate frequencies, becomes of a considerable importance for high frequencies. Actually, any *regular* computer algorithm will crash if the frequency of oscillation hits any eigenfrequency. With the frequency increasing *the probability* to be faced with irregular value ω_n increases, which makes numerical algorithm less stable.

$2°$. Asymptotic formula (4.86) can be inverted:

$$\lambda_n \sim \frac{4\pi}{S}n + o(n), \qquad n \to \infty. \tag{4.100}$$

This justifies what was noted about rate of increasing of λ_n following from Hilbert–Shmidt theorem for the 2D problem (see Eq. (4.25)). Indeed, from Eq. (4.100) it is clear that the sum of the series $\sum_{n=1}^{\infty}(1/\lambda_n^2) < \infty$.

4.5. Exact Explicit Results for Some Polygons

As can be seen from previous investigation, explicit analytical results can be obtained for domains where coordinates can be separated. A detailed study of this question was carried out by Kuznetsov (1966). Rectangular domain gives an example of the separation in the Cartesian coordinate system, and an explicit-form representation for eigenfunctions and eigenfrequencies was obtained in Section 4.4. In the polar coordinate system, the solution in the round disk is explicitly expressed in terms of Bessel functions (Section 2.2). On the other hand, Makai (1970) has derived exact analytical formulas for eigenfunctions and eigenvalues for three types of triangles, where variables do not separate.

At the present section we give a more complete solution for these triangles. First of all, we will understand why only these three types of triangles (among all others) admit exact analytical solution. Then we will construct an explicit-form representation of the Green's function for the Helmholtz operator in these domains as an exponentially convergent series (which has not been made until now). But what is much more important is that the method proposed here will allow us to extend this result to some 3D polyhedrons (see the next section), where there are no analogous results at all. For all that, we will use the well known virtual-image technique. To be more specific, we restrict the consideration by the Neumann boundary value problem (the case of acoustically hard boundary surfaces).

The basics of the virtual image method.

We have already been faced implicitly with application of the virtual image method in Section 3.5. Actually, the arisen series terms express the contribution of virtual (really not present in a layer) sources placed symmetrically to the respective point source related to a previous term. Thus, the next term of the series cancels the normal component of the velocity generated by previous virtual sources. As a result, we arrive at some infinite series.

In the investigation of the interior acoustic problem for polygons we start from the interior problem for a wedge with angle ϑ, $0 < \vartheta < 2\pi$. It can be shown (see Skudrzyk, 1971) that in the generic case the total wave field consists of the geometric-diffraction component (see below Section 6.3) and the wave diffracted by the wedge corner. The latter disappears when $\vartheta = \pi/n$, n is an arbitrary positive integer. Essentially, this property is connected with the evident statement that only for these values of ϑ the number of virtual sources outside the wedge is finite, being equal to $2n$ (including the real source). Obviously, the case $n = 2$ is related to the quarter-space.

Let us try to evaluate the class of polygons admitting exact analytical solution. The main statement of the present section is that we can construct an exact explicit solution for those polygonal spaces where only a geometric-diffraction wave component is present. In other words, each angle of the polygon has to be taken in the form π/n.

Let us study in more detail the set of such polygons. Let l denote the number of the sides; then the sum of the interior angles is $\pi(l-2)$. On the other hand, if each angle $\vartheta = \pi/n \le \pi/2$, then this sum is no greater than $\pi l/2$, which entails $\pi(l-2) \le \pi l/2$, or $l \le 4$. A trivial treatment shows that there are only four possible cases: 1) rectangle ($n = 2$ for each angle); 2) equilateral triangle ($n = 3$ for each angle); 3) isosceles right-angled triangle ($n = 2$ for one angle and $n = 4$ for others); 4) right-angled triangle ($\vartheta_1 = \pi/2$, $n = 2$) with $\vartheta_2 = \pi/3$ ($n = 3$) and $\vartheta_3 = \pi/6$ ($n = 6$).

An alternative representation for the rectangular space.

The main idea of the image method for the rectangular rigid-wall room is clear from Fig. 4.1. If there is a point source (x_0, y_0) : $p_0 = H_0^{(1)}\left[k\sqrt{(x_0 - x)^2 + (y_0 - y)^2}\right]$ placed inside the space, then to satisfy the trivial boundary condition for the normal derivative: $\partial p/\partial n = 0$, one may organize a two-dimensional array of virtual images, which provides an intrinsic symmetry with respect to arbitrary boundary point on the wall. The response at the receiving point is then given by a two-dimensional series.

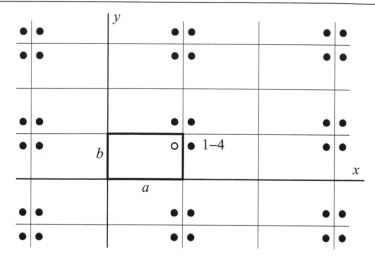

Figure 4.1. Arrangement of virtual sources for the rectangular domain

This technique was applied by many authors for the case of impulse source (Morse and Ingard, 1968), when this series is convergent, since only a finite number of virtual sources can contribute to the response at a certain moment. We can apply this approach to the harmonic process, which leads, in the 2D case, to the following result:

$$p(x_1, y_1) = s(x_0 - x_1, y_0 - y_1, a, b) + s(x_0 + x_1, y_0 - y_1, a, b)$$
$$+ \ s(x_0 - x_1, y_0 + y_1, a, b) + s(x_0 + x_1, y_0 + y_1, a, b), \tag{4.101}$$

where the function S is given by the double series

$$s(x, y, a, b) = \sum_{m,n=-\infty}^{\infty} H_0^{(1)}\left[k\sqrt{(x + 2am)^2 + (y + 2bn)^2}\right]. \tag{4.102}$$

Although the solution is expressed in explicit form (4.102), the series does not converge in any classical sense (see Section 1.3). Thus, we cannot use divergent series (4.102) for direct computations.

In order to correctly treat this series we may operate within the framework of the *extremely small attenuation* principle (see Section 3.2). If a small imaginary component is added to the wave number: $k_\varepsilon = k + i\varepsilon$, $0 < \varepsilon \ll 1$, then the series (4.102) becomes exponentially convergent, and it has a finite limit when $\varepsilon \to 0$, as in the Poisson–Abel method of generalized summation (see Section 1.3). However it is not clear how this approach can be treated directly. That is why we develop another idea. Since the function $s(x, y, a, b)$ is periodic with respect to both its arguments, let us suppose that $0 \le x < 2a$, $0 \le y < 2b$. Then we attract the integral representation of the Hankel function

$$H_0^{(1)}(k_\varepsilon \sqrt{x^2 + y^2}) = \frac{1}{\pi i} \int_{-\infty}^{\infty} e^{-\gamma(\alpha)x} \frac{e^{-i\alpha y}}{\gamma(\alpha)} d\alpha, \quad \gamma(\alpha) = \sqrt{\alpha^2 - k_\varepsilon^2}. \tag{4.103}$$

Thus, the series (4.102) may be rewritten as

$$S(x, y, a, b) = \frac{1}{\pi i} \sum_{n=-\infty}^{\infty} \int_{-\infty}^{\infty} \left[\sum_{m=0}^{\infty} e^{-\gamma(\alpha)(2am+x)} + \sum_{m=1}^{\infty} e^{-\gamma(\alpha)(2am-x)}\right] \frac{e^{-i\alpha|2bn+y|}}{\gamma(\alpha)} d\alpha. \tag{4.104}$$

The branching function $\gamma(\alpha)$ has a positive real part: $\text{Re}[\gamma(\alpha)] > 0$ when $\varepsilon > 0$ (see Section 1.1). Hence, the last series is a geometric progression, which yields

$$
s(x,y,a,b) = \frac{1}{\pi i} \sum_{n=-\infty}^{\infty} \int_{-\infty}^{\infty} \frac{\exp[-\gamma(\alpha)x] + \exp[\gamma(\alpha)(x-2a)]}{1 - \exp[-2a\gamma(\alpha)]} \frac{e^{-i\alpha|2bn+y|}}{\gamma(\alpha)} d\alpha
$$

$$
= \frac{1}{\pi i} \sum_{n=-\infty}^{\infty} \int_{-\infty}^{\infty} \frac{\cosh[(a-x)\gamma(\alpha)]}{\sinh[a\gamma(\alpha)]} \frac{e^{-i\alpha|2bn+y|}}{\gamma(\alpha)} d\alpha.
$$

(4.105)

There is no obstacle here to applying the standard expansion in residues at simple poles, because the integrand is an exponentially decaying meromorphic function with the integration contour removing down to infinity. When doing so, we arrive at the following result:

$$
s(x,y,a,b) = -i \sum_{m=0}^{\infty} \delta_m \frac{\cos(\pi m x/a)}{q_m} \sum_{n=-\infty}^{\infty} \exp\left(-|2bn+y|\,q_m/a\right),
$$

(4.106)

$$
q_m = \sqrt{(\pi m)^2 - (ak_\varepsilon)^2}, \qquad \delta_0 = 1, \qquad \delta_m = 2 \ (m = 1, 2, \dots).
$$

Since the real part of the quantities q_m is positive with $\varepsilon > 0$, the last series is again a geometric progression, with its sum being equal to

$$
\sum_{n=0}^{\infty} \exp\left[-\frac{(2bn+y)}{a}q_m\right] + \sum_{n=1}^{\infty} \exp\left[-\frac{(2bn-y)}{a}q_m\right]
$$

$$
= \frac{\exp[-yq_m/a] + \exp[(y-2b)q_m/a]}{1 - \exp[-2bq_m/a]} = \frac{\cosh[(b-y)q_m/a]}{\sinh[bq_m/a]}.
$$

(4.107)

Therefore, the final representation ($\varepsilon \to +0$) may be obtained from (4.106) and (4.107), as follows:

$$
s(x,y,a,b) = -i \sum_{m=0}^{\infty} \delta_m \frac{\cos(\pi m\, x/a)}{\sqrt{(\pi m)^2 - (ak)^2}} \frac{\cosh\left[\sqrt{(\pi m)^2 - (ak)^2}\,(b-y)/a\right]}{\sinh\left[\sqrt{(\pi m)^2 - (ak)^2}\,b/a\right]}.
$$

(4.108)

It should be noted that if $0 \le y < 2b$, then the last series converges exponentially, i.e., its terms decrease as in some geometric progression with a certain basis $q : |q| < 1$, when the number m infinitely grows.

Extension to more complex polygons.

It is not easy to extend the image method to more complex polygonal shapes. Irregular shapes involve a more complicated procedure for finding the positions of the virtual sources. It is important to clarify the intrinsic property of the exact solution for rectangular spaces. If we look at Fig. 4.1 again, it becomes clear that four virtual sources crowd around each corner of the main rectangle, symmetrically with respect to its faces, at equal distances from the corner. They correspond to virtual sources in the exact solution for a point source placed in a quarter-plane.

Analogous constructions may be applied to the discovered three triangle geometries, which admit exact analytical solution.

In the case of equilateral triangle, the corresponding array of virtual sources is shown in Fig. 4.2. It becomes clear from this figure that the structure of the solution is as follows:

$$
p(x_1, y_1) = \sum_{i=1}^{12} s(\xi_i - x_1, \eta_i - y_1, a, b), \qquad a = \frac{3}{2}c, \quad b = \frac{\sqrt{3}}{2}c,
$$

(4.109)

where c is the length of the side.

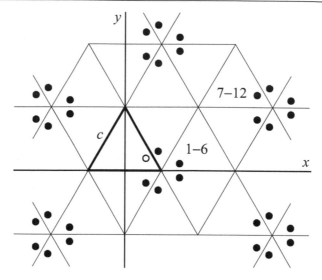

Figure 4.2. Arrangement of virtual sources for equilateral triangle

For isosceles right-angled triangle with the leg a, as it follows from Fig. 4.3, the solution can be expressed in the following way:

$$p(x_1, y_1) = \sum_{i=1}^{8} s(\xi_i - x_1, \eta_i - y_1, a, a). \tag{4.110}$$

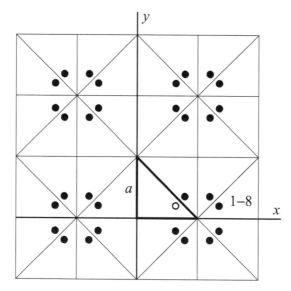

Figure 4.3. Virtual sources for isosceles right-angled triangle

At last, the structure of the solution for right-angled triangle with the acute angles $30°$ and $60°$ and the hypotenuse c becomes clear from Fig. 4.4:

$$p(x_1, y_1) = \sum_{i=1}^{24} s(\xi_i - x_1, \eta_i - y_1, a, b), \qquad a = \frac{\sqrt{3}}{2} c, \quad b = \frac{3}{2} c. \tag{4.111}$$

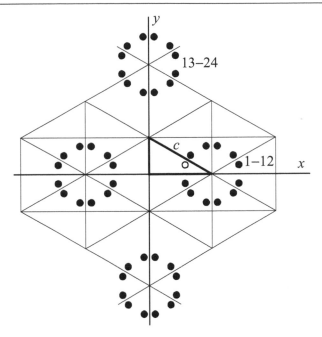

Figure 4.4. Virtual sources for right-angled triangle with the acute angles 30° and 60°

Estimate of the eigenfrequencies.

Eigenvalues of the wave number k are determined by Eq. (4.108) when the hyperbolic sine in the denominator becomes equal to zero:

$$\sinh\left[\sqrt{(\pi m)^2 - (ak)^2}\, b/a\right] = 0 \quad \sim \quad \sqrt{(\pi m)^2 - (ak)^2}\, b/a = -\pi n i$$

$$\sim \quad k_{mn} = \pi\sqrt{\frac{m^2}{a^2} + \frac{n^2}{b^2}} \tag{4.112}$$

for arbitrary integer values of m and n. For the rectangular domain with sides a and b this result is well known in the literature. For the triangular spaces this involves some new exact estimates:

1) Equilateral triangle with the side-length c:

$$k_{mn} = \frac{2\pi}{3c}\sqrt{m^2 + 3n^2}. \tag{4.113}$$

It can be easily seen that in the case of the considered triangle the first six terms $i = 1, 2, \ldots, 6$ in the sum (4.109) and the other six terms $i = 6, 7, \ldots, 12$ cancel each other if $n + m$ is odd. Indeed, the second crowd of the six virtual sources is obtained from the first six corresponding sources if we change (ξ_i, η_i) by $(\xi_i + a, \eta_i + b)$. Expression $T_{mn}(x, y) = \cos(\pi m\,x/a)\cosh\left[\sqrt{(\pi m)^2 - (ak)^2}\,(b-y)/a\right]$ in Eq. (4.108), under constraints (4.112), is $(-1)^{n+m}\, T_{mn}(x + a,\ y + b)$. Therefore, the contributions of the mentioned six pairs of virtual sources cancel each other if $n + m$ is odd, so that there is no singularity in the denominator of expression (4.108), and consequently no eigenvalues. From this consideration we can conclude that formula (4.113) should be applied only for such combinations of m and n, when $n + m$ is even. This is equivalent to

$$k_{mn} = \frac{2\pi}{3c}\sqrt{m^2 + 3n^2} = \frac{2\pi}{3c}\sqrt{(n + 2l)^2 + 3n^2} = \frac{4\pi}{3c}\sqrt{n^2 + nl + l^2}$$

$$\sim \quad k_{mn} = \frac{4\pi}{3c}\sqrt{m^2 + mn + n^2}, \qquad \forall m, n = 0, 1, 2, \ldots. \tag{4.114}$$

2) Isosceles right-angled triangle with the length of the leg equal to a:

$$k_{mn} = \frac{\pi}{a}\sqrt{m^2 + n^2}. \tag{4.115}$$

Of course, we should distinguish different eigenfunctions by their form. However, with the wave number defined by Eq. (4.115), the numerator in Eq. (4.108) becomes $\cos(\pi mx/a) \times \cos[\pi n(a-y)/a] = (-1)^n \cos(\pi mx/a)\cos(\pi ny/a)$ (recall that $b = a$). Since the sign "minus" does not influence the form of the eigenfunction, we can notice that the obtained structure is symmetric with respect to change of variable. So, Eq. (4.115) must be applied only under the following restriction:

$$k_{mn} = \frac{\pi}{a}\sqrt{m^2 + n^2}, \qquad m \geq n, \quad n = 0, 1, 2, \dots. \tag{4.116}$$

3) Right-angled triangle with the acute angles equal to $30°$ and $60°$ and the hypotenuse c:

$$k_{mn} = \frac{2\pi}{3c}\sqrt{3m^2 + n^2}. \tag{4.117}$$

For the same reason as in case 1, the real array of eigenvalues here is:

$$k_{mn} = \frac{4\pi}{3c}\sqrt{m^2 + mn + n^2} \qquad \forall m, n = 0, 1, 2, \dots. \tag{4.118}$$

All three derived formulas (4.114), (4.116), (4.118) coincide with results of Makai (1970) obtained by an absolutely different method.

Helpful remarks

$1°$. It should be noted that the representation (4.108) is also a mode expansion of the solution (compare with results of Section 3.1). Analogously to the case of a layer of constant thickness, a few first terms of the series (the number of which depends on the value of the wave number k) behave like an oscillating function. Then, beginning from $m = [ak/\pi] + 1$, all terms decay exponentially.

$2°$. It is noteworthy that the double divergent series (4.102) has been transformed during our transformations into a single series which converges like a geometric progression. The constructed solutions give a representation of the Green's function, which can be efficiently calculated without any problem.

$3°$. Formula (4.108) as representing the sum of divergent series (4.102) is worthy of special discussion. We can test the result obtained, for instance, by a boundary-element technique. In order to create wave field without singularities inside the closed space, we may remove the point-source, which leads to the nontrivial boundary condition:

$$\frac{\partial p}{\partial n} = -\frac{\partial H_0^{(1)}(kr)}{\partial n} = kH_1^{(1)}(kr)\frac{\partial r}{\partial n}, \qquad r = \left[(x-x_0)^2 + (y-y_0)^2\right]^{1/2}, \qquad (x,y) \in l. \tag{4.119}$$

Then the Kirchhoff–Helmholtz integral formula determines the wave field at arbitrary point (x_1, y_1) inside the boundary contour l, as follows:

$$p(x_1, y_1) = H_0^{(1))}(kr_0) + \frac{ki}{4}\int_l \left[H_0^{(1)}(kr_1)H_1^{(1)}(kr)\frac{\partial r}{\partial n} + u(x,y)H_1^{(1)}(kr_1)\frac{\partial r_1}{\partial n}\right] dl_{xy},$$

$$r_0 = \left[(x_1-x_0)^2 + (y_1-y_0)^2\right]^{1/2}, \qquad r_1 = \left[(x-x_1)^2 + (y-y_1)^2\right]^{1/2}, \tag{4.120}$$

which contains the unknown boundary value of the regular wave function $u(x, y) = p(x, y) - H_0^{(1)}(kr)$, $(x, y) \in l$. This function can be defined from the boundary integral equation given by the same Kirchhoff formula:

$$
\begin{aligned}
\frac{u(x_1, y_1)}{2} &- \frac{ki}{4} \int_l u(x, y) H_1^{(1)}(kr_1) \frac{\partial r_1}{\partial n} dl_{xy} \\
&= \frac{ki}{4} \int_l H_0(kr_1) H_1^{(1)}(kr) \frac{\partial r_1}{\partial n} dl_{xy}, \quad (x_1, y_1) \in l.
\end{aligned}
\tag{4.121}
$$

Recall that if using the BIE in such a form, we must set the mesh nodes so that no node coincides with any corner of the boundary contour l. The constructed algorithm for this regular Fredholm equation of the second kind was briefly described in Section 1.5.

An example of the comparison between the exact explicit solution given by (4.108), (4.109) and direct numerical computations in the case of equilateral triangle is reflected in the table below, where N designates the number of nodes over the boundary line. Note that explicit formula (4.108) predetermines the wave function to be imaginary. The results of the numerical simulation by BEM show a nontrivial but very small real part when compared with the imaginary one. The calculated results are related to the case when $k = 2$, $c = 1$, $x_0 = 0.25$, $y_0 = 0.25/\sqrt{3}$, $y = 0.4$.

x	(8), (9)	BEM ($N = 300$)	BEM ($N = 600$)
−0.25	$3.19\,i$	$-0.0374 + 3.15\,i$	$-0.0222 + 3.17\,i$
−0.20	$3.16\,i$	$-0.0378 + 3.12\,i$	$-0.0225 + 3.14\,i$
−0.15	$3.10\,i$	$-0.0380 + 3.06\,i$	$-0.0226 + 3.08\,i$
−0.10	$3.02\,i$	$-0.0381 + 2.97\,i$	$-0.0227 + 2.99\,i$
−0.05	$2.92\,i$	$-0.0381 + 2.85\,i$	$-0.0227 + 2.88\,i$
0.00	$2.78\,i$	$-0.0379 + 2.71\,i$	$-0.0227 + 2.73\,i$
0.05	$2.63\,i$	$-0.0376 + 2.55\,i$	$-0.0225 + 2.57\,i$
0.10	$2.44\,i$	$-0.0371 + 2.37\,i$	$-0.0222 + 2.39\,i$
0.15	$2.24\,i$	$-0.0365 + 2.18\,i$	$-0.0219 + 2.20\,i$
0.20	$2.03\,i$	$-0.0357 + 1.99\,i$	$-0.0214 + 2.01\,i$
0.25	$1.82\,i$	$-0.0347 + 1.80\,i$	$-0.0209 + 1.82\,i$

The results discussed in the present section are related to the author's work (Sumbatyan et al., 2000).

$4°$. It can directly be verified that the explicit formulas (4.115), (4.118) for the particular geometries studied in this section justify not only the asymptotic estimate (4.94) universally valid for all domains but also Weyl's hypothesis (4.87). The proof follows the one applied in Section 4.4 for rectangular domains (for more detail, see Makai, 1970).

4.6. Explicit Analytical Results for Some Polyhedra

The method proposed in the previous section can be extended to some polyhedra. Here we also consider the Neumann problem only.

In the class of polyhedral boundary surfaces S the only known classical geometry that gives an exact solution in explicit form is parallelepiped. The solution is well known and can be constructed by using a separation of variables, which involves again a modal representation. Since the latter does not admit extension to more complex geometries, we have developed an alternative approach.

If a point source is placed at an arbitrary interior point $(x_0, y_0, z_0) \in V : p_0 = \exp(ikr)/r$ with $r = [(x_0 - x)^2 + (y_0 - y)^2 + (z_0 - z)^2]^{1/2}$, then in order to satisfy the trivial boundary condition (1) for the normal derivative, one may arrange a three-dimensional periodic array of virtual sources (*mirror images*), which provides natural symmetry about the arbitrary boundary point. The response at the receiving point $(x_1, y_1, z_1) \in V$ is then given by a three-dimensional series ($0 < x < a$; $0 < y < b$; $0 < z < c$ is the size of the parallelepiped):

$$p(x_1, y_1, z_1) = \sum_{j=1}^{8} S(|\xi_j - x_1|, |\eta_j - y_1|, |\zeta_j - z_1|, a, b, c), \qquad (4.122)$$

where

$$
\begin{aligned}
&(\xi_1, \eta_1, \zeta_1) = (x_0, y_0, z_0); &&(\xi_5, \eta_5, \zeta_5) = (2a - x_0, 2b - y_0, z_0); \\
&(\xi_2, \eta_2, \zeta_2) = (2a - x_0, y_0, z_0); &&(\xi_6, \eta_6, \zeta_6) = (2a - x_0, y_0, 2c - z_0); \\
&(\xi_3, \eta_3, \zeta_3) = (x_0, 2b - y_0, z_0); &&(\xi_7, \eta_7, \zeta_7) = (x_0, 2b - y_0, 2c - z_0); \\
&(\xi_4, \eta_4, \zeta_4) = (x_0, y_0, 2c - z_0); &&(\xi_8, \eta_8, \zeta_8) = (2a - x_0, 2b - y_0, 2c - z_0);
\end{aligned}
\qquad (4.123)
$$

and the function S

$$S(x, y, z, a, b, c) = \sum_{m,n,l=-\infty}^{\infty} \frac{\exp\left[ik\sqrt{(x + 2am)^2 + (y + 2bn)^2 + (z + 2cl)^2}\right]}{\sqrt{(x + 2am)^2 + (y + 2bn)^2 + (z + 2cl)^2}} \qquad (4.124)$$

is given by a series, which, in the classical sense, does not converge. Therefore, the series (4.124) is absolutely unsuitable for any direct numerical treatment.

Rapidly convergent representation for the function S.

The series (4.124) can be regularized by means of Poisson–Abel summation (see Section 1.3). To follow his idea let us apply again the principle of extremely small attenuation. If a small imaginary component is added to the wave number: $k_\varepsilon = k + i\varepsilon$, $0 < \varepsilon \ll 1$, then the series (4.124) becomes exponentially convergent, and it has a finite limit when $\varepsilon \to 0$. Since the function $S(x, y, z, a, b, c)$ is periodic with respect to all its arguments, let us assume that $0 \le x < 2a$, $0 \le y < 2b$, $0 \le z < 2c$. Then we apply the Weyl integral representation

$$\frac{\exp(ik\sqrt{x^2 + y^2 + z^2})}{\sqrt{x^2 + y^2 + z^2}} = \frac{1}{2\pi} \iint_{-\infty}^{\infty} e^{-\gamma(\alpha)z} \frac{e^{-i(\alpha_1 x + \alpha_2 y)}}{\gamma(\alpha)} \, d\alpha_1 \, d\alpha_2,$$

$$\alpha = (\alpha_1, \alpha_2), \qquad \gamma(\alpha) = \sqrt{\alpha_1^2 + \alpha_2^2 - k_\varepsilon^2}. \qquad (4.125)$$

Thus, the series (4.124) may be rewritten as

$$
\begin{aligned}
S(x, y, z, a, b, c) = {}&\frac{1}{2\pi} \sum_{m,n=-\infty}^{\infty} \iint_{-\infty}^{\infty} \left[\sum_{l=0}^{\infty} e^{-\gamma(\alpha)(2cl+z)} + \sum_{l=1}^{\infty} e^{-\gamma(\alpha)(2cl-z)} \right] \\
&\times \frac{\exp[-i(\alpha_1|2am + x| + \alpha_2|2bn + y|)]}{\gamma(\alpha)} \, d\alpha_1 \, d\alpha_2.
\end{aligned}
\qquad (4.126)
$$

The branching function $\gamma(\alpha)$ (see Section 1.1) has a positive real part, $\mathrm{Re}[\gamma(\alpha)] > 0$, when $\varepsilon > 0$. Hence, both series in (4.126) are geometric progressions, which give

$$
\begin{aligned}
S(x, y, z, a, b, c) = {}&\frac{1}{2\pi} \sum_{m,n=-\infty}^{\infty} \iint_{-\infty}^{\infty} \frac{\exp[-\gamma(\alpha)] + \exp[\gamma(\alpha)(z - 2c)]}{1 - \exp[-2c\gamma(\alpha)]} \\
&\times \frac{\exp[-i(\alpha_1|2am + x| + \alpha_2|2bn + y|)]}{\gamma(\alpha)} \, d\alpha_1 \, d\alpha_2 \\
= {}&\frac{1}{2\pi} \sum_{m,n=-\infty}^{\infty} \iint_{-\infty}^{\infty} \frac{\cosh[(c - z)\gamma(\alpha)]}{\sinh[c\gamma(\alpha)]} \frac{\exp[-i(\alpha_1|2am + x| + \alpha_2|2bn + y|)]}{\gamma(\alpha)} \, d\alpha_1 \, d\alpha_2.
\end{aligned}
\qquad (4.127)
$$

Let us fix the value of the variable α_1. Then the integrand in Eq. (4.127) is a meromorphic function of α_2, exponentially decaying at infinity. Therefore, with the help of the Jordan lemma, one can apply a standard expansion in residues at simple poles, by removing the integration contour down to infinity. The simple poles are solutions of the following transcendental equation:

$$\sinh[c\gamma(\alpha_1, \alpha_2)] = 0 \quad \sim \quad \gamma_l = \frac{\pi l i}{c}$$

$$\sim \quad \alpha_{2l} = -iq(\alpha_1, l), \quad q(\alpha_1, l) = \sqrt{\alpha_1^2 + \left(\frac{\pi l}{c}\right)^2 - k_\varepsilon^2}, \quad l = 0, 1, \ldots . \tag{4.128}$$

Hence, integration over α_2 in (4.127) leads to the following representation:

$$S(x, y, z, a, b, c) = \sum_{l=0}^{\infty} \frac{\delta_l \cos(\pi l z / c)}{c\, q(\alpha_1, l)} \int_{-\infty}^{\infty} d\alpha_1 \sum_{m,n=-\infty}^{\infty} \exp(-i\alpha_1 |2am + x|)$$

$$\times \exp\left[-q(\alpha_1, l)|2bn + y|\right], \quad \delta_0 = \tfrac{1}{2}, \quad \delta_l = 1 \quad (l = 1, 2, \ldots). \tag{4.129}$$

Since the real part of the root square function $q(\alpha_1, l)$ in Eq. (4.129) is positive when $\varepsilon > 0$, the last series (taken over n) is again a geometric progression, its sum being equal to

$$\sum_{n=0}^{\infty} \exp[-q(\alpha_1, l)\,(2bn + y)] + \sum_{n=1}^{\infty} \exp[-q(\alpha_1, l)\,(2bn - y)]$$

$$= \frac{\cosh[(b-y)\,q(\alpha_1, l)]}{\sinh[b\,q(\alpha_1, l)]} = \frac{\cosh\left[(b-y)\,\sqrt{\alpha_1^2 + (\pi l/c)^2 - k_\varepsilon^2}\right]}{\sinh\left[b\,\sqrt{\alpha_1^2 + (\pi l/c)^2 - k_\varepsilon^2}\right]}, \tag{4.130}$$

which involves

$$S(x, y, z, a, b, c) = \sum_{l=0}^{\infty} \frac{\delta_l}{c} \cos(\pi l z / c) \sum_{m=-\infty}^{\infty} \int_{-\infty}^{\infty}$$

$$\times \frac{\cosh\left[(b-y)\,\sqrt{\alpha_1^2 + (\pi l/c)^2 - k_\varepsilon^2}\right] \exp(-i\alpha_1 |2am + x|)}{\sqrt{\alpha_1^2 + (\pi l/c)^2 - k_\varepsilon^2}\, \sinh\left[b\,\sqrt{\alpha_1^2 + (\pi l/c)^2 - k_\varepsilon^2}\right]}\, d\alpha_1. \tag{4.131}$$

The last integrand is a meromorphic function of the variable α_1, which satisfies the necessary conditions of the Jordan lemma. Therefore, by removing the contour of integration down to infinity, the integral in (4.131) can be rewritten as an expansion in residues at simple poles. The latter are given as follows:

$$\sinh\left[b\,\sqrt{\alpha_1^2 + (\pi l/c)^2 - k_\varepsilon^2}\right] = 0 \quad \sim \quad \sqrt{\alpha_1^2 + (\pi l/c)^2 - k_\varepsilon^2} = \frac{\pi n i}{b}$$

$$\sim \quad \alpha_1 = -iq_{nl}, \quad q_{nl} = \sqrt{\left(\frac{\pi n}{b}\right)^2 + \left(\frac{\pi l}{c}\right)^2 - k_\varepsilon^2}, \tag{4.132}$$

which yields the following representation in the form of a triple series:

$$\frac{S(x, y, z, a, b, c)}{2\pi} = \sum_{n=0}^{\infty} \sum_{l=0}^{\infty} \sum_{m=-\infty}^{\infty} \frac{\delta_n \delta_l \cos(\pi n y / b) \cos(\pi l z / c)}{b\, c\, q_{nl}} \exp(-q_{nl}|2am + x|). \tag{4.133}$$

Since $\mathrm{Re}(q_{nl}) > 0$ when $\varepsilon > 0$, the last series with parameter m is a geometric progression:

$$\sum_{m=0}^{\infty} \exp[-q_{n,l}\,(2am+x)] + \sum_{m=1}^{\infty} \exp[-q_{n,l}\,(2am-x)] = \frac{\cosh[(a-x)\,q_{n,l}]}{\sinh(a\,q_{n,l})}. \qquad (4.134)$$

Therefore, the final representation ($\varepsilon \to +0$) can be obtained from (4.133) and (4.134), as follows:

$$S(x,y,z,a,b,c) = 2\pi \sum_{n=0}^{\infty} \sum_{l=0}^{\infty} \frac{\delta_n \delta_l \cos(\pi ny/b)\cos(\pi lz/c)}{bc\,\sqrt{(\pi n/b)^2 + (\pi l/c)^2 - k^2}}$$
$$\times \frac{\cosh\left[(a-x)\,\sqrt{(\pi n/b)^2 + (\pi l/c)^2 - k^2}\right]}{\sinh\left[a\,\sqrt{(\pi n/b)^2 + (\pi l/c)^2 - k^2}\right]}. \qquad (4.135)$$

Note that with an arbitrary fixed wave parameter k the terms of the series (4.135) decay exponentially when n or/and l tend to infinity (recall that $0 < x < 2a \sim -a < a - x < a$). Thus, Eq. (4.135) is indeed an efficient rapidly convergent representation for the function S.

It should also be noted that, although the original formula (4.134) is absolutely symmetric with respect to the coordinates x, y, z, the representation (4.135) has an asymmetric form. It thus admits a standard cyclic substitution: $(x,a) \to (y,b) \to (z,c) \to (x,a)$, with a final form under arbitrary substitution of this type being absolutely equivalent to (4.135).

Exact solution for triangular prisms.

Apparently, in the case of a parallelepiped, the representation (1.435) can be directly obtained by using a separation of variables. However the application of this type of modal expansion is only possible for rectangular polyhedra. That is why in the previous section we developed a different approach, namely the virtual sources method, because this provides direct extension to more complex polyhedra, as soon as the complete disposition of the set of virtual sources is known.

An appropriate exact solution can be constructed for polyhedra, for which every interior dihedral angle possesses a finite set of virtual sources, all situated outside of the polyhedron. In such cases an explicit solution can be obtained in the form of a finite superposition (compare with mode expansion for parallelepiped) of functions S with certain values of (ξ_j, η_j, ζ_j) and (a, b, c). It is known that this holds only for wedges equal to π/n, n being an arbitrary positive integer. The case $n = 2$ for all dihedral angles has to be related to the above solution for the parallelepiped.

The main statement of the present section is that one can construct an exact explicit solution for polyhedral spaces where each dihedral angle of the polyhedron is π/n. We do not aim to give a complete description of the all such polyhedra and the relevant solutions, but give instead some explicit results for three right-angled triangular prisms.

If the base of the prism is a triangle, lying on the horizontal plane (see the previous section), then the corresponding sets of virtual sources are well known for 1) isosceles right-angled triangles; 2) equilateral triangles; and 3) right-angled triangle with acute angles $30°$ and $60°$. Complete three-dimensional sets of virtual sources for every case are obtained as a certain periodic continuation (certain reflections) of the corresponding horizontal system along the vertical z-axis.

Let (x_0, y_0, z_0) denote the real point source, and (x_1, y_1, z_1) the receiving point. Then for the three right-angled prisms we have:

1) The coordinates of the corners of this prism are $(0, 0, 0)$; $(a, 0, 0)$; $(0, a, 0)$; $(0, 0, h)$; $(a, 0, h)$; $(0, a, h)$ (a is a leg of the base). The exact solution is given as

follows:

$$p(x_1, y_1, z_1) = \sum_{j=1}^{16} S(|\xi_j - x_1|, |\eta_j - y_1|, |\zeta_j - z_1|, a, b, h), \qquad b = a, \qquad (4.136)$$

where

$$
\begin{array}{ll}
(\xi, \eta, \zeta)_1 = (x_0, y_0, z_0); & (\xi, \eta, \zeta)_9 = (x_0, y_0, 2h - z_0); \\
(\xi, \eta, \zeta)_2 = (a - y_0, a - x_0, z_0); & (\xi, \eta, \zeta)_{10} = (a - y_0, a - x_0, 2h - z_0); \\
(\xi, \eta, \zeta)_3 = (a + y_0, a - x_0, z_0); & (\xi, \eta, \zeta)_{11} = (a + y_0, a - x_0, 2h - z_0); \\
(\xi, \eta, \zeta)_4 = (2a - x_0, y_0, z_0); & (\xi, \eta, \zeta)_{12} = (2a - x_0, y_0, 2h - z_0); \\
(\xi, \eta, \zeta)_5 = (2a - x_0, -y_0, z_0); & (\xi, \eta, \zeta)_{13} = (2a - x_0, -y_0, 2h - z_0); \\
(\xi, \eta, \zeta)_6 = (a + y_0, x_0 - a, z_0); & (\xi, \eta, \zeta)_{14} = (a + y_0, x_0 - a, 2h - z_0); \\
(\xi, \eta, \zeta)_7 = (a - y_0, x_0 - a, z_0); & (\xi, \eta, \zeta)_{15} = (a - y_0, x_0 - a, 2h - z_0); \\
(\xi, \eta, \zeta)_8 = (x_0, -y_0, z_0); & (\xi, \eta, \zeta)_{16} = (x_0, -y_0, 2h - z_0);
\end{array}
\qquad (4.137)
$$

and the function S in (4.136) is given by Eq. (4.135).

2) If the side of the equilateral base is c, then the corners of this prism are at the points $(-\frac{1}{2}c, 0, 0)$; $(\frac{1}{2}c, 0, 0)$; $(0, \frac{1}{2}\sqrt{3}c, 0)$; $(-\frac{1}{2}c, 0, h)$; $(\frac{1}{2}c, 0, h)$; $(0, \frac{1}{2}\sqrt{3}c, h)$. The exact solution is as follows:

$$p(x_1, y_1, z_1) = \sum_{j=1}^{24} S(|\xi_j - x_1|, |\eta_j - y_1|, |\zeta_j - z_1|, a, b, h),$$

$$a = \tfrac{3}{2}c, \qquad b = \tfrac{1}{2}\sqrt{3}\, c, \qquad (4.138)$$

where

$$
\begin{aligned}
&(\xi, \eta, \zeta)_1 = (x_0, y_0, z_0); \\
&(\xi, \eta, \zeta)_2 = \left(\tfrac{3}{4}c - \tfrac{1}{2}x_0 - \tfrac{1}{2}\sqrt{3}y_0, \ \tfrac{1}{4}\sqrt{3}c - \tfrac{1}{2}\sqrt{3}x_0 + \tfrac{1}{2}y_0, \ z_0 \right); \\
&(\xi, \eta, \zeta)_3 = \left(\tfrac{3}{4}c - \tfrac{1}{2}x_0 + \tfrac{1}{2}\sqrt{3}y_0, \ \tfrac{1}{4}\sqrt{3}c - \tfrac{1}{2}\sqrt{3}x_0 - \tfrac{1}{2}y_0, \ z_0 \right); \\
&(\xi, \eta, \zeta)_4 = (\xi, -\eta, \zeta)_3; \quad (\xi, \eta, \zeta)_5 = (\xi, -\eta, \zeta)_2; \quad (\xi, \eta, \zeta)_6 = (\xi, -\eta, \zeta)_1; \\
&(\xi, \eta, \zeta)_{j+6} = \left(\xi_j + \tfrac{3}{2}c, \ \eta_j + \tfrac{1}{2}\sqrt{3}c, \ \zeta_j \right), \quad j = 1, \ldots, 6; \\
&(\xi, \eta, \zeta)_{j+12} = (\xi_j, \ \eta_j, \ 2h - \zeta_j), \quad j = 1, \ldots, 12.
\end{aligned}
\qquad (4.139)
$$

3) If the hypotenuse of the base is c, then the Cartesian coordinates of the prism corners are $(0, 0, 0,)$; $(0, \frac{1}{2}\sqrt{3}c, 0)$; $(0, \frac{1}{2}c, 0)$; $(0, 0, h)$; $(0, \frac{1}{2}\sqrt{3}c, h)$; $(0, \frac{1}{2}c, h)$. The explicit form of the solution is

$$p(x_1, y_1, z_1) = \sum_{j=1}^{48} S(|\xi_j - x_1|, |\eta_j - y_1|, |\zeta_j - z_1|, a, b, h), \qquad a = \tfrac{1}{2}\sqrt{3}c, \quad b = \tfrac{3}{2}c, \quad (4.140)$$

where

$$(\xi, \eta, \zeta)_1 = (x_0, \ y_0, \ z_0);$$

$$(\xi, \eta, \zeta)_2 = \left(\tfrac{1}{4}\sqrt{3}c + \tfrac{1}{2}x_0 - \tfrac{1}{2}\sqrt{3}y_0, \ \tfrac{3}{4}c - \tfrac{1}{2}\sqrt{3}x_0 - \tfrac{1}{2}y_0, \ z_0\right);$$

$$(\xi, \eta, \zeta)_3 = \left(\tfrac{1}{4}\sqrt{3}c + \tfrac{1}{2}x_0 + \tfrac{1}{2}\sqrt{3}y_0, \ \tfrac{3}{4}c - \tfrac{1}{2}\sqrt{3}x_0 + \tfrac{1}{2}y_0, \ z_0\right);$$

$$(\xi, \eta, \zeta)_4 = \left(\tfrac{3}{4}\sqrt{3}c - \tfrac{1}{2}x_0 - \tfrac{1}{2}\sqrt{3}y_0, \ \eta_3, \ z_0\right);$$

$$(\xi, \eta, \zeta)_5 = \left(\tfrac{3}{4}\sqrt{3}c - \tfrac{1}{2}x_0 + \tfrac{1}{2}\sqrt{3}y_0, \ \eta_2, \ z_0\right); \tag{4.141}$$

$$(\xi, \eta, \zeta)_6 = \left(\sqrt{3}c - x_0, \ y_0, \ z_0\right);$$

$$(\xi, \eta, \zeta)_j = (\xi, -\eta, \zeta)_{13-j}, \quad j = 7, \ldots, 12;$$

$$(\xi, \eta, \zeta)_{j+12} = (\xi_j + \tfrac{1}{2}\sqrt{3}c, \ \eta_j + \tfrac{3}{2}c, \ \zeta_j), \quad j = 1, \ldots, 12;$$

$$(\xi, \eta, \zeta)_{j+24} = (\xi_j, \ \eta_j, \ 2h - \zeta_j), \quad j = 1, \ldots, 24.$$

It should be noted that if any value of (ξ_j, η_j, ζ_j) in the above expressions appears to be outside the natural domain $(0 \le \xi_j \le 2a, \ 0 \le \eta_j \le 2b, \ 0 \le \zeta \le 2h)$, then one should use a periodicity of the function S, prior to applying formula (4.135).

Explicit formulas for eigenfrequencies.

The eigenvalues of the wave number k (related to resonance frequencies) can be obtained from Eq. (4.135) by setting its denominator equal to zero. Obviously, this holds for trivial values of the hyperbolic sine:

$$\sinh\left[a\sqrt{\left(\frac{\pi n}{b}\right)^2 + \left(\frac{\pi l}{c}\right)^2 - k^2}\right] = 0$$

$$\sim \quad a\sqrt{\left(\frac{\pi n}{b}\right)^2 + \left(\frac{\pi l}{c}\right)^2 - k^2} = \pi m i \quad \sim \quad k_{mnl} = \pi\sqrt{\frac{m^2}{a^2} + \frac{n^2}{b^2} + \frac{l^2}{c^2}}, \tag{4.142}$$

with arbitrary natural numbers m, n, l. For a parallelepiped domain with sides a, b, c this result is well known in the literature. For triangular prisms it involves some new exact estimates:

1) Right-angled prism with the base in the form of an isosceles right-angled triangle (the leg of the base is a, the height of the prism is h). The eigenvalues of the frequency parameter are

$$k_{mnl} = \pi\sqrt{\frac{m^2 + n^2}{a^2} + \frac{l^2}{h^2}}, \tag{4.143}$$

the same numbers as for a right-angled prism with a square base.

2) Right-angled prism with the base in the form of an equilateral triangle (the side of the base is c, the height of the prism is h). Following Eq. (4.142), we obtain $k_{mnl} = \pi\sqrt{4(m^2 + 3n^2)/(9c^2) + l^2/h^2}$.

Let us prove that this expression holds only if the sum $(m + n)$ is even (compare with Section 4.5). We start from the evident notice that Eqs. (4.138), (4.139) show: for every point $(\xi, \eta, \zeta)_j$, $j = 1, \ldots, 24$ in the sum (4.138) there is also a point $(\xi, \eta, \zeta)_{j\pm6}$ with $\xi_{j\pm6} = \xi_j \pm a$, $\eta_{j\pm6} = \eta_j \pm b$, $\zeta_{j\pm6} = \zeta_j$. Now if we take $\sqrt{(\pi n/b)^2 + (\pi l/c)^2 - k^2} = \pi m i/a$, as in Eq. (4.142), the following part of the numerator in Eq. (4.135) becomes equal to $\cos(\pi ny/b)\cosh\left[(a - x)\sqrt{(\pi n/b)^2 + (\pi l/c)^2 - k^2}\right] = \cos(\pi ny/b)\cos[\pi m(a - x)/a]$, the term which has opposite signs for $(\xi, \eta, \zeta)_j$ and $(\xi, \eta, \zeta)_{j\pm6}$ if $(m + n)$ is odd. Therefore, these terms in the sum (4.138) cancel each other out whenever $m + n$ is odd, which excludes this combination of n and m from the set of eigenvalues of the parameter k. The proof is complete. Further, in the same way as in the previous section, it can also be proved that using the obtained formula for k_{mnl} with $(m+n)$ even is equivalent to the following result:

$$k_{mnl} = \pi \sqrt{\frac{16(m^2 + mn + n^2)}{9c^2} + \frac{l^2}{h^2}}, \qquad (4.144)$$

if one applies this with arbitrary positive integers m, n, l.

3) Right-angled prism with a right-angled triangular base having the acute angles $30°$ and $60°$, the base hypotenuse is c and the height of the prism h. Equation (4.142) involves the following eigenvalues of the wave number $k_{mnl} = \pi \sqrt{4(3m^2 + n^2)/(9c^2) + l^2/h^2}$. However, arguments analogous to those given for case 2) prove that the correct result is (m, n, l are arbitrary natural numbers):

$$k_{mnl} = \pi \sqrt{\frac{16(m^2 + mn + n^2)}{9c^2} + \frac{l^2}{h^2}}, \qquad (4.145)$$

the same values as for the previous prism, given by (4.144).

The proposed results are connected with the author's works Sumbatyan (2000); Sumbatyan and Pompei (2001).

Helpful remarks

1°. The analytical results obtained are worthy of a special test to compare them with numerical calculations. To calculate the wave field at an arbitrary receiving point in the nonresonance case, one can apply, for instance, a finite-element or boundary-element technique. However, more precise results are obtained numerically for eigenfrequencies, since they are an *integral* measure of the polyhedron's geometry. We have used a finite element method for this purpose, and we have achieved the relative error in comparison between analytical and direct numerical calculations less than 1% for all considered geometries.

2°. Weyl–Carleman theory (see Section 4.4) can also be applied in the 3D case to a volume of arbitrary shape. It can be strictly proved that the leading asymptotic term, for both the Dirichlet and the Neumann problem, is expressed as follows (see Courant and Hilbert, 1953):

$$N(\lambda) = \frac{V}{6\pi^2} \lambda^{3/2} + O(\lambda \ln \lambda), \qquad \lambda \to \infty, \qquad (4.94)$$

where V is the measure of the volume.

Chapter 5

Integral Equations in Diffraction by Linear Obstacles

5.1. Integral Operators in Diffraction by Linear Screen and by a Gap in the Screen

As in wave problems for acoustic layer (Chapter 3), the problem considered here demonstrates the power of the Fourier transform.

Let a plane acoustic wave fall from below, propagating in direction of vertical axis y in the Cartesian coordinate system (x, y): $p^{\text{inc}}(x, y) = e^{iky}$. The problem is thus considered as two-dimensional. We will study here integral equations arising in diffraction by an acoustically hard linear obstacle placed over the axis x, i.e., somewhere at $y = 0$. If the obstacle occupies the whole axis x: $y = 0$, $-\infty < x < \infty$, then the problem is one-dimensional, and is reduced to a simple ordinary differential equation with constant coefficients. If there is a finite-length obstacle, then we consider simultaneously the two problems genetically related to each other:

α) There is a rigid screen placed on $|x| \leq a$, $y = 0$.

β) There is a gap (hole) in the infinite hard screen, occupying the interval $|x| \leq a$, $y = 0$. In Hönl et al. (1961) you can find some general results, which establish a relationship between these two problems.

For both problems we consider separately the lower ($y \leq 0$, all functions here are marked with the subscript 1) and the upper ($y \geq 0$, with the subscript 2) half-planes with appropriate (complex-valued) acoustic pressure $p_1(x, y)$ and $p_2(x, y)$, with both of them satisfying the Helmholtz equation

$$\Delta p_{1,2} + k^2 p_{1,2} = 0, \qquad k = \frac{\omega}{c}. \tag{5.1}$$

Normal incidence of a plane wave is considered here only for simplicity. Another form of the incident wave modifies the right-hand side of integral equations only.

If we apply the Fourier transform with respect to variable $-\infty < x < \infty$, then it can be easily seen that a solution whose diffracted component satisfies Sommerfeld's radiation condition is given by (cf. Chapter 3)

$$\alpha) \quad P_2(s, y) = A(s)e^{-\gamma(s)y} + P^{\text{inc}}(s, y), \quad y \geq 0,$$
$$P_1(s, y) = B(s)e^{\gamma(s)y} + P^{\text{inc}}(s, y), \quad y \leq 0, \quad \gamma(s) = \sqrt{\alpha^2 - s^2}, \tag{5.2}$$

$$\beta) \quad P_2(s, y) = C(s)e^{-\gamma(s)y}, \quad y \geq 0,$$
$$P_1(s, y) = D(s)e^{\gamma(s)y} + P^{\text{inc}}(s, y), \quad y \leq 0, \tag{5.3}$$

where $P^{\text{inc}}(s, y) = 2\pi\delta(s)e^{iky}$. The coefficients $A(s)$ and $B(s)$ are to be defined from the boundary conditions on the line $y = 0$, which are posed in accordance with the impenetrability of the hard screen, i.e., trivial normal component of the velocity. Since

$\mathbf{v} = (1/i\omega\rho_0)\,\mathrm{grad}\,p$, this involves the two problems in hand, for $y = 0$:

$$\alpha)\quad \frac{\partial p_1}{\partial y} = \frac{\partial p_2}{\partial y} = 0, \qquad |x| \le a, \qquad p_1 = p_2, \qquad \frac{\partial p_1}{\partial y} = \frac{\partial p_2}{\partial y}, \qquad |x| > a. \quad (5.4)$$

$$\beta)\quad p_1 = p_2, \qquad \frac{\partial p_1}{\partial y} = \frac{\partial p_2}{\partial y}, \qquad |x| \le a, \qquad \frac{\partial p_1}{\partial y} = \frac{\partial p_2}{\partial y} = 0, \qquad |x| > a, \quad (5.5)$$

where the second lines in Eqs. (5.4), (5.5) mean continuity of the wave field in the fluid.

We aim to reduce both problems $\alpha)$ and $\beta)$ to an integral equation over the interval $|x| \le a$. For this purpose, let us introduce the new unknown functions $g_\alpha(x)$ and $g_\beta(x)$ defined just on this interval:

$$\alpha)\ \ y = 0:\quad p_1 - p_2 = \begin{cases} 0, & |x| > a \\ g_\alpha(x), & |x| \le a. \end{cases} \quad (5.6)$$

$$\beta)\ \ y = 0:\quad \frac{\partial p_1}{\partial y} = \frac{\partial p_2}{\partial y} = \begin{cases} 0, & |x| > a \\ g_\beta(x), & |x| \le a. \end{cases} \quad (5.7)$$

The boundary conditions in Eqs. (5.4), (5.5) are written so that in both problem $\alpha)$ and $\beta)$ we have $\partial p_1/\partial y = \partial p_2/\partial y$, $y = 0$, $|x| < \infty$, which with the help of Eqs. (5.1), (5.2) yields

$$\begin{aligned} \alpha)&\quad -\gamma A = \gamma B \\ \beta)&\quad -\gamma C = \gamma D + 2\pi i k \delta(s). \end{aligned} \quad (5.8)$$

Finally, Eqs. (5.6), (5.7) in Fourier transforms imply

$$\begin{aligned} \alpha)&\quad B - A = G_\alpha(s) \\ \beta)&\quad -\gamma C = G_\beta(s), \end{aligned} \quad (5.9)$$

where

$$G_{\alpha,\beta} = \int_{-a}^{a} g_{\alpha,\beta}(\xi)\,e^{is\xi}\,d\xi. \quad (5.10)$$

Expressions (5.8)–(5.10) allow us to write out coefficients A, B and C, D in terms of G_α, G_β:

$$\begin{aligned} \alpha)&\quad B = \frac{G_\alpha(s)}{2}, \quad A = -\frac{G_\alpha(s)}{2}, \\ \beta)&\quad C = -\frac{G_\beta(s)}{\gamma(s)}, \quad D = \frac{G_\beta(s)}{\gamma(s)} - 2\pi\frac{ik}{\gamma(s)}\delta(s). \end{aligned} \quad (5.11)$$

Therefore by applying the convolution theorem (see Section 1.1) we arrive at the following representations:

$$\alpha)\quad p_1(x,y) = p^{\mathrm{inc}}(x,y) + \frac{1}{2\pi}\int_{-\infty}^{\infty} B(s)e^{\gamma y}e^{-isx}\,ds$$

$$= e^{iky} + \frac{1}{4\pi}\int_{-a}^{a} g_\alpha(\xi)\,d\xi \int_{-\infty}^{\infty} e^{\gamma y}e^{-is(x-\xi)}\,ds, \quad y \le 0, \quad |x| < \infty, \quad (5.12)$$

$$p_2(x,y) = e^{iky} - \frac{1}{4\pi}\int_{-a}^{a} g_\alpha(\xi)\,d\xi \int_{-\infty}^{\infty} e^{-\gamma y}e^{-is(x-\xi)}\,ds, \quad y \ge 0, \quad |x| < \infty.$$

$$\beta) \quad p_2(x,y) = \frac{1}{2\pi} \int_{-\infty}^{\infty} C(s) e^{-\gamma(s)y} e^{-isx} \, ds$$

$$= -\frac{1}{2\pi} \int_{-a}^{a} g_\beta(\xi) \, d\xi \int_{-\infty}^{\infty} \frac{e^{-\gamma(s)y}}{\gamma(s)} e^{-is(x-\xi)} \, ds, \quad y \geq 0, \quad |x| < \infty,$$

$$p_1(x,y) = p^{\text{inc}}(x,y) + \frac{1}{2\pi} \int_{-\infty}^{\infty} D(s) e^{\gamma(s)y} e^{-isx} \, ds \tag{5.13}$$

$$= e^{iky} + \frac{1}{2\pi} \int_{-a}^{a} g_\beta(\xi) \, d\xi \int_{-\infty}^{\infty} \frac{e^{\gamma y}}{\gamma(s)} e^{-is(x-\xi)} \, ds - ik \int_{-\infty}^{\infty} \frac{e^{\gamma y}}{\gamma(s)} \delta(s) e^{-isx} \, ds$$

$$= e^{iky} + \frac{1}{2\pi} \int_{-a}^{a} g_\beta(\xi) \, d\xi \int_{-\infty}^{\infty} \frac{e^{\gamma y}}{\gamma(s)} e^{-is(x-\xi)} \, ds + e^{-iky}, \quad y \leq 0, \quad |x| < \infty.$$

In the last transformations we have used the basic property of Dirac's delta (see Section 1.4), and the value $\gamma(0) = -ik$.

The required integral equations are obtained from (5.12), (5.13) if we attract the remaining boundary condition, which has not yet been used:

$$\alpha) \quad \left.\frac{\partial p_1}{\partial y}\right|_{y=0} = 0, \quad |x| \leq a \quad \sim \quad \int_{-a}^{a} g_\alpha(\xi) K_\alpha(x-\xi) \, d\xi = -2ik, \quad (x \leq a),$$

$$K_\alpha(x) = \frac{1}{2\pi} \int_{-\infty}^{\infty} L_\alpha(s) e^{isx} \, ds, \qquad L_\alpha(s) = \gamma(s), \tag{5.14}$$

$$\beta) \quad p_1 = p_2, \quad y = 0, \quad |x| \leq a \quad \sim \quad \int_{-a}^{a} g_\beta(\xi) K_\beta(x-\xi) \, d\xi = -1, \quad (|x| \leq a),$$

$$K_\beta(x) = \frac{1}{2\pi} \int_{-\infty}^{\infty} L_\beta(s) e^{isx} \, ds, \quad L_\beta(s) = \frac{1}{\gamma(s)}. \tag{5.15}$$

Properties of integral equations.
Since $L_\alpha(s)$ is unbounded for large s, the integral (5.14) for the kernel K_α should be treated in the generalized sense as a value of divergent integral. This can be calculated following to ideas of Section 1.3:

$$K_\alpha(x) = \frac{1}{\pi} \int_0^{\infty} s \cos(sx) \, ds + \frac{1}{\pi} \int_0^{\infty} \left(\sqrt{s^2 - k^2} - s \right) \cos(sx) \, ds$$

$$= \frac{1}{\pi} \lim_{\varepsilon \to +0} \int_0^{\infty} e^{-\varepsilon s} s \cos(sx) \, ds - \frac{k^2}{\pi} \int_0^{\infty} \frac{\cos(sx) \, ds}{\sqrt{s^2 - k^2} + s}$$

$$= \frac{1}{\pi} \lim_{\varepsilon \to +0} \frac{\varepsilon^2 - x^2}{(\varepsilon^2 + x^2)^2} - \frac{k^2}{\pi} \int_0^{\infty} \frac{\cos(kxs) \, ds}{\sqrt{s^2 - 1} + s} \tag{5.16}$$

$$= -\frac{1}{\pi x^2} - \frac{k^2 I_c(kx)}{\pi}, \qquad I_c(kx) = \int_0^{\infty} \frac{\cos(kxs) \, ds}{\sqrt{s^2 - 1} + s}.$$

The second term in (5.16) is bounded for finite x and k, because this integral is convergent in the classical sense if $k, x > 0$. For $kx \to 0$ its asymptotic behavior can be estimated from formula (1.83): $I_c(kx) \sim O(\ln(k|x|))$, $kx \to 0$. Therefore the kernel in problem $\alpha)$ is a characteristic hyper-singular kernel plus a kernel with a (weak) logarithmic singularity.

The kernel $K_\beta(x)$ can also be estimated in the same way; however it is more convenient to operate with the explicit representation

$$K_\beta(x) = \frac{i}{2} H_0^{(1)}(k|x|), \tag{5.17}$$

which is a kernel with a weak logarithmic singularity (cf. the asymptotic behavior of the Hankel function for small arguments, see Eq. (2.69)):

$$K_\beta(x) \sim -\frac{1}{\pi} \left[\ln(k|x|) + \gamma - \ln 2 \right] + \frac{i}{2} + O\left(x^2 \ln|x|\right), \qquad x \to 0. \tag{5.18}$$

THEOREM 1. *Integral equation (5.14) is uniquely solvable in the class of bounded functions at least for sufficiently small k.*

Proof. Let us extract, according to Eq. (5.16), characteristic hyper-singular part of the kernel

$$-\frac{1}{\pi} \int_{-a}^{a} \frac{g_\alpha(\xi)\,d\xi}{(x-\xi)^2} = -2ik + \frac{k^2}{\pi} \int_{-a}^{a} g_\alpha(\xi)\, I_c(k|x-\xi|)\,d\xi, \qquad |x| \le a, \tag{5.19}$$

and apply inversion formula (1.190) for the operator of the left-hand side.

$$g_\alpha(x) = 2ik \left[\frac{\sqrt{a^2 - x^2}}{\pi} \int_{-a}^{a} \frac{\xi\,d\xi}{\sqrt{a^2 - \xi^2}(x - \xi)} + \int_{-a}^{a} K_c(x, \xi)\, g_\alpha(\xi)\,d\xi \right],$$
$$K_c(x, \xi) = O(k \ln k), \qquad k \to 0, \tag{5.20}$$

where the estimate of the kernel K_c is uniform with respect to $(x, \xi) \in (-a, a)$. The first integral here can be calculated as follows

$$\int_{-a}^{a} \frac{\xi\,d\xi}{\sqrt{a^2 - \xi^2}(x - \xi)} = \int_{-a}^{a} \frac{(\xi - x) + x}{\sqrt{a^2 - \xi^2}(x - \xi)}\,d\xi = -\pi, \quad \text{since} \quad \int_{-a}^{a} \frac{d\xi}{\sqrt{a^2 - \xi^2}(x - \xi)} = 0, \tag{5.21}$$

as the Cauchy principal value. We thus have arrived at the Fredholm integral equation of the second kind, which, due to the remark right after Theorem 5, Section 1.5, is uniquely solvable. This completes the proof.

It should be noted that the solvability question for the arising equation (5.14) is a rather abstract issue. A much more practical question is related to the uniqueness, provided that correct conditions are set at infinity. If we have constructed a correct solution of this linear problem, it must be unique.

Note that a bounded solution vanishes near the edges of the interval $(-a, a)$ as, for example, $\sqrt{a \pm x}, x \to \mp a$.

THEOREM 2. *For all $0 < k < \infty$, the solution of equation (5.14) is unique in the class of differentiable functions: $g_\alpha(x) \in C_1(-a, a)$, with $g_\alpha(x) \sim \sqrt{a \pm x}, x \to \mp a$.*

Proof. This statement can be proved by the method of Section 1.6 adopted to the case when the symbolic function $L(s)$ is nonpositive (in our problem it is complex-valued). If there are two solutions $u_1(x)$ and $u_2(x)$ of the class described in the conditions of the theorem, then for the considered operator equation $G_\alpha g_\alpha = f_\alpha$ we have $G_\alpha(u_1 - u_2) = 0$. The scalar product of this relation with the function $(u_1 - u_2)$ results in

$$\int_{-\infty}^{\infty} L_\alpha(s)|U_1(s) - U_2(s)|^2\,ds = 0, \qquad L_\alpha(s) = \sqrt{s^2 - k^2}. \tag{5.22}$$

It is clear that under the conditions of this theorem $U_1(s), U_2(s) \sim |s|^{-3/2}, s \to \infty$ being continuous for finite s, the integral (5.22) is finite. Separation of real and imaginary parts in relation (5.22) results in

$$\left(\int_{-\infty}^{-k} + \int_{k}^{\infty} \right) \sqrt{s^2 - k^2}\, |U_1(s) - U_2(s)|^2\,ds - i \int_{-k}^{k} \sqrt{k^2 - s^2}\, |U_1(s) - U_2(s)|^2\,ds = 0. \tag{5.23}$$

Both the real and the imaginary terms in (5.23) should be equal to zero, which is possible only if $U_1(s) \equiv U_2(s), |s| < \infty \sim u_1(x) \equiv u_2(x), |x| \le a$. This proves the theorem.

Equation (5.15) can be studied by absolutely the same technique.

THEOREM 3. *1) The integral equation (5.15) is uniquely solvable in the class of continuous functions with a square root singularity at the ends of the interval at least for sufficiently small k, and 2) its solution is unique in the considered class for all k.*

Proof. The inversion of an integral operator with the characteristic logarithmic kernel (see Section 1.7) reduces equation (5.15) to a Fredholm equation of the second kind. Further procedure to prove the first statement of the theorem is absolutely the same as for the previous theorem. However, the second statement is also proved by analogy with Theorem 2. We have, instead of Eq. (5.22),

$$\int_{-\infty}^{\infty} L_\beta(s)|U_1(s) - U_2(s)|^2 \, ds = 0, \qquad L_\beta(s) = \frac{1}{\sqrt{s^2 - k^2}}. \tag{5.24}$$

For functions $u_1(x)$, $u_2(x)$ from the considered class the separation of the real and imaginary parts in (5.24),

$$\left(\int_{-\infty}^{-k} + \int_k^{\infty} \right) \frac{|U_1(s) - U_2(s)|^2}{\sqrt{s^2 - k^2}} \, ds + i \int_{-k}^k \frac{|U_1(s) - U_2(s)|^2}{\sqrt{k^2 - s^2}} \, ds = 0, \tag{5.25}$$

leads to some finite-valued integrals. If $u_1, u_2 \in C_1(-a, a)$ with behavior $(a\pm)^{-1/2}$, $x \to \mp a$, the asymptotics of U_1, U_2 is $U_1(s), U_2(s) \sim s^{-1/2}$, $s \to \infty$; therefore the first two integrals converge at infinity, which completes the proof.

Helpful remarks

For applications, the theory discussed in this section permits the calculation of the far-field scattering pattern. At low frequencies this can be expressed in explicit form (compare with the results of Chapter 2). For the sake of brevity, we give here only the sketch.

It is clear that for small frequencies ($k \to 0$) the kernels (5.16), (5.18) of the integral operators (5.14), (5.15) are reduced to their characteristic parts, which are characteristic hyper-singular kernel and characteristic logarithmic one. Both of them admit exact analytical inversion. As follows from Eq. (5.20) in the first problem we have $g_\alpha(x) = -2ik\sqrt{a^2 - x^2}$. The analogous low-frequency solution in the second problem, as a solution to a logarithmic characteristic equation, is $g_\beta(x) = B/\sqrt{a^2 - x^2}$, with some constant B. Then this function may be substituted into Eqs. (5.12), (5.13) giving wave field in all space. The far-field asymptotics can be extracted from there if we estimate integrals

$$I_\alpha = \int_{-\infty}^{\infty} e^{-\gamma|y|} e^{-is(x-\xi)} \, ds, \qquad I_\beta = \int_{-\infty}^{\infty} \frac{e^{-\gamma|y|}}{\gamma(s)} e^{-is(x-\xi)} \, ds, \tag{5.26}$$

for large $R = \sqrt{x^2 + y^2}$. This is given by the stationary phase method (see Eq. (1.88), (1.92)):

$$J = \int_{-\infty}^{\infty} F(s) \, e^{-isx} e^{-\gamma(s|y|} \, ds \sim \sqrt{\frac{2\pi k}{R}} \cos \alpha \, e^{i(kR-\pi/4)} \, F(k \sin \alpha), \quad R = \sqrt{x^2 + y^2} \to \infty, \tag{5.27}$$

hence

$$I_\alpha \sim \sqrt{\frac{2\pi k}{r}} \, e^{i(kr-\pi/4)} \cos \alpha, \quad r = \sqrt{(x-\xi)^2 + y^2} \to \infty, \tag{5.28}$$

$$I_\beta \sim \sqrt{\frac{2\pi k}{r}} \, e^{i(kr-\pi/4)} \frac{\cos \alpha}{\gamma(-k \sin \alpha)} = \sqrt{\frac{2\pi}{kr}} \, e^{i(kr+\pi/4)}, \quad r = \sqrt{(x-\xi)^2 + y^2} \to \infty. \tag{5.29}$$

Further treatment is based upon the classical far-field approximation for the quantity r in the argument of the exponential functions in (5.28), (5.29) (cf. Chapter 2):

$$r = \sqrt{x^2 + y^2 - 2x\xi + \xi^2} = R\sqrt{1 - \frac{2x\xi - \xi^2}{R^2}} \approx R\left[1 - \frac{x\xi}{R^2} + O\left(\frac{1}{R^2}\right)\right]$$

$$= R - \xi\cos\alpha + O\left(\frac{1}{R}\right), \quad R = \sqrt{x^2 + y^2} \to \infty. \tag{5.30}$$

5.2. Operator Equation in Diffraction Problem on a Crack in Unbounded Elastic Medium

Here we will study an integral operator arising in diffraction by a crack in elastic medium. Again, only for simplicity, we consider the 2D case of the normal incidence of a plane longitudinal wave.

Let a crack of length $2a$ be situated over the interval $|x| \le a$, $y = 0$, and the plane longitudinal wave fall to this crack from $y = -\infty$: $\varphi^{\text{inc}}(x, y) = e^{ik_p y}$, $\psi^{\text{inc}}(x, y) = 0$. The governing equations to this problem can be found in Section 1.9, and the application of the Fourier transform in the variable $-\infty < x < \infty$ is explained in Section 3.3.

Let us consider again the two half-planes $y \le 0$ (labeled with the superscript (1)) and $y \ge 0$ (superscript (2)) separately. Then in both half-planes the full wave field is the sum of the incident and scattered ones:

$$\varphi = \varphi^{\text{inc}} + \varphi^{\text{sc}}, \qquad \psi = \psi^{\text{sc}} \qquad (\psi^{\text{inc}} = 0). \tag{5.31}$$

In terms of Fourier images, a solution satisfying a correct radiation condition at infinity, which in the considered case of unbounded space can be applied simply as Sommerfeld's radiation condition, has the following form (compare with Section 3.3):

$$\Phi_{1,2}(s, y) = A_{1,2}(s)\, e^{-\gamma(s)|y|}, \qquad \gamma(s) = \sqrt{s^2 - k_p^2},$$

$$\Psi_{1,2}(s, y) = B_{1,2}(s)\, e^{-q(s)|y|}, \qquad q(s) = \sqrt{s^2 - k_s^2}, \tag{5.32}$$

where both Lamé potentials are taken for the scattered component of the elastic wave field.

Of course, the coefficients $A_{1,2}$, $B_{1,2}$ will be found from the boundary conditions at the line $y = 0$; in the case of load free crack faces, these conditions are written out as

$$y = 0: \quad \begin{aligned} &\sigma_{xy}^{(1,2)} = -\sigma_{xy}^{\text{inc}}, \quad \sigma_{yy}^{(1,2)} = -\sigma_{yy}^{\text{inc}}, \quad |x| \le a, \\ &\sigma_{xy}^{(1)} = \sigma_{xy}^{(2)}, \quad \sigma_{yy}^{(1)} = \sigma_{yy}^{(2)}, \quad u_x^{(1)} = u_x^{(2)}, \quad u_y^{(1)} = u_y^{(2)}, \quad |x| > a, \end{aligned} \tag{5.33}$$

where the equations of the second line here mean the continuity of stresses and displacements.

As usual, we notice that conditions (5.33) imply $\sigma_{xy}^{(1)} = \sigma_{xy}^{(2)}$, $\sigma_{yy}^{(1)} = \sigma_{yy}^{(2)}$, on the full line $y = 0$, $|x| < \infty$; hence

$$2\frac{\partial^2\varphi_1}{\partial x \partial y} + \frac{\partial^2\psi_1}{\partial y^2} - \frac{\partial^2\psi_1}{\partial x^2} = 2\frac{\partial^2\varphi_2}{\partial x \partial y} + \frac{\partial^2\psi_2}{\partial y^2} - \frac{\partial^2\psi_2}{\partial x^2},$$

$$2\left(\frac{\partial^2\varphi_1}{\partial x^2} + \frac{\partial^2\psi_1}{\partial x \partial y}\right) + k_s^2\varphi_1 = 2\left(\frac{\partial^2\varphi_2}{\partial x^2} + \frac{\partial^2\psi_2}{\partial x \partial y}\right) + k_s^2\varphi_2, \tag{5.34}$$

that is

$$-2is\Phi_1' + \Psi_1'' + s^2\Psi_1 = -2is\Phi_2' + \Psi_2'' + s^2\Psi_2,$$

$$2(-s^2\Phi_1 - is\Psi_1') + k_s^2\Phi_1 = 2(-s^2\Phi_2 - is\Psi_2') + k_s^2\Phi_2 \quad (y = 0), \tag{5.35}$$

where the sign of the derivative is related to y. Equations (5.34) imply

$$
\begin{aligned}
&-2is\gamma A_1 + (2s^2 - k_s^2)B_1 = 2is\gamma A_2 + (2s^2 - k_s^2)B_2,\\
&-(2s^2 - k_s^2)A_1 - 2isqB_1 = -(2s^2 - k_s^2)A_2 + 2isqB_2.
\end{aligned}
\tag{5.36}
$$

In order to reduce the problem to an integral equation, let us introduce the new unknown functions

$$
y = 0: \quad u_x^{(1)} - u_x^{(2)} = \begin{cases} 0, & |x| > a, \\ g_x(x), & |x| \le a, \end{cases} \qquad u_y^{(1)} - u_y^{(2)} = \begin{cases} 0, & |x| > a, \\ g_y(x), & |x| \le a. \end{cases}
\tag{5.37}
$$

Then in Fourier variables

$$
(-is\Phi_1 + \Psi_1') - (-is\Phi_2 + \Psi_2') = G_x(s), \qquad G_x(s) = \int_{-a}^{a} g_x(\xi)e^{is\xi}\, d\xi,
$$
$$
(\Phi_1' + is\Psi_1) - (-\Phi_2' + is\Psi_2) = G_y(s), \qquad G_y(s) = \int_{-a}^{a} g_y(\xi)e^{is\xi}\, d\xi,
\tag{5.38}
$$

which is equivalent to

$$
-is\, A_1 + qB_1 + isA_2 + qB_2 = G_x, \qquad \gamma A_1 + isB_1 + \gamma A_2 - isB_2 = G_y.
\tag{5.39}
$$

Now the solution of the system of equations (5.35), (5.38) allows us to express the four coefficients A_1, B_1, A_2, B_2 in terms of the functions G_x, G_y.

$$
A_1 = -A_2 = -\frac{2s^2 - k_s^2}{2\gamma k_s^2}\, G_y + \frac{is}{k_s^2}\, G_x, \qquad B_1 = B_2 = -\frac{is}{k_s^2}\, G_y - \frac{2s^2 - k_s^2}{2qk_s^2}\, G_x.
\tag{5.40}
$$

Then the Fourier images of the normal and tangential stress over the crack faces are

$$
\tilde{\sigma}_{yy}^{(1)}(s,0) = \frac{\Delta(s)}{2\gamma k_s^2}\, G_y, \qquad \tilde{\sigma}_{xy}^{(1)}(s,0) = -\frac{\Delta(s)}{2qk_s^2}\, G_x,
\tag{5.41}
$$

where

$$
\Delta(s) = (2s^2 - k_s^2)^2 - 4s^2\gamma q
\tag{5.42}
$$

is the well-known Rayleigh function.

Inversion of relations (5.41), with the help of the last remaining boundary conditions written in the first line of Eq. (5.33), leads to the following integral equations for the functions $g_x(x)$ and $g_y(x)$:

$$
\int_{-a}^{a} g_y(\xi)K_y(x - \xi)\, d\xi = -\frac{2\sigma_{yy}^{\text{inc}}(x,0)k_s^2}{\rho c_s^2} \qquad (|x| \le a),
$$
$$
K_y(x) = \frac{1}{2\pi}\int_{-\infty}^{\infty} L_y(s)e^{-isx}\, ds, \qquad L_y(s) = \frac{\Delta(s)}{\gamma(s)},
\tag{5.43}
$$

$$
\int_{-a}^{a} g_x(\xi)K_x(x - \xi)\, d\xi = \frac{2\sigma_{xy}^{\text{inc}}(x,0)k_s^2}{\rho c_s^2} \qquad (|x| \le a),
$$
$$
K_x(x) = \frac{1}{2\pi}\int_{-\infty}^{\infty} L_x(s)e^{-isx}\, ds, \qquad L_x(s) = \frac{\Delta(s)}{q(s)}.
\tag{5.44}
$$

Since $\sigma_{yy}^{\text{inc}}(x,0) = (k_s^2 - 2k_p^2)$, $\sigma_{xy}^{\text{inc}}(x,0) = 0$, so the function g_x is trivial: $g_x(x) \equiv 0$, $|x| \le a$. Thus the only nontrivial function here is the relative opening of the crack's faces in the vertical direction $g_y(x) = (u_y^{(1)} - u_y^{(2)})(x,0)$, $|x| \le a$, which after the change of variable $s = \tilde{s}k_s$ is reduced to the form (tildes are omitted below)

$$
\int_{-a}^{a} g_y(\xi)K_y(x-\xi)\, d\xi = -\frac{2(1-2\beta^2)}{\mu}, \qquad |x| \le a \qquad \left(\beta = \frac{k_p}{k_s} < 1\right),
$$
$$
K_y(x) = \frac{k_s^2}{2\pi}\int_{-\infty}^{\infty} L_y(s)\, e^{-isk_sx}\, ds, \qquad L_y = \frac{\Delta(s)}{\gamma(s)} = \frac{(2s^2-1)^2 - 4s^2\sqrt{s^2-1}\sqrt{s^2-\beta^2}}{\sqrt{s^2-\beta^2}},
\tag{5.45}
$$

where we have recalled that $c_s^2 = \mu/\rho$ (see Section 1.9).

Properties of the integral equation.

Let us start from the observation that the symbolic function $L_x(s)$ is integrable over any finite part of the axis $|s| < \infty$. Hence, the integral defining the kernel $K_y(x)$ in Eq. (5.43) diverges at infinity only, and this can be treated in a generalized sense. Notice that

$$L_y(s) = \frac{\Delta(s)}{\gamma(s)} = \frac{2(\beta^2-1)s^2[1+O(1/s^2)]}{s[1+O(1/s^2)]} = 2(\beta^2-1)s + O\left(\frac{1}{s}\right), \qquad s \to \infty, \quad (5.46)$$

hence

$$K_y(x) = \frac{2k_s^2}{\pi}(\beta^2-1)\int_0^\infty s\cos(sk_sx)\,ds + \frac{k_s^2}{\pi}\int_0^\infty \left[\frac{\Delta(s)}{\gamma(s)} - 2(\beta^2-1)s\right]\cos(sk_sx)\,ds$$

$$= -\frac{2(\beta^2-1)}{\pi x^2} + \frac{k_s^2}{\pi}I_y(k_sx), \quad I_y(k_sx) = \int_0^\infty \left[\frac{\Delta(s)}{\gamma(s)} - 2(\beta^2-1)s\right]\cos(sk_sx)\,ds.$$

$$(5.47)$$

Since expression in the square brackets in Eq. (5.47) is $O(1/s)$ as $s \to \infty$, the asymptotic estimate based on Eq. (1.83) results in $I_y(k_sx) \sim O[(k_s|x|)^{-2}]$ as $k_sx \to 0$. Therefore, we have arrived again at a hyper-singular integral equation.

In absolutely the same manner as in the previous section we can prove the following theorems.

THEOREM 1. *The integral equation (5.45) is uniquely solvable in the class of bounded functions at least for sufficiently small frequencies.*

The proof is based again on the equivalent representation (5.47) and inversion of the characteristic hyper-singular part of the kernel and the fact that the regular (more precisely, weakly singular) part of the kernel is uniformly small at $k_s \to 0$.

Let us note that a solution of equation (5.45), if exists, must be even for even right-hand side.

THEOREM 2. *The solution of equation (5.45) in the class of even bounded functions that behave as $\sqrt{a \pm x}$ for $x \to +a$ is unique for all $k > 0$.*

Proof. As usual, if there are two different solutions $u_1(x)$, $u_2(x)$ of the mentioned class, then $u(x) = u_1(x) - u_2(x)$ is a solution of the homogeneous equation (5.43), which is another form of (5.45). If we apply the scalar product of both sides of (5.43) with the same function $u(x)$, we arrive at equality

$$\int_{-\infty}^\infty L_y(s)|U(s)|^2\,ds = 0. \qquad (5.48)$$

The left-hand integral here is finite since the behavior of the integrand at infinity is $\sim s(s^{-3/2})^2 = O(1/s^2)$.

Further steps of the proof are based on the classical properties of the Rayleigh function $\Delta(s)$ (see, for example Achenbach, 1973). Usually, the Rayleigh equation

$$\Delta(s) = 0 \quad \sim \quad (2s^2 - k_s^2) - 4s^2\gamma q = 0 \qquad (5.49)$$

has one real root s_0 such that $s_0 > k_s > k_p$. Further, the function $L_y(s) = \Delta(s)/\gamma(s)$ of (5.45), present in (5.48), is complex-valued: $L_y(s) = \alpha + i\beta$, with

$$\begin{array}{llll} s > s_0: & \alpha < 0, & \beta = 0; & k_s < s < s_0: \quad \alpha > 0, \quad \beta = 0; \\ k_p < s < k_s: & \alpha > 0, & \beta > 0; & 0 < s < k_p: \quad \alpha = 0, \quad \beta > 0; \end{array} \qquad (5.50)$$

so separation of real and imaginary parts in (5.48) does not result in $U(s) \equiv 0$, because of negative sign of the symbolic function for $s > s_0$.

In order to provide that the Fourier image of the kernel is positive for $s > s_0$, keeping positive signs for its real and imaginary parts on all other intervals, we apply the idea of Vorovich and Babeshko (1979).

We develop below this approach to our case of hyper-singular equations.

Let us represent equation (5.43) in the equivalent form:

$$\frac{1}{2\pi} \int_{-a}^{a} g_y(\xi)\, d\xi \int_{-\infty}^{\infty} (s_0^2 - s^2) L_y(s) \frac{1}{s_0^2 - s^2}\, e^{-is(x-\xi)}\, ds = f,$$

$$|x| \le a, \quad f = -\frac{2(k_s^2 - 2k_p^2)}{\mu}\, k_s^2, \tag{5.51}$$

that is

$$\frac{1}{2\pi} \int_{\infty}^{\infty} v(\xi)\, d\xi \int_{\infty}^{\infty} (s_0^2 - s^2) L_y(s)\, e^{-is(x-\xi)}\, ds = f, \qquad |x| \le a,$$

$$v(x) = \frac{1}{2\pi} \int_{-a}^{a} g_y(\xi)\, d\xi \int_{-\infty}^{\infty} \frac{e^{-is(x-\xi)}}{s_0^2 - s^2}\, ds, \tag{5.52}$$

where the last integral is treated as a singular integral of the Cauchy type, which can be calculated as the sum of contributions by simple poles $s = \pm s_0$.

We thus notice that

$$V(s) = \frac{G_y(s)}{s_0^2 - s^2}, \tag{5.53}$$

and so $v(x)$ is even for all $|x| < \infty$ if $G_y(s)$ is even (the latter is even since $g_y(x)$ is even). Then

$$v(x) = -\frac{i}{2} \int_{-a}^{a} g_y(\xi) \left[\frac{e^{is_0|x-\xi|}}{2s_0} + \frac{e^{-is_0|x-\xi|}}{-2s_0} \right] d\xi = \frac{1}{2} \int_{-a}^{a} g_y(\xi) \sin(s_0|x - \xi|)\, d\xi. \tag{5.54}$$

Now we note that for $|x| > a$

$$v(x) = \pm\frac{1}{2} \int_{-a}^{a} g_y(\xi) \sin[s_0(x - \xi)]\, d\xi$$

$$= \pm\frac{1}{2} \left[\sin(s_0 x) \int_{-a}^{a} g_y(\xi) \cos(s_0 \xi)\, d\xi + \cos(s_0 x) \int_{-a}^{a} g_y(\xi) \sin(s_0 \xi)\, d\xi \right] = 0, \tag{5.55}$$

since the second integral here is zero due to evenness of $g_y(x)$, and the first one must be trivial due to evenness of $v(x)$ $\forall x \in (-\infty, \infty)$. This proves that Eq. (5.52) is a standard integral equation (not of Fredholm type), when there is the same interval of variation for interval (ξ) and external x variables.

After this auxiliary transformation the proof is completed by our usual reducing to relation of the type (5.48), which now reads

$$\int_{-\infty}^{\infty} (s_0^2 - s^2) L_y(s) |U(s)|^2\, ds = 0, \tag{5.56}$$

where the symbolic function $(s_0^2 - s^2) L_y(s)$ has positive real and imaginary parts over the total axis $|s| < \infty$. Then separation of the real and imaginary parts finally proves the Theorem.

Helpful remarks

As in the scalar problem of diffraction by a linear scatterer, the developed method allows us for low frequencies to obtain the far-field scattering pattern of the linear crack of finite length. Briefly speaking, the calculation scheme is as follows. For example, for the longitudinal scattered wave we have

$$\varphi_{1,2}(x,y) = \mp \frac{1}{2\pi k_s^2} \int_{-a}^{a} g_y(\xi)\,d\xi \int_{-\infty}^{\infty} \frac{2s^2 - k_s^2}{2\gamma}\,e^{-\gamma|y|}\,e^{-is(x-\xi)}\,ds$$

$$\sim \mp \frac{1}{4\pi k_s^2}\sqrt{\frac{2\pi}{k_p R}}\,e^{i(kR+\pi/4)}(2k_p^2\cos^2\alpha - k_s^2)\int_{-a}^{a} g_y(\xi)e^{-ik_p\xi\cos\alpha}\,d\xi,$$

(5.57)

and the problem is then reduced to the calculation of the arisen integral, which was discussed in detail above.

The scattering diagram of the transverse wave can be constructed in the same way.

5.3. High-Frequency Asymptotics in Diffraction by Linear Obstacles in Unbounded Medium

The integral equations (5.14), (5.15) arising in diffraction by linear obstacles, as a Fredholm equation of the first kind with a weak singularity in the kernel, and a hyper-singular equation, admit the application of direct numerical methods (see Chapter 9) for low and moderate frequencies. For extremely high frequencies, when the wavelength $\lambda = 2\pi/k$ becomes too short, the application of any numerical method requires very large number of nodes, since stable computations with reliable results need at least 10 nodes per wavelength. This is why the high-frequency (i.e., short-wave) regime requires a special analytical asymptotic approach. In our investigation we follow ideas suggested by Koiter (1954) and Aleksandrov (1968).

Let us rewrite equations (5.14), (5.15) in a uniform manner as follows (we apply here the change of variables $s = \tilde{s}k$, $x = \tilde{x}/k$, $\xi = \tilde{\xi}/k$, and tildes are omitted below)

$$\int_{-ak}^{ak} g(\xi)K(x-\xi)\,d\xi = f, \quad |x| \le ak \quad (f = \text{const}),$$

$$K(x) = \frac{1}{2\pi}\int_{-\infty}^{\infty} L(s)\,e^{isx}\,ds, \quad L(s) = \begin{cases} \alpha)\ L(s) = \gamma(s) = \sqrt{s^2-1}, & f = -2i, \\[2mm] \beta)\ L(s) = 1/\gamma(s) = 1/\sqrt{s^2-1}, & f = -k. \end{cases}$$

(5.58)

Let us represent Eq. (5.58) in equivalent form, where the key point is a representation of the integral over finite-length interval as a combination of several integrals, with each of them being taken over some semi-infinite integral. In order to derive this representation, let us consider the three equalities

$$\int_{-a}^{\infty} v(ak+\xi)K(x-\xi)\,d\xi = f + \int_{-\infty}^{-ak} [v(ak-\xi)-w]K(x-\xi)\,d\xi, \quad -ak \le x < \infty,$$

$$\int_{-\infty}^{a} v(ak-\xi)K(x-\xi)\,d\xi = f + \int_{ak}^{\infty} [v(ak+\xi)-w]K(x-\xi)\,d\xi, \quad -\infty < x \le ak,$$

$$\int_{-\infty}^{\infty} w(\xi)K(x-\xi)\,d\xi = f, \quad -\infty < x < \infty,$$

(5.59)

with each of them holding, among other domains, also on the basic interval $|x| \le ak$.

Note that a solution of the last equation is simply constructed by the direct application of the Fourier transform:

$$W(s)L(s) = 2\pi f \delta(s) \ \sim \ W(s) = 2\pi f \frac{\delta(s)}{L(s)} \ \sim \ w(x) = \frac{f}{L(0)} \equiv \text{const} = \begin{cases} \alpha) \ 2, \\ \beta) \ ik, \end{cases} \quad (5.60)$$

and for this reason we could write in Eqs. (5.59) the function $w(x)$ as a constant w.

After this remark, let us take the sum of the first two equations (5.59) and subtract this sum by the third equation (5.59). By cancelling some identical integrals, we can see that

$$g(x) = v(ak + x) + v(ak - x) - w \qquad (5.61)$$

satisfies initial equation (5.58). Note that the (constant) function w is already found in Eq. (5.60), so the only remaining unknown function to be found is $v(x)$. The latter satisfies Eq. (5.59) which in a rewritten form is

$$\int_0^\infty v(\xi)K(x - \xi)\,d\xi = f + \int_0^\infty [v(2ak + \xi) - w]K(x + \xi)\,d\xi, \quad 0 < x < \infty. \quad (5.62)$$

The right-hand integral here represents a *tail*, and we will prove that this is asymptotically small when $k \to \infty$, and so Eq. (5.62) becomes a Wiener–Hopf equation, which admits solution in explicit form (see Section 1.2).

A rather widespread mathematical approach consists in accepting first this statement as a hypothesis, to construct the full solution under this hypothesis, and finally to make sure that the obtained solution validates this hypothesis.

If we reject the tail integral in (5.62), then the solution of the Wiener–Hopf equation

$$\int_0^\infty v(\xi)K(x - \xi)\,d\xi = f, \qquad 0 \le x < \infty, \quad (5.63)$$

can be derived by the method described in Section 1.2 as follows:

$$\begin{aligned} \alpha) \ & V_+(s)\sqrt{s^2 - 1} = \left(\frac{f}{-is}\right)_+ + F_-(s) \\ \sim \ & V_+(s)(\sqrt{s+1})_+(\sqrt{s-1})_- = \left(\frac{f}{-is}\right)_+ + F_-(s) \\ \sim \ & V_+(s)(\sqrt{s+1})_+ = \left(\frac{f}{-is}\right)_+ + \left(\frac{1}{\sqrt{s-1}}\right)_- + \frac{F_-(s)}{(\sqrt{s-1})_-} \\ \sim \ & V_+(s)(\sqrt{s+1})_+ = \left(\frac{1}{-i}\frac{f}{-is}\right)_+ + N_-(s) \\ \sim \ & V_+(s)(\sqrt{s+1})_+ - \left(\frac{1}{-i}\frac{f}{-is}\right)_+ = N_-(s) \equiv 0 \\ \sim \ & V_+(s) = \frac{1}{-i}\frac{f}{(-is)\sqrt{s+1}} = \frac{1}{-i}\frac{f}{p\sqrt{1+ip}} = e^{\pi i/4}\frac{f}{p\sqrt{p-i}}, \qquad p = -is, \end{aligned} \quad (5.64)$$

where we have passed from the Fourier transform variable s to the Laplace transform variable p, in order to take advantage of the tables of inverse Laplace transforms (Bateman and Erdelyi, 1954):

$$v(x) = e^{\pi i/4}\frac{f}{\sqrt{-i}}\,\text{Erf}\left(\sqrt{-ix}\right) = e^{\pi i/4}f\sqrt{2}\,[C(x) + iS(x)], \quad (5.65)$$

where Erf(x) is the probability integral (error function) and $C(x), S(x)$ are the Fresnel integrals (Bateman and Erdelyi, 1953). Now, having constructed a solution of equation (5.63) in explicit form, we can prove that the right-hand integral in (5.62) is asymptotically small. Indeed, since

$$C(x), S(x) = \frac{1}{2} + O(x^{-1/2}), \qquad x \to +\infty, \tag{5.66}$$

we can see that

$$v(x) = if + O(x^{-1/2}) = 2 + O(x^{-1/2}) = w + O(x^{-1/2}), \qquad x \to +\infty, \tag{5.67}$$

so expression in the square brackets in (5.62) is

$$v(2ak + \xi) - w = O\big((ak)^{-1/2}\big), \qquad k \to \infty. \tag{5.68}$$

If the kernel $K(x)$ possesses integrable behavior at infinity, then the estimate (5.68) takes place for the full right-hand integral (5.62). We have

$$K(x) = \frac{1}{\pi} \int_0^\infty \sqrt{s^2 - 1} \cos(sx) \, ds = \frac{1}{\pi} \left(\int_0^1 + \int_1^\infty \right) \sqrt{s^2 - 1} \cos(sx) \, ds$$

$$= \frac{1}{\pi} \left\{ \int_0^1 \sqrt{s(2 - s)} \cos[(1 - s)x] \, ds + \int_0^\infty \sqrt{s(2 + s)} \cos[(1 + s)x] \, ds \right\}. \tag{5.69}$$

Now Erdelyi's lemma (see Section 1.4) shows that $K(x) \sim x^{-3/2}$, $x \to +\infty$, if we set there $\alpha = 1$, $\beta = 3/2$ and express cosines in Eq. (5.69) in terms of exponential functions of imaginary arguments with the help of Euler's formula.

β) By analogy with the problem α) we can find here that

$$V_+(s) = \frac{-if\sqrt{s + 1}}{-is} = \frac{-if\sqrt{1 + ip}}{p} = e^{-\pi i/4} f \frac{\sqrt{p - i}}{p}, \qquad p = -is, \tag{5.70}$$

hence (Bateman and Erdelyi, 1954)

$$v(x) = e^{-\pi i/4} f \left[\frac{e^{ix}}{\sqrt{\pi x}} + \sqrt{-i} \, \mathrm{Erf}(\sqrt{-ix}) \right] = e^{-\pi i/4} f \left\{ \frac{e^{ix}}{\sqrt{\pi x}} - \sqrt{2} \, i[C(x) + iS(x)] \right\}. \tag{5.71}$$

Now, having derived explicit expression for $v(x)$, estimate of the right-hand integral is based upon the following asymptotic behavior at infinity:

$$v(x) \sim -if - \frac{e^{ix}}{\sqrt{\pi} x^{3/2}} + O\left(x^{-3/2}\right) = ik - \frac{e^{ix}}{\sqrt{\pi} x^{3/2}} + O\left(x^{-3/2}\right)$$

$$= w - \frac{e^{ix}}{\sqrt{\pi} x^{3/2}} + O\left(x^{-3/2}\right), \qquad x \to +\infty. \tag{5.72}$$

If we add to this estimate also the asymptotics of the kernel

$$K(x) = \frac{1}{2\pi} \int_{-\infty}^\infty \frac{e^{isx}}{\sqrt{s^2 - 1}} \, ds = \frac{i}{2} H_0^{(1)}(|x|) \sim \frac{e^{i(x + \pi/4)}}{\sqrt{2\pi x}}, \qquad x \to +\infty, \tag{5.73}$$

it becomes clear that the tail in (5) is asymptotically small:

$$\int_0^\infty [v(2ak + \xi) - w] K(x + \xi) \, d\xi$$

$$\sim -\frac{e^{i(2ak + x - \pi/4)}}{\sqrt{\pi} (2ak)^{3/2}} \int_0^\infty \frac{e^{2i\xi}}{\sqrt{x + \xi}} \, d\xi = O\left[(ak)^{-3/2}\right], \qquad ak \to \infty. \tag{5.74}$$

Helpful remarks

1°. The structure of the derived solution is such that there is a *degenerate* solution w, which asymptotically represents the actual solution over the total interval $|x| \leq ak$, except small *boundary layers* near the screen (or hole) ends $x \sim \pm ak$, where the structure of the solution is affected by these sharp edges. This idea is quite clear since if $|x \pm ak| = O(1)$, then $v(ak \pm x) \to w$, $k \to \infty$, and hence in (5.61) we have $g(x) \to w$, $k \to \infty$. We thus can conclude that the degenerate solution, which is obtained from the integral equation over infinite axis $|x| < \infty$, is valid over the considered interval. This should be corrected only in a small neighborhood of the edges, and correct structure of the solution in these small boundary layers can be determined by solving the corresponding Wiener–Hopf equation.

2°. Heuristically, the degenerate solution, which is valid outside small boundary layers, must be for high frequencies in accordance with Kirchhoff's physical diffraction theory. We can verify that these heuristic ideas are correct. Indeed, in problem α) the pressure "on the light" is, according to Kirchhoff's theory, $p(x,-0) = p_1(x) = 2p^{\text{inc}} = 2$, $|x| \leq a$, and solution "in the shadow" is $p(x,+0) = p_2(x) = 0$. Then $g_2(x) = p_1(x) - p_2(x) = 2$, which coincides with our obtained value $w = 2$. In problem β) the physical diffraction theory may be interpreted so that in the case, when the width of the opening is much longer than the wavelength, the structure of the solution over the opening is as the one in the incident wave, i.e., the edges of the screen modify the latter very poorly. Therefore, in this problem $g_\beta(x) = (\partial p_1/\partial x)(x,0) = (\partial p_2/\partial x)(x,0) = (\partial p^{\text{inc}}/\partial x)(x,0) = ik$, which coincides with the degenerate solution $w = ik$.

3°. In the literature, a standard approach to improve simple ideas of physical (and geometrical, see below Section 6.3) diffraction theory is called the method of *boundary* (or *edge*) *waves* (see Keller, 1962; Hönl et al., 1961). This is based on some nonstrict physical observations, and is also reduced to a Wiener–Hopf equation. The method described here gives a direct mathematical instrument, which admits a clear mathematical justification.

4°. From results of Section 5.1 we could come to an observation that the behavior of the solution near the interval ends is $\sim (ak \pm x)^{1/2}$ $(x \to \mp ak)$ in problem α). It is interesting to check if this qualitative property is provided by the constructed solutions. As follows from Eq. (5.61), (5.65), if $x \to ak$, then $g(x) \to v(2ak) - w + v(ak - x) = O\left((ak - x)^{1/2}\right)$, since $\text{Erf}(ax) = O(x)$, $x \to 0$. By analogy, it can be easily seen from formula (5.71) that this possesses the required square root singularity near the screen edges in problem β).

5.4. High-Frequency Asymptotics for Diffraction by Linear Obstacles in Open Waveguides

Many efforts of researchers were devoted to study wave propagation in waveguides (see, for example, Chapter 3; Jones, 1986; Achenbach, 1980; Mittra and Lee, 1971). From the theoretical point of view the case of a homogeneous infinite waveguide of constant width has been studied by means of the Fourier transform and the obtained results have now become classical (Collin, 1960). Through the application of the Wiener–Hopf method (Section 1.2), many problems for open-ended waveguides have been successfully investigated (see Vajnshtejn, 1969). It was clarified that the wave field in elongated waveguides is locally similar to that in infinite ones (results contiguous to what is obtained in the previous section).

   The *edge waves* method developed in the previous section cannot be transferred directly to the diffraction by a finite-length waveguide, due to presence of the propagating mode waves that lead to a strong interaction between the ends of the waveguide. Nevertheless, for specific values of the frequency, close to these mode values, Vajnshtejn (1969) first proposed, on the basis of some heuristic physical ideas, an approach similar to the method of *edge waves*. He discovered that at these frequencies the wave propagating along the waveguide does not permit any mode distinct from the incident one, when reflecting from the open edge. Hence, it is a transformation of a one-type wave, as in diffraction by the single slit (see Section 5.3), and so successive reflections from the edges correctly describe the wave picture. Nobody knows if this idea is correct for arbitrarily high frequency, different from these critical mode values. Here we develop an asymptotic approach valid for high frequencies antipodal to those of Vajnshtejn.

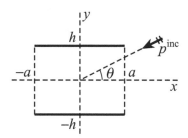

Figure 5.1. High-frequency diffraction by the open waveguide

Let the plane incident acoustic wave

$$p^{\text{inc}}(x, y) = e^{-ik(x \cos \theta + y \sin \theta)} \tag{5.75}$$

be scattered by a finite-length waveguide with incident angle θ, as illustrated in Fig. 5.1. The full wave field, as usual, consists of the incident and the scattered one: $p(x, y) = p^{\text{inc}}(x, y) + p^{\text{sc}}(x, y)$ and all three functions satisfy the Helmholtz equation

$$\Delta p + k^2 p = 0, \tag{5.76}$$

everywhere outside the two finite parallel plates of length $2a$, with the distance between them being equal to $2h$. Assume, for instance, that the plates are acoustically soft:

$$p(x, y) = 0: \qquad p^{\text{sc}}(x, y) = -p^{\text{inc}}(x, y), \quad |x| \le a, \ y = \pm h. \tag{5.77}$$

By applying the Fourier transform along the x-axis to the Helmholtz equation (5.76), we obtain the following representation for the function $P^{\text{sc}}(s, y)$ (cf. Sections 3.1, 5.1) in the form satisfying radiation condition at infinity:

$$P^{\text{sc}}(s, y) = \begin{cases} A(s) \, e^{-\gamma y}, & y > h, \\ C(s) \, e^{-\gamma y} + D(s) \, e^{\gamma y}, & |y| \le h, \\ B(s) \, e^{\gamma y}, & y < -h, \end{cases} \tag{5.78}$$

where $A(s), B(s), C(s), D(s)$ are unknown functions which should be defined from the boundary conditions over the screens, and $\gamma = \gamma(s) = \sqrt{s^2 - k^2}$. In the definition of the branching function $\gamma(s)$ we use, as usual, the branch cut such that $\text{Re}(\gamma) \ge 0$ for arbitrary complex-valued s.

It is evident that the boundary condition (5.77) provides continuity of the potential $p(x, y)$ over the plates:

$$\begin{aligned} p^{\text{sc}}(x, h + 0) &= p^{\text{sc}}(x, h - 0) = -p^{\text{inc}}(x, h); \\ p^{\text{sc}}(x, -h + 0) &= p^{\text{sc}}(x, -h - 0) = -p^{\text{inc}}(x, -h), \quad |x| \le a. \end{aligned} \tag{5.79}$$

Together with the continuity of the wave field outside the plates, this involves the equality $p(x, h + 0) = p(x, h - 0)$; $p(x, -h + 0) = p(x, -h - 0)$, $-\infty < x < \infty$, which gives the two relations

$$A(s) e^{-\gamma h} = C(s) e^{-\gamma h} + D(s) e^{\gamma h},$$
$$B(s) e^{-\gamma h} = C(s) e^{\gamma h} + D(s) e^{-\gamma h}. \tag{5.80}$$

Let us introduce the new unknown functions $u(x)$ and $v(x)$ as follows:

$$\left.\frac{\partial p}{\partial y}\right|_{y=h+0} - \left.\frac{\partial p}{\partial y}\right|_{y=h-0} = \begin{cases} 0, & |x| > a, \\ u(x), & |x| \leq a, \end{cases}$$
$$\left.\frac{\partial p}{\partial y}\right|_{y=-h-0} - \left.\frac{\partial p}{\partial y}\right|_{y=-h+0} = \begin{cases} 0, & |x| > a, \\ v(x), & |x| \leq a. \end{cases} \tag{5.81}$$

The differences vanish here along $|x| > a$, due to continuity of the wave field.

In terms of Fourier images the last equations are equivalent to

$$\gamma \left[A(s) e^{-\gamma h} + D(s) e^{\gamma h} - C(s) e^{-\gamma h} \right] = -\int_{-a}^{a} u(\xi) e^{is\xi} d\xi,$$
$$\gamma \left[B(s) e^{-\gamma h} + C(s) e^{\gamma h} - D(s) e^{-\gamma h} \right] = \int_{-a}^{a} v(\xi) e^{is\xi} d\xi. \tag{5.82}$$

It follows from (5.80), (5.82) that

$$C(s) = \frac{e^{-\gamma h}}{2\gamma} \int_{-a}^{a} v(\xi) e^{is\xi} d\xi, \qquad D(s) = -\frac{e^{-\gamma h}}{2\gamma} \int_{-a}^{a} u(\xi) e^{is\xi} d\xi, \tag{5.83}$$

with $A(s)$ and $B(s)$ being defined from (5.80). Lastly, the boundary condition (5.79) yields the following system of two integral equations ($|x| \leq a$):

$$\frac{1}{2\pi} \int_{-a}^{a} u(\xi) d\xi \int_{-\infty}^{\infty} \frac{e^{is(\xi-x)}}{2\gamma} ds - \frac{1}{2\pi} \int_{-a}^{a} v(\xi) d\xi \int_{-\infty}^{\infty} \frac{e^{is(\xi-x)-2\gamma h}}{2\gamma} ds = e^{-ik(h\sin\theta + x\cos\theta)},$$
$$\frac{1}{2\pi} \int_{-a}^{a} v(\xi) d\xi \int_{-\infty}^{\infty} \frac{e^{is(\xi-x)}}{2\gamma} ds - \frac{1}{2\pi} \int_{-a}^{a} u(\xi) d\xi \int_{-\infty}^{\infty} \frac{e^{is(\xi-x)-2\gamma h}}{2\gamma} ds = -e^{ik(h\sin\theta - x\cos\theta)}. \tag{5.84}$$

By applying summation and subtraction to (5.84), we can reduce the last system to a pair of two independent integral equations:

$$\int_{-a}^{a} [u(\xi) + v(\xi)] K_1(x - \xi) d\xi = -4i \sin(kh\sin\theta) e^{-ikx\cos\theta}, \qquad |x| \leq a, \tag{5.85}$$

$$\int_{-a}^{a} [u(\xi) - v(\xi)] K_2(x - \xi) d\xi = 4 \cos(kh\sin\theta) e^{-ikx\cos\theta}, \qquad |x| \leq a, \tag{5.86}$$

where

$$K_{1,2}(x) = \frac{1}{2\pi} \int_{-\infty}^{\infty} G_{1,2}(s) e^{-isx} ds, \qquad G_{1,2}(s) = \frac{1 \mp e^{-2\gamma h}}{\gamma}. \tag{5.87}$$

Note that the symbolic functions $G_1(s)$ and $G_2(s)$ have the branch points $s = \pm k$, and there are no singularities, except these two points, which is an intrinsic property of open structures. The closed ones may possess a countable set of simple poles, with a finite number of them being distributed on the real axis (compare with the results of Chapter 3).

When any pole coincides with the origin, we have a resonant case (see Vorovich and Babeshko, 1979). As shown below, the distribution of the zeros of the symbol $G_{1,2}(\alpha)$ is important too. To treat correctly disposition of (possibly existing) real zeros, let us apply the principle of extremely low absorption (Section 3.2). If a small attenuation is added to the medium, then the positive zeros and the branch point $s = k$ move upwards, and the negative ones downwards. This implies that for an ideal medium, for which these points are situated on the real axis, the integration contour in (5.87) should bend around the positive points from below and the negative ones from above.

It is conventionally recognized that presence of the branch points in the symbolic function complicates analytical treatment. The algorithm developed below is insensitive to the influence of the branching effect.

In order to solve equations (5.85) and (5.86), it suffices to treat the following equation written in a nondimensional form:

$$\int_{-b}^{b} w(\xi)K(x-\xi)\,d\xi = f(x), \qquad |x| \le b, \quad \chi = kh, \quad b = a/h,$$

$$u(x) \pm v(x) = \left\{\begin{array}{c} -4i\sin(\chi\sin\theta) \\ 4\cos(\chi\sin\theta) \end{array}\right\} \frac{w(x)}{h}, \qquad f(x) = e^{-i\chi x\cos\theta}, \tag{5.88}$$

$$K(x) = \frac{1}{2\pi}\int_{-\infty}^{\infty} G_{1,2}(s)\,e^{-isx}\,ds, \qquad G_{1,2}(s) = \frac{1 \mp e^{-2\gamma}}{\gamma}, \quad \gamma = \sqrt{s^2 - \chi^2}.$$

There are two independent dimensionless parameters in this problem: the frequency parameter χ and the relative length of the waveguide $b = a/h$. The asymptotic analysis undertaken by Jones (1952) operates with a one-mode regime ($\pi < \chi < 2\pi$) (see Section 3.1) when $b \to \infty$. We study the opposite case: the high-frequency regime $\chi = kh \to \infty$ with bounded b (so $ka \to \infty$ too), which is more complex, since the number of propagating modes grows with the frequency increasing.

To attract the powerful Wiener–Hopf technique, we represent again the basic equation (5.88) as a superposition of the three functions:

$$w(x) = w_1(b+x) + w_2(b-x) - w_0(x), \tag{5.89}$$

with the new ones satisfying the following system equivalent to the initial equation (5.88):

$$\int_{-b}^{\infty} w_1(b+\xi)K(x-\xi)\,d\xi = f(x) + \int_{-\infty}^{-b}[w_2(b-\xi)-w_0(\xi)]K(x-\xi)\,d\xi, \quad -b < x < \infty,$$

$$\int_{-\infty}^{b} w_2(b-\xi)K(x-\xi)\,d\xi = f(x) + \int_{b}^{\infty}[w_1(b+\xi)-w_0(\xi)]K(x-\xi)\,d\xi, \quad -\infty < x < b,$$

$$\int_{-\infty}^{\infty} w_0(\xi)\,K(x-\xi)\,d\xi = f(x), \quad -\infty < x < \infty.$$

$$\tag{5.90}$$

Here again, from the physical point of view the integral operators on the left-hand sides of (5.90) are related to one infinite waveguide and two semi-infinite ones. As we could see in previous section, for a wide class of physical problems additional integrals on the right-hand side of the first two equations (5.90), the so-called *tails*, appear to be small in the asymptotic sense. For instance, if the frequency is low enough, so that there is no real zero of the symbol $G_{1,2}(s)$ (k is less than the first critical value), then for a long waveguide ($b = a/h \gg 1$ is the asymptotic case considered by Jones, 1952) the differences in the square brackets vanish when $\xi \to \pm\infty$. Thus, it can be proved that the integrals on the right tend to zero with $a \to \infty$. Physically, this means that the edges of the narrow open-ended waveguide do not influence each other in the asymptotic sense. A similar approach is possible for the short-wave diffraction problem for a single isolated plate (as well as for a single slot in the infinite plate). The symbolic function $G(\alpha)$ there may be formally obtained from (5.87) at $h \to \infty$ to be $G(\alpha) \sim 1/\gamma$; thus there are no zeros of the function $G(\alpha)$. It involves again an asymptotic disappearance of the right-hand integrals in (5.90), with the same physical meaning: the edges of the plate influence each other weakly when $a/\lambda \to \infty$ ($\lambda = 2\pi/k$ is the wavelength). The described approach generates again the *edge-waves* method (see Keller, 1962), which was developed by means of a different mathematical technique.

In the problem at hand the zeros $\pm s_{1m}, \pm s_{2m}$ of the function $G_{1,2}(s)$:

$$s_{1m} = \sqrt{\chi^2 - (\pi m)^2}, \qquad s_{2m} = \sqrt{\chi^2 - \left(\pi m - \frac{\pi}{2}\right)^2}, \qquad m = 1, 2, \dots, \qquad (5.91)$$

play the most important role.

The number of positive zeros (5.91) increases with the growth of the frequency parameter $\chi = kh$. It generates more and more propagating *mode waves* inside the waveguide. Thus, the ends of the open finite-length waveguide just affect each other by these propagating waves, and the internal wave process differs considerably from that in semi-infinite structures. As a result, the differences $[w_2(b-\xi) - w_0(\xi)]$, $[w_1(b+\xi) - w_0(\xi)]$ do not vanish as $\chi \to \infty$, in contrast to case of diffraction in unbounded medium, Section 5.3.

In spite of the finite-length structure cannot be represented as a composition of two semi-infinite ones, there is a classical idea that diffraction by the open end of the finite-length waveguide is similar to diffraction by the edge of a semi-infinite one, when the wave number approaches any mode value. We mean the optical range of electromagnetism, where both geometric sizes of the structure (a and h) are several thousand times larger than the wavelength (see Vajnshtejn, 1969). Below we determine in which sense this point of view is correct.

To begin with, let us write out the explicit representation for the kernels

$$K_{1,2}(x) = \frac{i}{2} \left\{ H_0^{(1)}(\chi|x - \xi|) \mp H_0^{(1)}\left[\chi\sqrt{4 + (x - \xi)^2}\right] \right\}, \qquad (5.92)$$

where the asymptotic behavior of the Hankel function $H_0^{(1)}$ is given by Eq. (3.115). It is clear from (5.92) that at high frequencies ($\chi \gg 1$) the kernel $K(x)$ on the right-hand sides of (5.90) is strongly oscillating. So the right-hand integrals in (5.90) may be small if the differences in square brackets are asymptotically bounded.

This heuristic idea should be verified. Let us assume, as usual, that these integrals are asymptotically small indeed (correctness of this assumption should be verified afterwards). Then we can construct the solution of the equations (5.90) over the semi-infinite intervals by using the Wiener–Hopf method (Section 1.2). It gives the following result:

$$w_1(x) = \frac{e^{-i\chi(x-b)\cos\theta}}{G(\chi\cos\theta)} - \frac{1}{2G_+(-\chi\cos\theta)} \sum_{m=1}^{\infty} \frac{(\pi m)^2}{s_m(s_m + \chi\cos\theta)} G_+(s_m) e^{i(s_m x + \chi b \cos\theta)},$$

$$w_2(x) = \frac{e^{i\chi(x-b)\cos\theta}}{G(\chi\cos\theta)} - \frac{1}{2G_+(\chi\cos\theta)} \sum_{m=1}^{\infty} \frac{(\pi m)^2}{s_m(s_m - \chi\cos\theta)} G_+(s_m) e^{i(s_m x - \chi b \cos\theta)},$$

$$(5.93)$$

where $G(s) = G_+(s)\, G_-(s)$ is a factorization of the function $G(\alpha)$.

A solution of the third equation (5.90) is obtained by the application of the Fourier transform as follows:

$$w_0(x) = \frac{e^{-i\chi x \cos \theta}}{G(\chi \cos \theta)}, \tag{5.94}$$

hence we can give the final representation for $w(x)$, under the accepted assumption, as

$$
\begin{aligned}
w(x) = \frac{e^{-i\chi x \cos \theta}}{G(\chi \cos \theta)} &- \frac{1}{2} \sum_{m=1}^{\infty} (\pi m)^2 \frac{G_+(s_m)}{s_m} \\
&\times \left\{ \frac{e^{i[s_m(b-x)-\chi b \cos \theta]}}{(s_m - \chi \cos \theta)\, G_+(\chi \cos \theta)} + \frac{e^{i[s_m(b+x)+\chi b \cos \theta]}}{(s_m + \chi \cos \theta)\, G_+(-\chi \cos \theta)} \right\} + O\left(\chi^{-1/2}\right).
\end{aligned} \tag{5.95}
$$

The asymptotic estimate of the error of this formula is proved below.

The structure of the obtained solution (5.95) is the following. The first term corresponds to the case of an infinite waveguide and has the order $O(\chi)$, $\chi \to \infty$ (so far as $G(\chi \cos \theta) = i(1 \mp \exp(2\chi i \sin \theta))/(\chi \sin \theta)$). Several oscillating terms under the sum sign may appear to have the same order $O(\chi)$ with respect to the parameter χ (if $\alpha_m \sim \chi \cos \theta$). Thus, the behavior of the solution is also highly oscillating.

When estimating the value of the right-hand side integrals in (5.90), let us note that the argument $(x - \xi)$ in the kernel $K(x - \xi)$ is of a constant sign for all x and ξ there, so $K(x - \xi) \sim O(1/\sqrt{\chi})$ uniformly on x and ξ. Thus, the integrand is of the order of $O(\chi^{1/2})$, $\chi \to \infty$. Besides, there is no stationary point, in contrast to the left-hand side kernel with its stationary point $x - \xi = 0 \sim \xi = x$. Therefore, it can be shown by integration by parts that, owing to the oscillating structure of the integrand, the value of the integrals has the order $O(\chi^{1/2})/\chi = O(\chi^{-1/2})$, $\chi \to \infty$. This proves the correctness of the method when the right-hand integrals in (5.90) are neglected. The only breakdown takes place when the value of the variable x coincides with the edge point ($x = -b$ in (5.90)$_1$ and $x = b$ in (5.90)$_2$). For these values the argument $(x - \xi)$ may be small, and the asymptotic estimate (5.93) for the kernel is not valid. Thus, the developed approach gives the correct representation (5.95) all over the interval $-b < x < b$, excluding the *boundary layers* where $b \pm x \sim 1/\chi$. Another restriction is related to the cases when the incident angle θ approaches those values where the incident wave generates standing modes. In these cases $G(\chi \cos \theta)$ tends to zero, $w_1(x)$ and $w_2(x)$ grow to infinity, and the right-hand integrals in (5.90) become unbounded. Thus, for some particular values of the frequency and the angle of incidence the main asymptotic result (5.95) fails.

Helpful remarks

1°. The main obtained formula (5.95) is worthy of special discussion. An interesting fact is that the mode waves do not decay with distance; however some integrals of these wave functions decay, so they could be neglected in (5.90). The nondecaying structure of the wave field in (5.95) shows that the solution has no relation to the problem of semi-infinite waveguide where only one type of edge-wave, instead of two, is present.

It should also be noted that the asymptotic contribution to the sum in (5.95) is given only by nondecaying terms in the sum (when s_m is real). Therefore, in practice the summation can be restricted by $m = n$ if the frequency parameter is $\chi = \pi(n + \delta)$ ($0 < \delta < 1/2$ or $1/2 < \delta < 1$), where n is a large positive integer. Thus, formula (5.95) is quite appropriate to evaluate the split function $w(x)$ at high frequencies, because efficient high-frequency representations for the coefficients $G_+(s_m)$ are proposed in Mittra and Lee (1971).

$2°$. Whenever a solution of the main integral equation is determined by (5.95), the scattered wave field $p^{sc}(x, y)$ can be obtained directly from (5.78) and (5.83)

$$
\begin{aligned}
p^{sc}(x, y) = \ &\frac{i}{4} \int_{-a}^{a} v(\xi) H_0^{(1)} \left[k \sqrt{(x - \xi)^2 + (y + h)^2} \right] d\xi \\
&- \frac{i}{4} \int_{-a}^{a} u(\xi) H_0^{(1)} \left[k \sqrt{(x - \xi)^2 + (y - h)^2} \right] d\xi,
\end{aligned}
\tag{5.96}
$$

uniformly over (x, y).

5.5. High-Frequency Diffraction by a Linear Discontinuity in the Waveguide

Here we consider a problem contiguous to that studied in the previous section. We will see that qualitative properties of the main symbolic function (i.e., the Fourier transform of the kernel of the basic integral equation) for this closed waveguide are absolutely different from the one for open structure studied in Section 5.4.

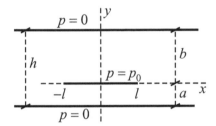

Figure 5.2. Oscillating linear plate in a closed infinite waveguide

To be more specific, we consider acoustic waveguide (i.e., a layer of constant thickness) with acoustically soft faces (see Fig. 5.2). Let a plane rigid finite-length plate be located inside a layer. Let its faces $y = \pm 0$, $|x| < l$ oscillate with an amplitude p_0, which is assumed, for definiteness, to be uniform, $p_0 = \text{const}$, without any restriction. As usual, we study the wave process harmonic in time, where the wave pressure $p(x, y)$ satisfies the Helmholtz equation.

By applying the Fourier transform along the x axis to the wave (Helmholtz) equation, we arrive at the following representation for the function $P(\alpha, y)$, which is chosen so that the boundary conditions $p(x, b) = p(x, -a) = 0$ on the free boundary surfaces are automatically satisfied:

$$
P(s, y) = \begin{cases} A(s) \sinh[\gamma(y + a)], & -a < y < 0, \\ B(s) \sinh[\gamma(y - b)], & 0 < y < b, \end{cases}
\tag{5.97}
$$

where $A(s)$ and $B(s)$ are unknown functions to be defined from the boundary conditions, and $\gamma = \sqrt{s^2 - k^2}$.

Obviously, the wave field is continuous at $|x| > l$ along the line $y = 0$. On the other hand, the boundary condition $p = p_0$, $y = \pm 0$, $|x| \le l$ implies its continuity on the interval $|x| \le l$ also. Thus we have $p(x, -0) = p(x, +0)$, $|x| < \infty$, and hence

$$
A(s) \sinh(a\gamma) = -B(s) \sinh(b\gamma) \quad \sim \quad B(s) = -A(s) \frac{\sinh(a\gamma)}{\sinh(b\gamma)}.
\tag{5.98}
$$

Let us introduce a new unknown function $u(x)$, $|x| \le l$, so that

$$
\left. \frac{\partial p}{\partial y} \right|_{y=-0} - \left. \frac{\partial p}{\partial y} \right|_{y=+0} = \begin{cases} 0, & |x| > l, \\ u(x), & |x| \le l. \end{cases}
\tag{5.99}
$$

Then the function $A(s)$ can be expressed in terms of $u(x)$:

$$A(s) = \frac{\sinh(b\gamma)}{\sinh(h\gamma)} \int_{-l}^{l} u(\xi)\, e^{is\xi}\, d\xi, \quad h = a + b. \tag{5.100}$$

At last, the boundary condition $p = p_0$, $|x| \leq l$, with the help of Eqs. (5.97)–(5.100), yields the following integral equation for the function $u(x)$:

$$\int_{-l}^{l} u(\xi)K(x - \xi)\, d\xi = p_0, \quad (|x| \leq l),$$

$$K(x) = \frac{1}{2\pi} \int_{-\infty}^{\infty} G(s)e^{-isx}\, ds, \qquad G(s) = \frac{\sinh(a\gamma)\sinh(b\gamma)}{\gamma \sinh(h\gamma)}. \tag{5.101}$$

The symbolic function $G(s)$, given by (5.101), is well known in the theory of branching electromagnetic waveguides. However there is no mathematical theory appropriate for Eq. (5.101) at high frequencies.

For the sake of brevity, further consideration is restricted, without any loss of generality, to a particular case $a = b$, i.e., for the discontinuity positioned parallel to and midway between the faces of the waveguide. Let us rewrite equation (5.101) in a nondimensional form ($h = 2a = 2b$)

$$\int_{-d}^{d} v(\xi)K(x - \xi)\, d\xi = 1, \quad |x| \leq d, \quad \chi = ka, \quad d = l/a, \quad v(x) = \frac{u(x)}{2p_0},$$

$$K(x) = \frac{1}{2\pi} \int_{\sigma} G(\alpha)e^{-i\alpha x}\, d\alpha, \quad G(\alpha) = \frac{\tanh \gamma}{\gamma}, \quad \gamma = \sqrt{\alpha^2 - \chi^2}. \tag{5.102}$$

The short-wave range implies that the parameter d is fixed, and the frequency parameter $\chi \gg 1$.

The function $G(s)$ in Eq. (5.102) is a meromorphic function of the complex variable s (see Section 1.1). There is a countable set of its zeros $\pm s_m$:

$$\alpha_m = \sqrt{\chi^2 - (\pi m)^2}, \qquad m = 1, 2, \ldots \tag{5.103}$$

and a countable set of its poles $\pm \beta_m$:

$$\beta_m = \sqrt{\chi^2 - [\pi(m - 1/2)]^2}, \qquad m = 1, 2, \ldots, \tag{5.104}$$

so the kernel (5.102) can be explicitly represented as a series in residues at simple poles (Section 1.1):

$$K(x) = i \sum_{m=1}^{\infty} \frac{\exp(i\beta_m x)}{\beta_m}. \tag{5.105}$$

Such a representation permits reducing the problem to an infinite system of linear algebraic equations (see Mittra and Lee, 1971).

For arbitrary large value of the frequency parameter χ there is a finite number of real zeros and poles. When a pole approaches the origin with the frequency change, then we encounter a resonance. If a small attenuation is added to the medium, then the positive poles move upwards, and the negative ones move downwards. It implies that for the ideal medium, when these singular points are situated on the real axis, the integration contour σ in (5.102) should bend round the positive poles from below and the negative ones from above.

It should be noted that with the frequency increasing (the parameter χ tends to infinity) more and more zeros and poles appear on the real axis. This essential feature complicates the high-frequency analysis. Let

$$\chi = \pi(n + \delta), \qquad 0 \le \delta < 1, \qquad \delta \ne 1/2, \tag{5.106}$$

with n being a large positive integer: $n \gg 1$. The critical value $\delta = 1/2$, related to the resonance case, should be excluded from the consideration.

As in the previous sections, it is easily seen that the basic equation (5.102) is equivalent to a pair of equations for the new unknown functions $p(x)$ and $w(x)$, which hold, respectively, on semi-infinite and infinite interval:

$$\int_0^\infty w_1(\xi)K(x - \xi)\, d\xi = 1 + \int_0^\infty [w_1(2d + \xi) - w]K(x + \xi)\, d\xi, \quad 0 < x < \infty,$$
$$\int_{-\infty}^\infty w(\xi)K(x - \xi)\, d\xi = 1, \qquad -\infty < x < \infty, \tag{5.107}$$

if

$$v(x) = w_1(d + x) + w_1(d - x) - w, \qquad |x| \le d. \tag{5.108}$$

The main result of the present work is given by the following

THEOREM. *If $\delta \ne 0$, $1/2$, 1 and $x > 0$, then the right-hand integral in Eq. (5.107) is asymptotically small:*

$$\int_0^\infty [w_1(2d + \xi) - w]K(x + \xi)\, d\xi \to 0, \quad n \to \infty, \tag{5.109}$$

and so can be rejected.

The proof is based again on the asymptotic properties of the kernel $K(x)$. Representation (5.105) is not suitable for this purpose, and we use here a different expansion. The function $G(s)$ in Eq. (5.102) can be rewritten as

$$G(x) = \frac{1 - \exp(-2\gamma)}{\gamma[1 + \exp(-2\gamma)]} = \frac{1}{\gamma} - \frac{2\exp(-2\gamma)}{\gamma[1 + \exp(-2\gamma)]} = \frac{1}{\gamma} - 2\sum_{m=1}^\infty (-1)^{m-1} \frac{\exp(-2m\gamma)}{\gamma}. \tag{5.110}$$

Taking into account the integral identity (see Eq. (1.17))

$$\int_\sigma \frac{\exp(-|z|\gamma)}{\gamma} e^{-isx}\, ds = \pi i H_0^{(1)}\left(\chi\sqrt{z^2 + x^2}\right), \tag{5.111}$$

we then arrive at the following expression for the kernel $K(x)$:

$$K(x) = \frac{i}{2} H_0^{(1)}(\chi|x|) - i\sum_{m=1}^\infty (-1)^{m-1} H_0^{(1)}\left(\chi\sqrt{4m^2 + x^2}\right), \tag{5.112}$$

where $H_0^{(1)}$ is the Hankel function.

It is evident that with $\chi \to \infty$ the argument of each term under the sum in Eq. (5.112) increases infinitely, uniformly with respect to x. Hence, with the help of the asymptotic formula (cf. Eq. (3.115))

$$H_0^{(1)}(z) \sim \sqrt{\frac{2}{\pi z}} e^{i(z - \pi/4)}, \qquad z \to \infty, \tag{5.113}$$

the uniform high-frequency representation for the kernel can be finally obtained as follows:

$$K(x) \sim \frac{i}{2} H_0^{(1)}(\chi|x|) - \sqrt{\frac{2}{\pi\chi}} e^{\pi i/4} \sum_{m=1}^{\infty} (-1)^{m-1} \frac{\exp(i\chi\sqrt{4m^2 + x^2})}{(4m^2 + x^2)^{1/4}}. \tag{5.114}$$

For the open finite-length waveguide (cf. Section 5.4) there is only the first term of the infinite series (5.114). It should also be noted that the term $(i/2)H_0^{(1)}(\chi|x|)$ cannot be simplified asymptotically at $\chi \to \infty$, since the argument of the kernel $K(x + \xi)$ in the considered right-hand integral (5.107) may be arbitrarily small when $\xi \sim x \sim 0$.

To prove the theorem let us suppose, as usual, that its statement is valid. Then the function $w_1(x)$ is a solution of the Wiener–Hopf integral equation (5.107) holding over a semi-infinite interval. Omitting some routine transformations, one comes to the following expression for its solution:

$$\begin{aligned} w_1(x) &= \frac{1}{G(0)} - \frac{1}{2G_+(0)} \sum_{m=1}^{\infty} \frac{(\pi m)^2 G_+(\alpha_m)}{\alpha_m^2} e^{i\alpha_m x} \\ &= \frac{\chi}{\tan\delta} - \frac{1}{2} \sum_{m=1}^{\infty} \frac{G_+(\alpha_m)}{G_+(0)} \frac{(\pi m)^2}{\alpha_m^2} e^{i\alpha_m x}, \quad x > 0, \end{aligned} \tag{5.115}$$

where

$$G(s) = G_+(s)\, G_-(s), \qquad G_-(-s) = G_+(s), \tag{5.116}$$

is a result of the factorization of the symbolic function $G(s)$ given by Eq. (5.102).

The solution of the convolution equation (5.107) can be trivially obtained by using the Fourier transform, as follows:

$$w(x) \equiv w = \frac{1}{G(0)} = \frac{\chi}{\tan\delta}. \tag{5.117}$$

It now becomes clear from Eqs. (5.115), (5.117) that the values $\delta = 0$ and $\delta = 1$ (as well as the resonance value $\delta = 1/2$) should be excluded from the consideration, which is stated in the body of the theorem. If δ does not coincide with any of those critical values, then $w_1(x)$ and $w(x)$ are uniformly of the order of $O(\chi)$, $\chi \to \infty$. It is interesting to note that a few first terms in the series (5.115), in addition to the first one, have the same order $O(\chi)$, $\chi \to \infty$ (for m such that $\pi m \sim \chi$). Note also that the summation in the series (5.115) may be taken over $1 \le m \le n$, because, in the asymptotic sense, only for these values of m the quantities α_m are real. All other terms are exponentially small ($i\alpha_m \sim -\chi^{1/2}, m > n$), since the argument of the function $w_1(2d + \xi), \xi > 0$ in Eq. (5.107)$_1$ is greater than $2d$ (i.e., $x > 2d$ in (5.115)).

Let us estimate the proper asymptotic order of the right-hand integral in Eq. (5.107):

$$\begin{aligned} \int_0^{\infty} [w_1(2d + \xi) - w] K(x + \xi)\, d\xi &= \frac{i}{2} \int_0^{\infty} [w_1(2d + \xi) - w] H_0^{(1)}[\chi(x + \xi)]\, d\xi \\ &- \sqrt{\frac{2}{\pi\chi}} e^{\pi i/4} \sum_{m=1}^{\infty} (-1)^{m-1} \int_0^{\infty} [w_1(2d + \xi) - w] \frac{\exp[i\chi\sqrt{4m^2 + (x + \xi)^2}]}{[4m^2 + (x + \xi)^2]^{1/4}}\, d\xi. \end{aligned} \tag{5.118}$$

The right-hand integrals in Eq. (5.118) should be estimated by different ways for the free term and the terms in the series. For the last ones we introduce the phase function

$$S(x) = \sqrt{4m^2 + (x + \xi)^2}, \tag{5.119}$$

which determines highly oscillating nature of the exponential function $\exp[i\chi S(x)]$ in (5.118). It is evident that $S'(x) = (x + \xi)/S(x) > 0$, when $x > 0$, $\xi \geq 0$, hence this function has no stationary point. Thus a standard multiple integration by parts proves that the order of each term under the sum sign in Eq. (5.118) is less than $O(\chi^{-\beta})$, with β being an arbitrarily large positive number: $\beta \gg 1$.

Further, the free right-hand side term in Eq. (5.118) can be estimated as

$$
\int_0^\infty [w_1(2d + \xi) - w] H_0^{(1)}[\chi(x + \xi)] \, d\xi = \int_0^\infty \frac{w_1(2d + \xi/\chi) - w}{\chi} H_0^{(1)}(\chi x + \xi) \, d\xi
$$

$$
\sim \int_0^\infty \frac{w_1(2d + \xi/\chi) - w}{\chi} e^{i\xi} \, d\xi \, \frac{e^{i\chi x}}{\sqrt{\chi x}} = O\left(\frac{1}{\sqrt{\chi x}}\right),
$$

$$(5.120)$$

if the value x is outside of a *boundary layer* (where $x \sim 1/\chi$). Thus the theorem is proved.

The problem in hand seems to be a typical asymptotic problem with singular perturbations. The structure of the solution outside of the boundary layer is absolutely different from that valid near the *boundary point* $x = 0$. As can be seen, the theorem indicates how to construct an *external* solution only. Indeed, by substituting (5.115) and (5.117) into Eq. (5.108), one comes to the following high-frequency representation:

$$
v(x) = \frac{\chi}{\tan \delta} - \frac{1}{2} \sum_{m=1}^n \frac{G_+(\alpha_m)}{G_+(0)} \frac{(\pi m)^2}{\alpha^2} \left[e^{i\alpha_m(d+x)} + e^{i\alpha_m(d-x)} \right], \qquad |x| < d. \quad (5.121)
$$

The solution in the boundary layer $d \pm x \sim 1/\chi$ cannot be obtained by this method. Note that a finite number of terms $1 \leq m \leq n$ are taken in the series (5.121), since all other terms are exponentially small outside of the boundary layer (α_m becomes imaginary).

Efficient high-frequency representation for $G_+(\alpha)$.

The main asymptotic formula (5.121) contains some values of the function $G_+(s)$ (s is real) which is a result of the factorization of the main symbolic function $G(s)$ (5.102). It can be written as the product of three functions: $G = G_1 G_2^{-1} G_3$, where $G_1 = 1 - \exp(-2\gamma)$, $G_2 = 1 + \exp(-2\gamma)$, $G_3 = (\gamma)^{-1}$, and $\gamma = \sqrt{s^2 - \chi^2}$, with the trivial factorization of the third of them: $G_3(s) = (s + \chi)^{-1/2}(s - \chi)^{-1/2}$. For the first two functions one can deal with the general representation (see Section 1.2; Noble, 1962)

$$
G_{1,2}^+(s) = \exp\left\{ \frac{1}{2\pi i} \int_\sigma \frac{\ln\left(1 \mp e^{-2\sqrt{u^2 - \chi^2}}\right)}{u - s} \, du \right\}, \quad (5.122)
$$

which has too complex form.

A good high-frequency approximation proposed by Vajnshtejn (1969) is known:

$$
G_1^+ G_3^+(s) \approx 2\sqrt{\frac{\delta}{n}} \left(1 + \frac{s}{\sqrt{2\pi\delta\chi}}\right) \exp\left[-0.824 \frac{1-i}{\sqrt{2\chi}} s\right], \quad \chi = \pi(n+\delta), \quad 0 < \delta < 1. \quad (5.123)
$$

The last approximation is correct when $|\alpha| \ll \chi$, and $0 < \delta \ll 1$, i.e., for high frequencies near eigen-modes. Hence, that is of no interest for the present investigation since the asymptotic approach proposed fails for $\delta \to 0$, as discussed above.

We start from the following approximate equality:

$$
G_1^+(s) \approx \frac{e^{\pi i/4}}{\sqrt{s + \chi}} \exp\left\langle \frac{s_*}{\pi} \sqrt{\frac{\chi}{2}} \int_0^\infty \frac{\ln\{1 - \exp[2i(\pi\delta + t)]\}}{2t + \chi s_*^2} \frac{dt}{\sqrt{t}} \right\rangle \quad (s_* = s/\chi), \quad (5.124)
$$

which is known to be valid at $\chi \gg 1$ and arbitrary $\mathrm{Im}(s) \geq 0$ (see Mittra and Lee, 1971). By the application of the Taylor expansion to the logarithmic function, the last integral can be rewritten as

$$-\sum_{j=1}^{\infty} \frac{e^{2\pi \delta ji}}{j} \int_0^{\infty} \frac{e^{2tji}}{2t + \chi s_*^2} \frac{dt}{\sqrt{t}}$$

$$= \frac{\pi}{s_*\sqrt{2\chi}} \sum_{j=1}^{\infty} \frac{e^{2\pi \delta ji}}{j} \left\{ \sqrt{2} \left[S(\chi s_*^2 j)i + C(\chi s_*^2 j) \right] e^{-i\pi/4} - 1 \right\} e^{-i\chi s_*^2 j}, \tag{5.125}$$

where $S(x)$ and $C(s)$ are the Fresnel integrals. These possess some rational approximation (see Abramowitz and Stegun, 1965) so that the integral in (5.124) can be reduced to the form

$$= -\frac{\pi}{s_*\sqrt{\chi}} e^{-\pi i/4} \sum_{j=1}^{\infty} \frac{e^{2\pi \delta ji}}{j} \left[if\left(\alpha_* \sqrt{\frac{2}{\pi}\chi j} \right) + g\left(\alpha_* \sqrt{\frac{2}{\pi}\chi j} \right) \right], \tag{5.126}$$

where

$$f(x) \approx \frac{1 + 1.22x}{2 + 2.17x + 1.22\pi x^2}, \qquad g(x) \approx \frac{1}{2 + 4.14x + 3.49x^2 + 6.67x^3}, \tag{5.127}$$

uniformly over $0 < x < \infty$.

If we extract the most slowly decaying term in the sum (5.126), $\sum_1^{\infty}[\exp(2\pi \delta ji)/j^{3/2}]$, then the series becomes rapidly convergent. The extracted series can be calculated by using our efficient approximation for the functions $C_\beta(x)$, $S_\beta(x)$ (see Section 3.4). Finally, we arrive at the following high-frequency representation:

$$G_1^+(s) \approx \frac{e^{\pi i/4}}{\sqrt{s+\chi}} \exp\left\langle \frac{i-1}{2} \left\{ \frac{iC_{3/2}(s_*\sqrt{2\chi/\pi}) - S_{3/2}(s_*\sqrt{2\chi/\pi})}{s_*\sqrt{2\pi\chi}} \right. \right.$$

$$\left. \left. + \sum_{j=1}^{\infty} \frac{e^{2\pi \delta ji}}{j} \left[if\left(s_*\sqrt{\frac{2}{\pi}\chi j} \right) + g\left(s_*\sqrt{\frac{2}{\pi}\chi j} \right) - \frac{i}{s_*\sqrt{2\pi\chi j}} \right] \right\} \right\rangle, \tag{5.128}$$

where the functions f, g are determined by (5.127) and $C_{3/2}$, $S_{3/2}$ in Section 3.4. The corresponding high-frequency approximation for $G_2^+(s)$ can be derived likewise.

The common term of the series (5.128) decreases as $O(j^{-5/2})$, so this formula permits efficient calculations indeed. The convergence of the series in (5.128) becomes slower when $\delta \to 0$, because there is a number of small $s_{*m} = \sqrt{(\pi m)^2 - \chi^2}/\chi \sim \sqrt{\delta}$, so the acceleration procedure in this range becomes inefficient.

The discussion in the last two sections follows the authors' works (Scalia and Sumbatyan, 1999, 2001).

Helpful remarks

As we have already discussed, the problems considered in the two last sections lead to integral equations containing two independent dimensionless parameters. The first of them is purely geometric, and this characterizes relative length of the slits, that is, $b = a/h$ (Section 5.4) or $d = l/a$ (this section). The second parameter χ is related to the frequency, being large for high frequencies. This type of high-frequency problem gives a bright example when heuristic ideas may lead to mistaken conclusions. Indeed, in the considered case of high-frequency diffraction the wavelength becomes asymptotically small when compared with the slits length or the width of the structure. So, there are two very widespread (mistaken) heuristic opinions. The first of them is based on the fact that if the width (thickness) of the waveguide is much greater than the wavelength, then the leading asymptotic term of the solution on the slit coincides with that in the problem for a half-plane. Another (mistaken) point of view declares that the solution for the finite-length slit is asymptotically the same as for semi-infinite one, since the slit is much longer than the wavelength. Both viewpoints ignore the fact that with frequency increasing both dimensional sizes become large with respect to the wavelength simultaneously. Formulas derived in the present chapter represent mathematical structure, which provides actual behavior with multiple re-reflections of waves between the faces of the waveguide.

5.6. Waves in Elastic Half-Space. Factorization of the Rayleigh Function

Let us consider vertical harmonic oscillations of a rigid punch coupled without friction with an elastic half-plane (2D problem). For this problem, as in many problems studied above, the choice of methods of investigation depends on the value of frequency. As usual, the case of low and moderate frequencies can be investigated by using regular (analytical or numerical) methods. With the frequency increasing regular methods lead to a loss of the computational stability. That is why we undertake here again an asymptotic analysis appropriate just for the high frequency case.

In order to derive the basic integral equation of the considered contact problem of dynamic elasticity, let us come back to equalities (5.41), which establish relations between some components of the stress tensor and those of the displacement vector. In the case when there is known a normal component of the stress, we arrived at the integral equation (5.45) of the dynamic cracks theory. If, in the opposite case, we know the amplitude of normal displacement, then it can be easily seen from Eq. (5.41) that we arrive at the following integral equation of the first kind:

$$\int_{-1}^{1} p(\xi) K\left(\chi|x - t|\right) d\xi = \frac{\mu}{a} w, \quad |x| \le 1, \quad K(x) = \frac{1}{2\pi} \int_{-\infty}^{\infty} L_*(s) e^{-isx} ds,$$

$$L_*(u) = \frac{\sqrt{u^2 - \beta^2}}{4u^2 \sqrt{u^2 - \beta^2} \sqrt{u^2 - 1} - (2u^2 - 1)^2}, \quad \chi = ak_s, \quad \beta = k_p/k_s = c_s/c_p < 1.$$

$$(5.129)$$

In Eq. (5.129), $p(x)$ is the amplitude of the normal contact pressure (the contact area is free of friction), w is the amplitude of the punch oscillations, χ is a parameter, which is large at high frequencies, μ is the shear elastic modulus, and a is the semi-width of the punch.

As in the previous sections, we will apply here our idea based upon the fact that initial equation (5.129) is (exactly) equivalent to the system:

$$\int_{0}^{\infty} \varphi(\xi) K(x - \xi) d\xi = \chi + \int_{0}^{\infty} [\varphi(2\chi + \tau) - v(\tau)] K(x + \tau) d\tau, \quad x \ge 0,$$

$$\int_{-\infty}^{\infty} v(t) K\left[\chi|x - \xi|\right] d\xi = 1, \quad |x| < \infty,$$

$$(5.130)$$

if

$$p(x) = \frac{\mu}{a} w \left\{ \varphi\left[\chi(1 + x)\right] + \varphi\left[\chi(1 - x)\right] - v(x) \right\}, \quad |x| \le 1. \quad (5.131)$$

As we could see above, in many problems the last integral in (5.130) may be rejected, and equation (5.129) becomes asymptotically as a Wiener–Hopf equation on the semi-infinite interval. Usually, this brings the error of the order of $\exp(-\chi\varepsilon)$ $(\varepsilon > 0)$, uniformly with respect to x. Sometimes this error has a power character, i.e., becomes more essential. Below we will show that in the present problem the *tail* on the right-hand side of Eq. (5.129) is an (asymptotically) small quantity just of some power degree. Namely, the estimate

$$\int_0^\infty [\varphi(2\chi+\tau)-v(\tau)]\,K(x+\tau)\,d\tau = O(\chi^{-1/2}), \qquad \chi \to \infty, \tag{5.132}$$

holds uniformly over $x \geq 0$.

As in the previous sections we aim at construction of the leading asymptotic term. Heuristically, the estimate (5.132) means again that the leading asymptotic term (5.131) is determined from the solution for a pair of semi-infinite punches (5.130) and one infinite punch. Physically, this means that the high-frequency oscillations generate the waves of so small length that perturbations from the right edge of the punch do practically not influence the wave process near the left edge, and vice versa.

A solution of the convolution equation (5.130) is constructed immediately by applying the Fourier transform, and is given as follows:

$$v(x) = -\frac{i}{\beta\chi} \equiv v. \tag{5.133}$$

In order to successfully solve the equation on the semi-axis

$$\int_0^\infty \varphi(\xi)K(x-\xi)\,d\xi = \chi, \tag{5.134}$$

it is necessary to arrange factorization of the symbolic function (Fourier image of the kernel). As known (see Babeshko, 1971; Vorovich and Babeshko, 1979), the solution of equation (5.134) is stable with respect to small perturbations of the symbol on the real axis, so let us apply an approximate factorization (cf. Section 1.2).

The symbolic function $L_*(u)$ is a combination of the four root squares $\sqrt{s+\beta}$, $\sqrt{s+1}$, $\sqrt{s-\beta}$, $\sqrt{s-1}$, with each of them having branching points. Let us draw in the complex plane s the cuts (cf. Section 1.1) that join the points $s = -\beta$ and $s = -1$ with infinity in the lower half-plane, and the points $s = \beta$ and $s = 1$ with infinity in the upper half-plane. In addition to the branching points, on the real axis $\text{Im}(s) = 0$ there are two Rayleigh poles of the symbolic function: $a = \pm s_1$, $s_1 > 1$. According to the principle of extremely small absorption (see Section 3.2), the integration contour in the integral representation of the kernel in Eq. (5.129) coincides with the real axis, bending around positive singularities from below, and negative ones from above.

The function $L_*(s)$ has qualitatively different behavior on different intervals of the real axis. For $|s| \geq 1$ it is real-valued, for $\beta < |s| < 1$ it is complex-valued, and for $|s| \leq \beta$ it is imaginary-valued.

Let us approximate the symbol $L_*(s)$ by the expression

$$L_*(s) = \sqrt{s^2-\beta^2}\,\frac{4s^2\sqrt{s^2-\beta^2}\sqrt{s^2-1}+(2s^2-1)^2}{16s^4(s^2-\beta^2)(s^2-1)-(2s^2-1)^4}$$

$$\approx L(s) = \frac{A\sqrt{s^2-\beta^2}}{(s^2-s_1^2)(s^2-z^2)(s^2-\bar{z}^2)}M_+(s)M_-(s), \tag{5.135}$$

$$M_\pm(s) = Bs\sqrt{s\pm\beta}\sqrt{s\pm1}+\left(\sqrt{2}\,s\pm1\right)^2 \qquad (A,B,\text{Im}\,z>0).$$

This function $L(s)$, like $L_*(s)$, is even, has two Rayleigh poles $s = \pm s_1$ and possesses the same qualitative behavior on various intervals of the real axis. Moreover, it catches the behavior at the origin and at the infinity exactly. Besides, on all these intervals it has a true sign of the imaginary part, which is important for the uniqueness theorem to be valid. The expression $M_+(s)$ has a zero in the upper half-plane, which should be liquidated by the zero of the denominator $s = -\bar{z}$. Let us note that the point $s = -z$ is another zero.

With the help of this approximation factorization of the function $L(s)$ can be made very easily, in the following way:

$$L_+(s) = \frac{A\sqrt{s+\beta}}{(s+s_1)(s+z)(s+\bar{z})}\, M_+(s). \tag{5.136}$$

The realization of the Wiener–Hopf method for equation (5.134), with the help of factorization (5.136), leads to the following expression for the Fourier image of the function $\varphi(x)$:

$$
\begin{aligned}
\Phi_+(s) &= C\frac{(s+s_1)(s+z)(s+\bar{z})}{s\sqrt{s+\beta}\, M_+(s)} \\
&= C\frac{(s+s_1)(s+z)(s+\bar{z})}{s\sqrt{s+\beta}\, \Delta(s)}\left[Bs\sqrt{s+\beta}\sqrt{s+1}-(\sqrt{2}s+1)^2\right], \\
\Delta(u) &= B^2 u^2(u+\beta)(u+1)-(\sqrt{2}u+1)^4 = d(s+z)(s+\bar{z})(s+\eta)(s+\bar{\eta}),
\end{aligned}
\tag{5.137}
$$

$$d = B^2 - 4, \qquad C = \frac{s_1|z|^2}{A\sqrt{\beta}}\chi.$$

Therefore,

$$\Phi_+(s) = \frac{C}{d}\left[a_1\frac{\sqrt{s+1}}{s+\eta} + a_2\frac{\sqrt{s+1}}{s+\bar{\eta}} + \frac{a_3}{s\sqrt{s+\beta}} + \frac{a_4}{(s+\eta)\sqrt{s+\beta}} + \frac{a_5}{(s+\bar{\eta})\sqrt{u+\beta}}\right],$$

$$a_1 = B\frac{u_1-\eta}{\bar{\eta}-\eta}, \quad a_2 = \bar{a}_1, \quad a_3 = -\frac{u_1}{|\eta|^2}, \quad a_4 = \frac{(u_1-\eta)(\sqrt{2}\eta-1)^2}{\eta(\bar{\eta}-\eta)}, \quad a_5 = \bar{a}_4.$$

$$\tag{5.138}$$

This permits explicit representation for the function $\varphi(x)$, since the following inversion formulas hold (see Bateman and Erdelyi, 1954):

$$
\begin{aligned}
\frac{1}{(s+\eta)\sqrt{s+\beta}} &\Longleftarrow -i\frac{e^{i\eta x}}{\sqrt{\beta-\eta}}\operatorname{Erf}\sqrt{-ix(\beta-\eta)}, \qquad x > 0, \\
\frac{\sqrt{s+1}}{s+\eta} &\Longleftarrow \frac{e^{-i(\pi/4-x)}}{\sqrt{\pi x}} - ie^{i\eta x}\sqrt{1-\eta}\operatorname{Erf}\sqrt{-ix(1-\eta)}, \quad x > 0.
\end{aligned}
\tag{5.139}
$$

Note that this solution can be expressed in terms of the Fresnel integrals (compare with Section 5.3).

From explicit form of the function $\varphi(x)$, as well as directly from Eq. (5.138), we can obtain the estimate

$$\varphi(x) \sim v + a\,e^{i\beta x}x^{-1/2}, \qquad x \to +\infty, \tag{5.140}$$

which indicates that the *boundary layer* solution, as we did expect, turns in the *outer* zone into a solution for an infinite punch (5.133). However, we can see that this process is very slow.

If we add to the estimate (5.140) also the evident estimate

$$K(x) \sim c_1\,e^{iu_1 x} + c_2\,e^{i\beta x}x^{-3/2} + c_3\,e^{ix}x^{-3/2}, \qquad x \to +\infty, \tag{5.141}$$

then we justify the asymptotic relation (5.132).

It is interesting for many applications to define the relation between the applied force and the settlement of the punch base (we assume the punch to be weightless)

$$P = b \int_{-1}^{1} p(x)\,dx = \mu\,w \int_{-1}^{1} \{\varphi\,[\chi(1+x)] + \varphi\,[\chi(1-x)] - v\}\;dx$$
$$= \mu\,w \left\langle 2v + 2 \int_{-1}^{\infty} \{\varphi\,[\chi(1+x)] - v\}\;dx \right\rangle. \tag{5.142}$$

The last equality holds, due to the estimate (5.140), with the error $O(\chi^{-1/2})$, which is equal to the error of the constructed solution of the equation (5.129). Further, since

$$\int_{-1}^{\infty} \{\varphi\,[\chi(1+x)] - v\}\;dx = \frac{1}{\chi} \int_{-\infty}^{\infty} [\varphi(\xi) - v\,H(\xi)]\;dt = \frac{1}{\chi}[\Phi_{+}(s) - V_{+}(s)]_{s=0} \tag{5.143}$$

($H(\xi)$ is the Heaviside function), then finally we obtain the following expression for the pliability of the foundation:

$$w = \frac{P}{2\mu}\,\frac{\beta}{CD\beta - i\chi}, \qquad D = \frac{1}{d}\left\{ 2\,\mathrm{Re}\left[\frac{1}{\eta}\left(a_1 + \frac{a_4}{\sqrt{\beta}} \right) \right] - \frac{a_3}{2\beta^{3/2}} \right\}. \tag{5.144}$$

Let us note that the constants C and D in Eq. (5.144) are real-valued, so, for instance, the shift in phases between the settlement and the applied force is given by the formula

$$\theta = -\arctan\left(\frac{\chi}{CD\beta} \right). \tag{5.145}$$

Helpful remarks

For a long time the problem of factorization of the classical Rayleigh function remained unsolved, and in many problems where there was a need of this factorization some authors applied exact factorization formulas. This leads to extremely cumbersome calculations with very complex integrals. Instead, we use here an approximate factorization (5.135), which keeps all qualitative properties of the (complex-valued) initial Rayleigh function (compare with what is discussed in Section 1.2). Quantitatively, an arbitrarily precise approximation can be achieved by a regular factor representing a combination of some polynomials (as discussed in Section 1.2). Our calculations show that the proposed approximation, as follows from Eq. (5.135), provides an accuracy of few percent.

5.7. Integral Equation of the Mixed Boundary Value Problem for Elastic Layer

In this section we will demonstrate that the asymptotic approach developed in Section 5.5 for scalar problems in acoustic layer with mixed boundary conditions can be expanded to the case of elastic layer (2D in-plane problem).

Let a rigid punch of the width $2b$ be undertaken by vertical harmonic oscillations being placed on the free surface of elastic strip. Let the latter, to be more specific, be placed on an absolutely rigid basis. We assume that the friction in the contact zone, as well as between the basis and the strip, is absent. As follows from results of Section 3.3, the considered problem can be reduced to the integral equation for the unknown normal contact stress $p(x)$, related to the quantity $\mu w / h$:

$$\int_{-a}^{a} p(\xi) K(x - \xi)\, d\xi = 1, \quad |x| < a, \quad (a = b/h), \quad K(x) = \frac{1}{2\pi} \int_{-\infty}^{\infty} L(s)\, e^{-i\chi x u}\, ds,$$

$$L(s) = L_1(s) - L_2(s), \qquad L_1(s) = \sigma_1 / \Delta(s), \qquad L_2(s) = \sigma_1 P_1(s) / \Delta(s),$$

$$P_1(s) = e^{-2\chi \sigma_1} + e^{-2\chi \sigma_2} - e^{-2\chi(\sigma_1 + \sigma_2)}, \quad \chi = h k_s, \quad \sigma_1 = \sqrt{s^2 - \beta^2}, \quad \sigma_2 = \sqrt{s^2 - 1},$$

$$\Delta(s) = 4 s^2 \sigma_1 \sigma_2 G_1(s) F_2(s) - (2 s^2 - 1)^2 G_2(s) F_1(s), \qquad \beta = k_p / k_s = c_s / c_p < 1,$$

$$G_j(s) = 1 - e^{-2\chi \sigma_j}, \qquad F_j(s) = 1 + e^{-2\chi \sigma_j}, \qquad j = 1, 2,$$

$$\tag{5.146}$$

where h is the thickness of the layer. Equation (5.146) is written again in a dimensionless form, which is more convenient to be represented in a little different manner compared with the previous section. Note that the function $\Delta(s)$ represents again the Rayleigh–Lamb function, written in a slightly different manner.

Similarly to the scalar problem (Section 5.5) with increasing frequency (i.e., with the growth of dimensionless parameter χ) the number of zeros and poles of the symbolic function $L(s)$, situated on the real axis, unboundedly grows. Location of real zeros of the symbol $\pm \alpha_k$ and $\pm \beta_k$:

$$\alpha_k = \sqrt{1 - \left(\frac{\pi k}{\chi}\right)^2} \ (k = 1, \ldots, n_\alpha), \qquad \beta_k = \sqrt{\beta^2 - \left(\frac{\pi k}{\chi}\right)^2} \ (k = 1, \ldots, n_\beta). \tag{5.147}$$

In principle, the method of Sections 5.4 and 5.5 can be directly applied to the integral equation (5.146). However, justification of the using approach will be more complex and difficult. So we apply a slightly different idea, which leads to the leading asymptotic term more directly and more rapidly. Let us extract from the kernel $K(x)$ the function $K_1(x)$ with the symbol $L_1(s)$, which does not have any zeros on the real axis and whose behavior is algebraic at $u \to \pm \infty$:

$$\int_{-a}^{a} p(\xi) K_1(x - \xi)\, d\xi = 1 + \int_{-a}^{a} p(\xi) K_2(x - \xi)\, d\xi. \tag{5.148}$$

So far as symbolic the function $L_1(s)$ of the kernel $K_1(x)$ has no real pole, our standard representation of initial equation holding over the finite-length interval as a combination of three equations allows us to construct the *outer* (i.e., outside of the asymptotically small boundary layers near the punch edges) solution at $\chi \to \infty$ by extending the integral operator with kernel K_1 to the full real axis. We thus arrive at a convolution equation whose solution is easily constructed by the application of the Fourier transform. This technique reduces the equation to the simpler form:

$$\int_{-a}^{a} p(\xi) Q(x - \xi)\, d\xi = l_0, \quad |x| < a, \qquad Q(x) = \frac{1}{2\pi} \int_{-\infty}^{\infty} G(s) e^{-i\chi s x}\, ds,$$

$$G(s) = G_1(s) G_2(s) = \left(1 - e^{-2\chi \sigma_1}\right) \left(1 - e^{-2\chi \sigma_2}\right), \tag{5.149}$$

$$l_0 = \left. \frac{\Delta}{\sigma_1} \right|_{s=0} = -\frac{i}{\beta} (1 + e^{2i\chi\beta})(1 - e^{2i\chi}).$$

Taking into account the relation

$$\frac{1}{2\pi} \int_{-\infty}^{\infty} e^{-i\chi s x}\, ds = \frac{1}{\chi} \delta(x), \tag{5.150}$$

Eq. (5.149) can be reduced to the following form:

$$p(x) - \chi \int_{-a}^{a} p(\xi) Q_1(x - \xi)\, d\xi = l_0 \chi, \qquad |x| < a,$$

$$Q_1(x) = \frac{1}{2\pi} \int_{-\infty}^{\infty} P_1(s) e^{-i\chi sx}\, ds, \qquad P_1(s) = 1 - G(s). \tag{5.151}$$

Obviously, the kernel $Q_1(x)$ is continuous (moreover, it is even infinitely differentiable).

Now, by analogy with integral equations of the first kind, here we will apply our standard idea that equation (5.151) is equivalent to the pair of equations for two new unknown functions $\varphi(x)$ and $v(x)$:

$$\varphi(x) - \chi \int_0^{\infty} \varphi(\xi) Q_1(x-\xi)\, d\xi = l_0\chi - \chi \int_0^{\infty} [\varphi(2a+\xi) - v] Q_1(x+\xi)\, d\xi, \qquad x > 0, \tag{5.152}$$

$$v(x) - \chi \int_0^{\infty} v(\xi) Q_1(x - \xi)\, d\xi = l_0\chi, \qquad |x| < \infty, \tag{5.153}$$

if

$$p(x) = \varphi(a + x) + \varphi(a - x) - v, \qquad |x| < a. \tag{5.154}$$

Let us prove, as usual, that integral on the right-hand side of Eq. (5.152) is asymptotically small, starting again from the properties of the kernel $Q_1(x)$. Similarly to the previous sections, we first obtain an asymptotic expression for $Q_1(x)$ at $\chi \to \infty$. The principal trouble is connected with the estimate of the integral

$$J = \frac{1}{2\pi} \int_{-\infty}^{\infty} e^{-i\chi sx} e^{-2\chi(\sigma_1 + \sigma_2)}\, ds. \tag{5.155}$$

It can be simply shown that the main contribution to J is given by a neighborhood of the stationary point s_* of the phase F:

$$F(s, x) = sx + 2\sqrt{1 - s^2} + 2\sqrt{\beta^2 - s^2}, \tag{5.156}$$

where $0 < s_* < \beta$. It can also be shown that F_s' is a monotonically decreasing continuous function for any fixed $x > 0$, where $F_s' > 0$ for $u = 0$ and $F_s' < 0$ for $s = \beta - 0$. Therefore, equation $F_s' = 0$, which determines the stationary point s_*, always has a unique solution. As a result, we obtain

$$J \sim \frac{\exp(-i\pi/4)}{\sqrt{2\pi\chi}} \frac{\exp(i\chi F(s_*, x))}{\sqrt{|F_{ss}''(s_*, x)|}}, \qquad \chi \to \infty. \tag{5.157}$$

Asymptotic estimates for two other integrals present in the kernel $Q_1(x)$ can be derived by using the asymptotic behavior of the Hankel function for large arguments (compare with Section 5.5). Then we finally arrive at the following estimate:

$$Q_1 \sim \chi^{-1/2} \left[A_1(x) e^{i\chi F(s_*, x)} + A_2(x) e^{i\chi\beta\sqrt{4+x^2}} + A_3(x) e^{i\chi\sqrt{4+x^2}} \right], \tag{5.158}$$

where $A_1(x)$, $A_2(x)$, $A_3(x)$ are some smooth functions free of the parameter χ.

Further estimate of the integral on the right-hand side (5.152) will be based on the application of the method of integration by parts (see Section 1.4) well known in such cases. Since for $x > 0$ the phase function $(4 + x^2)^{1/2}$ has no stationary point, such integration by parts shows that contribution to the considered *tail* from the terms related to the functions A_2

and A_3 has the order $\chi^{-1/2}$. In order to establish the asymptotic behavior of the term related to A_1, let us first clarify that the function $F(s_*, x)$ has no stationary point (as a function of x) for $x > 0$. Indeed, the equation $F_s'(s, x) = 0$ defines the stationary point $s_* = s_*(x)$, so $F(s_*, x) = F[s_*(x), x]$. It follows from this relation that $dF/dx = F_s' s_{*x} + F_x' = F_x' = s = s_*(x)$.

We thus have discovered that $dF/dx = 0$ only for $s_* = 0$, which is possible only at $x = 0$, which can be proved directly. It follows from the last fact that the contribution of the term with A_1 to the right-hand side integral in Eq. (5.152) is also of the order of $\chi^{-1/2}$.

Omitting details of the Wiener–Hopf method for equation (5.152) and trivial treatment of the convolution equation (5.153), we can write out the final result

$$
\begin{aligned}
p(x) &= \frac{\chi}{\beta \tan \chi\beta} - \frac{i\pi l_0 \chi}{2G_-(0)} \\
&\times \left\{ \sum_{j=1}^{n_\alpha} \frac{j\, G_+(\alpha_j) H_j(x,1)}{1 - \exp[-2\chi\sigma_1(\alpha_j)]} + \sum_{j=1}^{n_\beta} \frac{j\, G_+(\beta_j) H_j(x,\beta)}{1 - \exp[-2\chi\sigma_2(\beta_j)]} \right\},
\end{aligned}
\tag{5.159}
$$

$$
H_j(x,\beta) = \frac{\exp\left[i\sqrt{(\chi\beta)^2 - (\pi j)^2}\,(a+x)\right] + \exp\left[i\sqrt{(\chi\beta)^2 - (\pi j)^2}\,(a-x)\right]}{(\chi\beta)^2 - (\pi j)^2},
$$

where $G(s) = G_+(s)\, G_-(s)$ is a factorization of the function $G(s)$.

Helpful remarks

Direct numerical solution of the initial equation (5.146) is faced with essential difficulties. This is caused by the evident remark that for $\chi \gg 1$ the kernel represents the sum of a delta-like and several strongly oscillating functions, which makes the calculation procedure very unstable.

With increasing of the frequency parameter χ the contact pressure diagram becomes *wave-like*. This phenomenon, just as in the scalar acoustic case, can be explained by multiple re-reflections of rays from the bottom of the layer and, as noted in the previous sections, differs from analogous problems for the half-plane (Section 5.6), where the asymptotics of the contact pressure tends with the frequency increasing to a constant distribution.

Chapter 6

Short-Wave Asymptotic Methods on the Basis of Multiple Integrals

6.1. Schoch's Method: Exact Representation of 3D Wave Fields by One-Dimensional Quadratures

It is conventionally recognized that modern computers permit, in principle, efficient calculation of wave fields by direct numerical methods for arbitrary boundary value problems. This is especially true for the cases of low and moderate frequencies. In the Preface to this book it was outlined why the high-frequency regime is so distinct from ordinary cases, when implemented on computers. Briefly speaking, this requires too large number of mesh nodes with the frequency increasing.

This chapter will be devoted to various methods to achieve efficient calculation of high-frequency wave fields. In the previous chapters we could see a number of unexpected features of high-frequency wave processes. Another interesting property is also rather unexpected: some exact solutions written in explicit form are absolutely unserviceable for computer calculations.

Example. Let us consider an elastic half-space $z \geq 0$, whose boundary plane surface $z = 0$ is free of tangential load, and an oscillating (with angular frequency ω) normal pressure of the amplitude p_0 is applied to this surface over the rectangular domain $|x| \leq a$, $|y| \leq b$. Equations of motion (see Section 1.9) in this 3D problem have the following form:

$$c_p^2 \frac{\partial^2 u_x}{\partial x^2} + c_s^2 \left(\frac{\partial^2 u_x}{\partial y^2} + \frac{\partial^2 u_x}{\partial z^2} \right) + \left(c_p^2 - c_s^2 \right) \frac{\partial}{\partial x} \left(\frac{\partial u_y}{\partial y} + \frac{\partial u_z}{\partial z} \right) + \omega^2 u_x = 0,$$

$$c_p^2 \frac{\partial^2 u_y}{\partial y^2} + c_s^2 \left(\frac{\partial^2 u_y}{\partial z^2} + \frac{\partial^2 u_y}{\partial x^2} \right) + \left(c_p^2 - c_s^2 \right) \frac{\partial}{\partial y} \left(\frac{\partial u_z}{\partial z} + \frac{\partial u_x}{\partial x} \right) + \omega^2 u_y = 0, \qquad (6.1)$$

$$c_p^2 \frac{\partial^2 u_z}{\partial z^2} + c_s^2 \left(\frac{\partial^2 u_z}{\partial x^2} + \frac{\partial^2 u_z}{\partial y^2} \right) + \left(c_p^2 - c_s^2 \right) \frac{\partial}{\partial z} \left(\frac{\partial u_x}{\partial x} + \frac{\partial u_y}{\partial y} \right) + \omega^2 u_z = 0.$$

The boundary conditions on the plane boundary are ($z = 0$):

$$\sigma_{xz} = \sigma_{yz} = 0, \qquad \sigma_{zz} = p_0(x,y) = \begin{cases} p_0, & (x,y) \in S = \{|x| \leq a\} \cap \{|y| \leq b\}, \\ 0, & (x,y) \in \mathbb{R}^2 \setminus S. \end{cases} \qquad (6.2)$$

The double Fourier transform with respect to the variables x, y, $|x| < \infty$, $|y| < \infty$, applied to these equations, leads to the Fourier images of the displacement components to the system of ordinary differential equations of the second order with constant coefficients

$$c_s^2 U_x'' + \left[\omega^2 - \left(c_p^2 s_1^2 + c_s^2 s_2^2 \right) \right] U_x + \left(c_p^2 - c_s^2 \right) (-is_1) \left(-is_2 U_y + U_z' \right) = 0,$$

$$c_s^2 U_y'' + \left[\omega^2 - \left(c_p^2 s_2^2 + c_s^2 s_1^2 \right) \right] U_y + \left(c_p^2 - c_s^2 \right) (-is_2) \left(-is_1 U_x + U_z' \right) = 0, \qquad (6.3)$$

$$c_p^2 U_z'' + \left(\omega^2 - c_s^2 s^2 \right) U_z + \left(c_p^2 - c_s^2 \right) \left(-is_1 U_x' - is_2 U_y' \right) = 0, \qquad s^2 = s_1^2 + s_2^2,$$

where $U_x = U_x(s_1, s_2, z)$, $U_y = U_y(s_1, s_2, z)$, $U_x = U_x(s_1, s_2, z)$ are the Fourier transforms of respective components of the displacement vector, and the sign of the derivative is of course related to the variable z. A solution of this system can be found by a standard operation with the characteristic polynomial. To this end, we should seek the solution in the form

$$\begin{pmatrix} U_x \\ U_y \\ U_z \end{pmatrix} = \begin{pmatrix} A_x \\ A_y \\ A_z \end{pmatrix} e^{\alpha z}, \tag{6.4}$$

where the characteristic parameter α should be determined as a root of the characteristic polynomial. The latter is obtained by the substitution of the representation (6.4) into system (6.3), which leads to a 3×3 homogeneous linear algebraic system. Since we seek a nontrivial solution of this system, its principal determinant must be trivial. This yields the characteristic equation, which in the considered problem after some transformations can be reduced to the following simple form:

$$(s^2 - \alpha^2 - k_s)^2(s^2 - \alpha^2 - k_p) = 0, \tag{6.5}$$

where $k_p = \omega/c_p$ and $k_s = \omega/c_s$ are longitudinal and transverse wave numbers, respectively.

As can be seen from Eq. (6.5), there are two simple α_1, α_2 and two double α_3, α_4 roots of this bi-cubic equation:

$$\alpha_{1,2} = \pm\gamma(s), \qquad \alpha_{3,4} = \pm q(s), \qquad \gamma(s) = \sqrt{s^2 - k_p^2}, \quad q(s) = \sqrt{s^2 - k_s^2}. \tag{6.6}$$

It is easily seen from Eq. (6.6) that the general solution is a combination of some structures of the type (6.4), and in our case can be expressed as follows:

$$\begin{aligned} U_x &= iAs_1\, e^{-\gamma(s)z} - Bq(s)\, e^{-q(s)z}, \\ U_y &= iAs_2\, e^{-\gamma(s)z} - Cq(s)\, e^{-q(s)z}, \\ U_z &= A\gamma(s)\, e^{-\gamma(s)z} + (is_1 B + is_2 C)\, e^{-q(s)z}, \end{aligned} \tag{6.7}$$

where we keep only those characteristic values of α that provide the satisfaction of the radiation boundary conditions at $z \to +\infty$, that is, we take the sign minus in all pairs $\pm\gamma(s)$, $\pm q(s)$.

The remaining unknown constants A, B, C also depend on the Fourier transform parameters (s_1, s_2) and should be determined by satisfying the boundary conditions (6.2) over the boundary plane $z = 0$. Note that the components of the stress tensor participating in Eqs. (6.2) are expressed in terms of components of the displacement vector in the following way:

$$\begin{aligned} \frac{\sigma_{xz}}{\rho} &= c_s^2 \left(\frac{\partial u_x}{\partial z} + \frac{\partial u_z}{\partial x} \right), \qquad \frac{\sigma_{yz}}{\rho} = c_s^2 \left(\frac{\partial u_y}{\partial z} + \frac{\partial u_z}{\partial y} \right), \\ \frac{\sigma_{zz}}{\rho} &= c_p^2 \frac{\partial u_z}{\partial z} + \left(c_p^2 - 2c_s^2 \right) \left(\frac{\partial u_x}{\partial x} + \frac{\partial u_y}{\partial y} \right). \end{aligned} \tag{6.8}$$

After some transformations, expressions (6.8) can be rewritten in Fourier images as (a tilde here denotes the corresponding Fourier transform of the stress tensor components)

$$\begin{aligned} \frac{\tilde{\sigma}_{xz}}{\rho c_s^2} &= U_x' - is_1 U_z = -2is_1\gamma A\, e^{-\gamma(s)z} + \left[(s_1^2 + q^2)\, B + s_1 s_2 C \right]\, e^{-q(s)z}, \\ \frac{\tilde{\sigma}_{yz}}{\rho c_s^2} &= U_y' - is_2 U_z = -2is_2\gamma A\, e^{-\gamma(s)z} + \left[s_1 s_2 B + (s_2^2 + q^2)\, C \right]\, e^{-q(s)z}, \\ \frac{\tilde{\sigma}_{zz}}{\rho c_s^2} &= \frac{k_s^2}{k_p^2} U_z' + \left(2 - \frac{k_s^2}{k_p^2} \right) (is_1\, U_x + is_2\, U_y) \\ &= (k_s^2 - 2s^2)\, A\, e^{-\gamma z} - 2iq(s_1 B + s_2 C)\, e^{-qz}. \end{aligned} \tag{6.9}$$

Now, satisfaction by these expressions to the boundary conditions (6.2), which, in Fourier transforms, become

$$\tilde{\sigma}_{xz}(s_1, s_2, 0) = 0, \qquad \tilde{\sigma}_{yz}(s_1, s_2, 0) = 0, \qquad \tilde{\sigma}_{zz}(s_1, s_2, 0) = P_0(s_1, s_2), \qquad (6.10)$$

leads to a 3×3 linear algebraic system to find coefficients A, B, C. By solving this system and taking into account that for the uniformly distributed applied constant normal oscillating pressure $p_0 \equiv \mathrm{const}$

$$P_0(s_1, s_2) = \int_{-a}^{a} \int_{-b}^{b} p_0 e^{i(s_1 x + s_2 y)} \, dx \, dy = 4p_0 \frac{\sin(a s_1)\sin(b s_2)}{s_1 s_2}, \qquad (6.11)$$

we then finally arrive at the following exact analytical representations for the most interesting physical quantities, the components of the stress tensor given explicitly by

$$\sigma_{xz}(x, y, z) = \frac{2ip_0}{\pi^2} \int_{-\infty}^{\infty} \int_{-\infty}^{\infty} \gamma(2s^2 - k_s^2) \frac{(e^{-\gamma z} - e^{-qz})\sin(a s_1)\sin(b s_2)}{s_2 \, \Delta(s)} e^{-i(x s_1 + y s_2)} \, ds_1 \, ds_2,$$

$$\sigma_{yz}(x, y, z) = \frac{2ip_0}{\pi^2} \int_{-\infty}^{\infty} \int_{-\infty}^{\infty} \gamma(2s^2 - k_s^2) \frac{(e^{-\gamma z} - e^{-qz})\sin(a s_1)\sin(b s_2)}{s_1 \, \Delta(s)} e^{-i(x s_1 + y s_2)} \, ds_1 \, ds_2, \qquad (6.12)$$

$$\sigma_{zz}(x, y, z) = \frac{p_0}{\pi^2} \int_{-\infty}^{\infty} \int_{-\infty}^{\infty} \frac{[(2s^2 - k_s^2)^2 e^{-\gamma z} - 4s^2 \gamma q e^{-qz}]\sin(a s_1)\sin(b s_2)}{s_1 s_2 \, \Delta(s)} e^{-i(x s_1 + y s_2)} \, ds_1 \, ds_2,$$

where $\Delta(s) = (2s^2 - k_s^2)^2 - 4s^2 \gamma q$ is again the classical Rayleigh function.

Calculation of integrals (6.12) is not a simple task. In the high-frequency range, where $k_p a \gg 1$, $k_p b \gg 1$, this requires too long computation time, even on modern powerful computers.

In order to overcome the troubles of such sort Schoch (1941) proposed a method that allows, for a domain S of arbitrary shape, the reduction of the considered problem without any approximation to the calculation of a single integral (instead of the double one) over a finite-length interval—a fantastic result! His idea is as follows. Let us first study small-amplitude oscillations (linear theory) of a rigid plate on the free surface of the scalar acoustic half-space $z = 0$ (see Fig. 6.1).

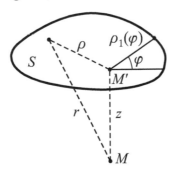

Figure 6.1. On the calculation of the wave field generated by a plane oscillator

If the amplitude of oscillations of the plane vibrator is $v_0 \equiv \mathrm{const}$, $(x, y) \in S$ then the problem is reduced to the Helmholtz equation for acoustic pressure

$$\Delta p(x, y, z) + k^2 p(x, y, z) = 0, \qquad z \geq 0. \qquad (6.13)$$

If the oscillator is fastened with a rigid screen occupying domain $z = 0$, $(x, y) \in \mathbb{R}^2 \setminus S$, then the boundary condition is

$$z = 0: \qquad \frac{\partial p}{\partial z} = g(x, y) = \begin{cases} i\omega \rho_0 v_0, & (x, y) \in S, \\ 0, & (x, y) \in \mathbb{R}^2 \setminus S. \end{cases} \qquad (6.14)$$

Let us denote, as in Chapter 2, the Green's function for full 3D acoustic space by

$$\Phi(r) = \frac{e^{ikr}}{4\pi r}, \qquad r = \sqrt{(x - x_0)^2 + (y - y_0)^2 + z^2}, \tag{6.15}$$

where the point M with Cartesian coordinates (x_0, y_0, z) is the observation point. Then, recalling the boundary property of the normal derivative of the single-layer potential (see Section 2.1), we can see that the exact solution of the posed problem can be expressed by the following formula:

$$p(M) = -2 \iint_{-\infty}^{\infty} g(x, y)\, \Phi(r)\, dx\, dy = -\frac{i\omega\rho_0}{2\pi} v_0 \iint_S \frac{e^{ikr}}{r}, \tag{6.16}$$

because the boundary value of the normal derivative of this expression is

$$-2 \iint_{-\infty}^{\infty} g(x, y) \frac{\partial \Phi(\rho, 0)}{\partial z}\, dx\, dy + g(x_0, y_0), \qquad \rho = \sqrt{(x - x_0)^2 + (y - y_0)^2}. \tag{6.17}$$

The last integral is equal to zero since $\partial \Phi(r)/\partial z = \big(\partial \Phi(r)/\partial r\big)\, (r \cdot n(x_0, y_0)) = 0$ and because $r \perp n$ when $z = 0$.

Schoch's idea works in the following manner. Let us place the origin of the cylindrical coordinate system at the vertical projection of the observation point, i.e., at the point $M'(x_0, y_0, 0)$ (see Fig. 6.1). For simplicity, we consider here only the case when $M' \in S$. If M' lies outside the domain S all transformations can be repeated in a similar way. Now, formula (6.16) reads

$$p(M) = -\frac{i\omega\rho_0}{2\pi} v_0 \iint_S \frac{e^{ikr}}{r} = -\frac{i\omega\rho_0}{2\pi} v_0 \int_0^{2\pi} d\varphi \int_0^{\rho_1(\varphi)} \frac{e^{ik\sqrt{\rho^2 + z^2}}}{\sqrt{\rho^2 + z^2}} \rho\, d\rho. \tag{6.18}$$

Now we may apply the change of variable (see Fig. 6.1): $r^2 = \rho^2 + z^2$, $dx\, dy = \rho\, d\rho\, d\varphi$, $r\, dr = \rho\, d\rho$. Then Eq. (6.18) becomes equal to

$$p(M) = -\frac{i\omega\rho_0}{2\pi} v_0 \int_0^{2\pi} d\varphi \int_z^{r_1(\varphi)} e^{ikr}\, dr = \frac{\omega\rho_0}{k} v_0 \left[e^{ikz} - \frac{1}{2\pi} \int_0^{2\pi} e^{ikr_1(\varphi)}\, d\varphi \right], \tag{6.19}$$

where $r_1(\varphi) = (z^2 + \rho_1^2)^{1/2}$.

Our purpose is achieved: the 3D wave field is (absolutely exactly) expressed in terms of a one-dimensional integral over the interval of finite length.

In some simple cases Eq. (6.19) permits analytical calculation of the full wave field. For example, if S is a disk of radius a, then

$$p(M) = \frac{\omega\rho_0}{k} v_0 \left(e^{ikz} - e^{ik\sqrt{a^2 + z^2}} \right). \tag{6.20}$$

Helpful remarks

We remarked above that the calculation of integrals of the type (6.12) for high frequencies is a hard problem, even when implemented on computers. In some cases this can be treated approximately. For instance, in the far field such integrals can be estimated asymptotically (the so-called *Fraunhofer approximation*). However, for small distances it is very difficult to obtain any result with acceptable accuracy by direct numerical treatment of the integral (6.12). This trouble arises even in the scalar acoustic problem, but in elastic case the integrand becomes more complex and the problem becomes more difficult.

6.2. High-Frequency Wave Fields in Elastic Half-Space

Schoch's approach can be spread to more complex problems. First of all, let us expand it to the scalar problem when on the boundary plane surface of the acoustic half-space there is applied a uniformly distributed pressure $p(x, y, 0) = p_0(x, y)$, $(x, y) \in S$ over the domain S, so that $p(x, y, 0) = 0$, $(x, y) \in \mathbb{R}^2 \setminus S$. Then we can see that the exact solution here can be represented by the double-layer potential (see Section 2.1)

$$p(M) = 2 \iint_{-\infty}^{\infty} p(x, y) \frac{\partial \Phi(r)}{\partial z} \, dx \, dy = \frac{p_0}{2\pi} \iint_S \frac{\partial}{\partial z} \left(\frac{e^{ikr}}{r} \right) dx \, dy, \tag{6.21}$$

$$r = \sqrt{(x - x_0)^2 + (y - y_0)^2 + z^2},$$

where $M(x_0, y_0, z)$ is the observation point. Correctness of the representation (6.21) is due to the boundary property of the potential of double layer (Section 2.1), which is

$$2 \iint_{-\infty}^{\infty} p(x, y) \frac{\partial \Phi(\rho, 0)}{\partial z} \, dx \, dy + p(x_0, y_0), \quad \rho = \sqrt{(x - x_0)^2 + (y - y_0)^2}. \tag{6.22}$$

The last integral is equal to zero since $\partial \Phi(r)/\partial z = \left(\partial \Phi(r)/\partial r \right) (r \cdot n(x, y)) = 0$ and because $r \perp n$ when $z = 0$.

Equation (6.21) yields

$$p(M) = \frac{p_0}{2\pi} \frac{\partial}{\partial z} \iint_S \frac{e^{ikr}}{r} \, dx \, dy = \frac{p_0}{2\pi} \frac{\partial}{\partial z} \int_0^{2\pi} d\varphi \int_z^{r_1(\varphi)} e^{ikr} \, dr$$

$$= -\frac{p_0}{ik} \frac{\partial}{\partial z} \left[e^{ikz} - \frac{1}{2\pi} \int_0^{2\pi} e^{ikr_1(\varphi)} \, d\varphi \right] = -p_0 \left[e^{ikz} - \frac{z}{2\pi} \int_0^{2\pi} \frac{e^{ikr_1(\varphi)}}{r_1(\varphi)} \, d\varphi \right]. \tag{6.23}$$

The problem becomes much more complex if the medium cannot be considered as an acoustic one. Thus, the solution for the normal point force applied to the free surface of elastic half-space cannot be expressed in elementary form, but is expressed in terms of the Fourier–Bessel integral. In order to arrive at this representation, we can apply the same technique as in the previous section, in the case $p_0(x, y) = p_0\delta(x - x_0, y - y_0)$. Then $P_0(s_1, s_2) = p_0 e^{i(x_0 s_1 + y_0 s_2)}$, so for example, for the σ_{zz} stress component we have

$$\sigma_{zz}^0(x, y, z) = \frac{p_0}{4\pi^2} \iint_{-\infty}^{\infty} \frac{(2s^2 - k_s^2)^2 e^{-\gamma z} - 4s^2 \gamma q e^{-qz}}{\Delta(s)} e^{-i[(x-x_0)s_1 + (y-y_0)s_2]} \, ds_1 \, ds_2. \tag{6.24}$$

If we apply here the change of variables $x - x_0 = \rho \cos\psi$, $x - x_0 = \rho \sin\psi$, $s_1 = s\cos\mu$, $s_2 = s\sin\mu$, and pass from the Cartesian to a cylindrical coordinate system, then we can see that expressions under the double integral in the polar coordinate system become $(x - x_0)s_1 + (y - y_0)s_2 = s\rho\cos(\mu - \psi)$. Since all other terms in the integrand depend only on $s = (s_1^2 + s_2^2)^{1/2}$, it can easily be seen that the integration with respect to s leads to

$$\sigma_{zz}^0(\rho, z) = \frac{p_0}{4\pi^2} \int_0^{\infty} \frac{(2s^2 - k_s^2)^2 e^{-\gamma z} - 4s^2 \gamma q e^{-qz}}{\Delta(s)} s \, ds \int_0^{2\pi} e^{-is\rho\cos(\mu-\psi)} \, d\mu. \tag{6.25}$$

Since for arbitrary periodic integrand the integral may be taken over any interval of the length equal to the period of this function, it can be directly proved that expression (6.25) is free of the polar angle ψ:

$$\int_0^{2\pi} e^{-is\rho\cos(\mu-\psi)} \, d\mu = \int_0^{2\pi} e^{-is\rho\cos\mu} \, d\mu = 2\pi J_0(s\rho) \tag{6.26}$$

(J_0 is the Bessel function), so

$$\sigma_{zz}^0(\rho, z) = \frac{p_0}{2\pi} \int_0^\infty \frac{(2s^2 - k_s^2)^2\, e^{-\gamma z} - 4s^2\gamma q\, e^{-qz}}{\Delta(s)}\, s\, J_0(s\rho)\, ds. \tag{6.27}$$

Our basic idea is the following. We can estimate this integral asymptotically for high frequencies. The stationary phase method (see Section 1.4) can be applied to this integral, after the change of variable $s = k_p\tilde{s}$ (the first exponent in (6.27)) or $s = k_p\tilde{s}$ (the second exponent), when we arrive at some exponentially oscillating function with a phase, containing a large frequency parameter. This also appears in the argument of the Bessel function, and an appropriate asymptotic formula should be taken to reduce the integral to the form containing only purely exponential oscillating functions. The principal contribution of the considered integral is given by the stationary point $s = \rho/r = \left[(x-x_0)^2 + (y-y_0)^2\right]^{1/2} / \left[(x-x_0)^2 + (y-y_0)^2 + z^2\right]^{1/2}$, which permits the following asymptotic estimate:

$$\sigma_{zz}^0(\rho, z) \sim -\frac{ip_0 z}{2\pi r^2}\left[k_p\, e^{ik_p r} f_1\left(\frac{\rho}{r}\right) + k_s\, e^{ik_s r} f_2\left(\frac{\rho}{r}\right)\right],$$

$$f_1(s) = \frac{(2s^2 - \delta^2)^2}{(2s^2 - \delta^2)^2 + 4s^2\sqrt{1 - s^2}\sqrt{\delta^2 - s^2}}, \qquad \beta = k_p/k_s = c_s/c_p < 1, \tag{6.28}$$

$$f_2(s) = \frac{4s^2\sqrt{1 - s^2}\sqrt{\beta^2 - s^2}}{(2s^2 - 1)^2 + 4s^2\sqrt{1 - s^2}\sqrt{\beta^2 - s^2}} \qquad \delta = 1/\beta.$$

If now the oscillator generates a uniformly distributed normal load p_0 applied to the free plane boundary surface over the domain S, then Eq. (6.28) allows us to operate with the Schoch method. Indeed, due to the linearity of the problem, the total wave field at an arbitrary point inside elastic half-space can be calculated as a superposition of the contributions from all point elementary forces of the type (6.28). This results in the following asymptotic formula:

$$\sigma_{zz}^0(\rho, z) = -\frac{ip_0 z}{2\pi}\int_0^{2\pi} d\varphi \int_z^{r_1(\varphi)}\left[k_p\, e^{ik_p r} f_1\left(\frac{\sqrt{r^2 - z^2}}{r}\right) + k_s\, e^{ik_s r} f_2\left(\frac{\sqrt{r^2 - z^2}}{r}\right)\right]\frac{dr}{r}. \tag{6.29}$$

In some sense this integral is an alternative representation of the integral given by Eq. (6.12). Unfortunately, in contrast with scalar acoustic case internal integral in (6.29) cannot be calculated analytically, so we would like to estimate it asymptotically. Such an approach is not equivalent to asymptotic approximation of integrals of the type (6.12), since here we have the only large dimensionless parameter $k_p r$ (or $k_s r$), which determines how many times the distance from the current applied point source is greater than the wavelength. By contrast, the asymptotic estimate of the integral (6.12) will encounter the two large dimensionless geometric parameters: the relative (with respect to the wavelength) distance of the observation point from the applied load, and relative (again compared with the wavelength) geometric size of the domain S. With the frequency increasing these parameters compete with each other. Domination of one of them relative to the other generates the far-field (Fraunhofer) approximation. In the opposite case we arrive at the so-called *Fresnel approximation*. Our approach, based upon brilliant ideas of Schoch, will lead us to a uniform high-frequency approximation. Besides, as we will see below, in the scalar acoustic case this approach gives exact formulas for the wave field.

An efficient high-frequency representation can be obtained if we apply to the integral (6.29) a standard integration by parts. If we restrict ourselves by the leading asymptotic term only, then we arrive at the following asymptotic representation:

$$
\sigma_{zz}^0(\rho, z) = -\frac{p_0 z}{2\pi} \int_0^{2\pi} \frac{d\varphi}{r} \left[e^{ik_p r} f_1 \left(\frac{\sqrt{r^2 - z^2}}{r} \right) + e^{ik_s r} f_2 \left(\frac{\sqrt{r^2 - z^2}}{r} \right) \right]_{z=r}^{r_1(\varphi)}
$$

$$
= p_0 \left\{ e^{ik_p z} - \frac{z}{2\pi} \int_0^{2\pi} \left[e^{ik_p r} f_1 \left(\frac{\sqrt{r^2 - z^2}}{r} \right) + e^{ik_s r} f_2 \left(\frac{\sqrt{r^2 - z^2}}{r} \right) \right] \frac{d\varphi}{r_1(\varphi)} \right\}.
$$

(6.30)

The problem of efficient calculation of the high-frequency wave field in elastic half-space is thus reduced, by the Schoch method to a one-dimensional integral over finite integral. In some simple cases this again can be calculated analytically. For instance, for round disk oscillator of radius a we obtain from (6.30)

$$
\sigma_{zz} = p_0 \left\{ e^{ik_p z} - \frac{z}{\sqrt{a^2 + z^2}} \left[e^{ik_p \sqrt{a^2 + z^2}} f_1 \left(\frac{a}{\sqrt{a^2 + z^2}} \right) + e^{ik_s \sqrt{a^2 + z^2}} f_2 \left(\frac{a}{\sqrt{a^2 + z^2}} \right) \right] \right\}.
$$

(6.31)

Let us note that in the far zone, as it follows from the structure of the functions f_1 and f_2, the last expressions (6.30) and (6.31) degenerate into the corresponding formulas for acoustic half-space. In the near-field they provide a high accuracy of the order of only a few percent, as was tested by numerous numerical simulations.

Helpful remarks

Some examples on the application of the method proposed in this section are quoted in Sumbatyan (1988). The accuracy of our approach is around 1% of relative error, and this works in real time when implemented on PC. From our experience, direct numerical treatment of integrals of the type (6.12) requires approximately 10^2 to 10^3 times more computational time than the approach proposed here.

A survey of existing methods of elastic wave fields calculation can be found in Achenbach (1973).

6.3. Asymptotic Nature of the Geometrical Diffraction Theory

As in Kirchhoff's physical diffraction theory you can find various formulations of the geometrical diffraction theory treated by different authors. A good survey can be found in Hönl et al. (1961, more physical treatment) and in Babich and Buldyrev (1989, more mathematical treatment). In two words, Kirchhoff's theory gives a leading high-frequency asymptotic term of the solution on the boundary of the obstacle. Analogously, the geometrical diffraction theory gives a leading high-frequency asymptotic term of the solution at an arbitrary point of observation. The foundations of this theory were established by Keller (1962), but in this section we give an alternative derivation of some of his classical results, starting (more naturally) from Kirchhoff's solution.

To be more specific, let us consider a point source x_0 of a (2D) acoustic wave which is incident on an acoustically hard obstacle D with the boundary line l. The Kirchhoff–Helmholtz integral formula gives an exact representation for the solution at arbitrary observation point x (see Section 2.2)

$$
p^{sc}(x) = \int_l p(y) \frac{\partial \Phi(kr)}{\partial n_y} \, dl_y, \quad \Phi(kr) = \frac{i}{4} H_0^{(1)}(kr), \quad r = |x - y|, \quad x \in \mathbb{R}^2 \setminus D, \quad (6.32)
$$

where $p^{sc}(x)$ is a scattered wave field at the point x, and $p(y)$, $y \in l$, is a full wave pressure on the boundary contour l.

According to Kirchhoff's physical diffraction theory, the leading asymptotic term for $p(y)$, $y \in l$ is given as follows (see Section 2.6):

$$p(y) = \begin{cases} 2p^{inc}(y), & y \in l^+, \\ 0, & y \in l^-, \end{cases} \qquad k \to \infty. \qquad (6.33)$$

Since the incident wave, given by the same (point) Green's function expressed again in terms of the Hankel function, has also an asymptotic representation, we may assume that

$$p^{inc}(y) = \frac{e^{ikr_0}}{\sqrt{r_0}}, \qquad k \to \infty, \qquad r_0 = |x_0 - y|. \qquad (6.34)$$

Now, taking into account that

$$2\frac{\partial \Phi(kr)}{\partial n_y} = e^{-\pi i/4}\sqrt{\frac{k}{2\pi}} \frac{e^{ikr}}{\sqrt{r}} \cos(n_y, r), \qquad k \to \infty, \qquad (6.35)$$

and collecting together Eqs. (6.32)–(6.35), we can see that at $k \to \infty$:

$$p^{sc}(x) = e^{-\pi i/4}\sqrt{\frac{k}{2\pi}} \int_{l^+} \frac{\cos(n_y, r)}{\sqrt{rr_0}} e^{ikS(y)} \, dl_y, \quad S(y) = r_0 + r = |x_0 - y| + |x - y|. \quad (6.36)$$

THEOREM 1. *Any point $y_0 \in l^+$ with $\widehat{n_y\,r} = \widehat{n_y\,r_0} = \theta$ supplies a stationary point for the phase function $S(y)$.*

Proof. Let such a point $y_0 \in l^+$ exist, and we consider the current point $y \in l^+$ from a small neighborhood of the point y_0. Then we will prove that y_0 is the only stationary point among all y from this neighborhood. To this end, we consider the distance $|x_0 - y|$ as a side of the triangle formed by the three sides $L_0 = |x_0 - y_0|$, $r = |x_0 - y|$, $\Delta s = |y - y_0|$, and use a cosine theorem:

$$|x_0 - y|^2 = L_0^2 + (\Delta s)^2 - 2L_0(\Delta s)\cos\alpha, \qquad \alpha = \widehat{L_0\,\Delta s}. \qquad (6.37)$$

Let us recall some classical results from differential geometry (see, for example, Pogorelov, 1978).

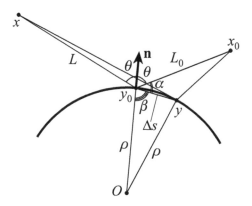

Figure 6.2. Single ray reflection from obstacle with a perfect boundary

For the pair of the points y and y_0 (see Fig. 6.2) situated on the smooth curve l^+ close to each other, there is an osculating circle whose radius ρ determines the curvature of the curve at the point y_0. Then geometry of the triangle Oyy_0 determines the angle β as follows: $\beta = \pi/2 - \arcsin[(\Delta s/2)/\rho] = \pi/2 - \Delta s/(2\rho) + O[(\Delta s)^3]$. As follows from Fig. 6.2, $\alpha = \pi - \theta - \beta = \pi/2 - \theta + \Delta s/(2\rho)$, so

$$\cos\alpha = \sin\left(\theta - \frac{\Delta s}{2\rho} + O[(\Delta s)^3]\right) = \sin\theta - \frac{\Delta s}{2\rho}\cos\theta + O((\Delta s)^2). \tag{6.38}$$

The last relation allows us to express the distance $|x_0 - y|^2$ in Eq. (6.37) in terms of (Δs):

$$|x_0 - y|^2 = L_0^2 + (\Delta s)^2 + 2L_0(\Delta s)\left(\frac{\Delta s}{2\rho}\cos\theta - \sin\theta\right) + O((\Delta s)^3). \tag{6.39}$$

By extracting the square root from both sides (6.39), we arrive at the following relation:

$$|x_0 - y| = L_0 - \Delta s\sin\theta + \frac{(\Delta s)^2}{2}\left(\frac{\cos^2\theta}{L_0} + \frac{\cos\theta}{\rho}\right) + O((\Delta s)^3). \tag{6.40}$$

By analogy, we can find that a similar formula holds for the second term in the phase function (6.36):

$$|x - y| = L + \Delta s\sin\theta + \frac{(\Delta s)^2}{2}\left(\frac{\cos^2\theta}{L} + \frac{\cos\theta}{\rho}\right) + O((\Delta s)^3), \tag{6.41}$$

so the phase function is

$$S(y) = |x_0 - y| + |x - y| = L_0 + L + \frac{(\Delta s)^2}{2}\cos^2\theta\left(\frac{1}{L_0} + \frac{1}{L} + \frac{2}{\rho\cos\theta}\right) + O((\Delta s)^3). \tag{6.42}$$

Note that linear term (Δs) is absent here, so the point of specular reflection is indeed the stationary point of the phase function $S(y)$ (see Section 1.4). The theorem is proved.

THEOREM 2. *If there is the only point $y_0 \in l^+$ of the specular reflection on the contour l^+ and expression in the parentheses in Eq. (6.42) is not equal to zero, then the leading asymptotic term of the scattered wave $p^{sc}(x)$ in Eq. (6.36) is*

$$p^{sc}(x) = \frac{\exp\{i[k(L_0 + L) + (\pi/4)(\delta - 1)]\}}{\sqrt{|L_0 + L + 2L_0 L/(\rho\cos\theta)|}}, \quad \delta = \mathrm{sign}[L_0 + L + 2L_0 L/(\rho\cos\theta)]. \tag{6.43}$$

The proof directly follows from the stationary phase method (see Section 1.4), since at the stationary point we have $\widehat{n_y r} = \widehat{n_y r_0} = \theta$, $r = L$, $r_0 = L_0$, $S(y) = r_0 + r = L_0 + L$.

The solution expressed by formula (6.43) is called a *geometrical diffraction theory solution*.

It is interesting to note that the obtained results implicitly contain solutions not only for a point-source incident wave but also for a plane incident wave. This case can be obtained from our solution (6.43) by putting $L_0 \gg L$, so the scattered wave pressure is

$$p^{sc}(x) = \frac{\exp\{i[k(L_0 + L) + (\pi/4)(\delta - 1)]\}}{\sqrt{L_0}\sqrt{|1 + 2L/(\rho\cos\theta)|}}, \quad \delta = \mathrm{sign}[1 + 2L/(\rho\cos\theta)]. \tag{6.44}$$

If, in addition to the listed assumptions, we consider also a far-field scattered wave field with $L_0, L \gg \rho$, then Eq. (6.43) reads

$$p^{sc}(x) = \frac{\exp\{i[k(L_0 + L) + (\pi/4)(\delta - 1)]\}}{\sqrt{2L_0 L}}\sqrt{\rho\cos\theta}, \quad \delta = \mathrm{sign}(\rho). \tag{6.45}$$

We thus can conclude that the amplitude in the far-field reflected plane acoustic wave is proportional to the square root of the curvature radius.

Helpful remarks

$1°$. The proved Theorem 1 asserts that the first reflection law of the linear wave theory, which states that the reflection angle is equal to the angle of incidence, well known for plane reflectors is also valid for arbitrary smooth curvilinear boundary.

$2°$. If there are several stationary points, i.e., several points of specular reflection on the reflecting boundary, then the total contribution to the scattered wave field is given as the sum of individual contributions of these (stationary) points.

$3°$. The leading asymptotic term for the scattered wave field derived in the present section is valid only if the stationary point is regular, i.e., if expression in the parentheses in Eq. (6.42) does not vanish and so the term with the small length increment of the second order $(\Delta s)^2$ is not trivial. Otherwise, when the stationary point is degenerated and this factor is zero, expression in the denominator vanishes and more detailed analysis shows that in such cases the leading asymptotic term is of a higher order with $k \to \infty$.

$4°$. It perhaps seems to be a little strange that the denominator of Eq. (6.43) may vanish in some cases. Note that the curvature radius ρ at the reflection point should be taken with its actual sign—positive for convex obstacles and negative for concave ones. In the case of nonconvex boundary with a negative curvature the considered expression may turn out zero. A bright example is given by illumination of a circle with radius a from a point source located in its center, when the observation point x and source point x_0 are taken at the same place. In this case we have $L = L_0 = a$, $\rho = -a$, $\theta = 0$, $\cos \theta = 1$, and hence $L_0 + L + 2L_0 L/(\rho \cos \theta) = 0$.

$5°$. Results of this section can explain why Kirchhoff's physical diffraction theory gives correct prediction only for convex obstacles. Indeed, from the point of view of geometrical diffraction theory the physical diffraction theory is related to a direct incidence of the incoming falling wave. If the boundary of the obstacle is convex, then this can guarantee that the only wave falling to a small neighborhood of the point $y_0 \in l^+$ is a (directly) incident one. If the boundary is not convex, then in addition to a directly falling incident wave there may exist also waves re-reflected from other boundary points and arriving, after these re-reflections, to the same neighborhood of y_0.

$6°$. Formula (6.43) admits a simple testing in the case of plane reflecting boundary. Heuristically, if a wave radiated from a point x_0 and represented by formula (6.34) propagates in the space and then encounters any plane reflecting acoustically hard surface, then simple *mirror* ideas allow us to suppose that the amplitude of this wave behaves as in the case where there is no reflector and the amplitude of re-reflected wave depends on the total flight of the acoustic ray. We thus can predict from heuristic considerations that the solution must in this case be of the following form (compare with the form of the radiated pressure (3)):

$$p(x) = \frac{e^{ik(L_0+L)}}{\sqrt{L_0 + L}}. \tag{6.46}$$

This result directly follows from our formula (6.43), since in the case of plane reflector $\rho = \infty$, $\delta = \mathrm{sign}(L_0 + L) = 1$.

$7°$. It should be noted that the results obtained in this section, in principle, can be obtained within the framework of Keller's approach to the geometrical diffraction theory (see Keller, 1962; Borovikov and Kinber, 1994; McNamara et al., 1990). However, because of the number of re-reflections, it is too hard to trace the acoustic beams history. We believe that our approach leads to the final results more directly and more clearly.

6.4. High-Frequency Diffraction with Re-Reflections

Let us consider repeated (double) reflection from acoustically hard boundary (see Fig. 6.3).

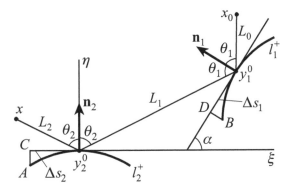

Figure 6.3. Repeated ray reflection from obstacle with a perfect boundary

A scalar acoustic wave is emitted from the point source x_0 and then received at the point x after re-reflections from arcs l_1^+ and l_2^+ of obstacle's boundary. According to Kirchhoff–Helmholtz integral formula, the received scattered pressure is given by

$$p^{sc}(x) = \int_{l_2^+} p(y_2)\frac{\partial\Phi(kr)}{\partial n_2}\, dl_2, \qquad r = |x - y_2|, \tag{6.47}$$

where $p(y_2)$ is the value of acoustic pressure on the arc l_2^+. For high frequencies, according to physical diffraction theory, $p(y_2) = 2p_2^{inc}(y_2)$, where $p_2^{inc}(y_2)$ is a wave field falling to the points $y_2 \in l_2^+$ of the contour l_2^+ after the first scattering happened on the contour l_1^+. The latter is determined again by the Kirchhoff–Helmholtz integral formula as follows:

$$p_2^{inc}(y_2) = \int_{l_1^+} p(y_1)\frac{\partial\Phi(kr_1)}{\partial n_1}\, dl_1, \qquad r_1 = |y_2 - y_1|, \qquad y_2 \in l_2^+, \tag{6.48}$$

where $p(y_1)$ is the value of acoustic pressure on the contour l_1^+. According to the physical diffraction theory, $p(y_1) = 2p^{inc}(y_1)$, where $p^{inc}(y_1)$ is a wave field incident on the contour l_1^+. Taking into account that $p^{inc}(y_1)$ is expressed for high frequencies again by the formula

$$p^{inc}(y) = \frac{e^{ikr_0}}{\sqrt{r_0}}, \qquad k \to \infty, \qquad r_0 = |x_0 - y_1|, \tag{6.49}$$

and collecting together all relations (6.47)–(6.49), we can write out the following basic representation valid at $k \to \infty$:

$$p^{sc}(x) = -\frac{ik}{2\pi}\int_{l_1^+}\int_{l_2^+} \frac{\cos(n_1, r_1)\cos(n_2, r)}{\sqrt{r_0 r_1 r}}\, e^{ikS(y_1,y_2)}\, dl_1\, dl_2, \tag{6.50}$$

$$S(y_1, y_2) = r_0 + r_1 + r = |x_0 - y_1| + |y_2 - y_1| + |x - y_2|.$$

THEOREM. *Any pair of points $y_1^0 \in l_1^+$ and $y_2^0 \in l_2^+$ with $\widehat{n_1 r_1} = \widehat{n_1 r_0} = \theta_1$ and $\widehat{n_2 r} = \widehat{n_2 r_1} = \theta_2$ supplies a stationary value for the phase function $S(y_1, y_2)$.*

Proof. Let such a pair exist and for these values of $y_1 = y_1^0$, $y_1 = y_2^0$ the quantities participating in the phase function take the following values: $r_0 = L_0$, $r_1 = L_1$, $r = L_2$, $S(y_1, y_2) = L_0 + L_1 + L_2$. Then we consider the current point $A = y_1 \in l_1^+$ from a small neighborhood of the point y_1^0 and analogous point $B = y_2 \in l_2^+$ from a small neighborhood of the point y_2^0. Similarly to the derivation procedure of the previous section we can easily obtain

$$|x_0 - y_1| = L_0 + \Delta s_1 \sin \theta_1 + \frac{(\Delta s_1)^2}{2} \left(\frac{\cos^2 \theta_1}{L_0} + \frac{\cos \theta_1}{\rho_1} \right) + O((\Delta s_1)^3),$$

$$|x - y_2| = L_2 - \Delta s_2 \sin \theta_2 + \frac{(\Delta s_2)^2}{2} \left(\frac{\cos^2 \theta_2}{L_2} + \frac{\cos \theta_2}{\rho_2} \right) + O((\Delta s_2)^3).$$

(6.51)

The key problem is to evaluate the third distance $|y_1 - y_2|$. To this end, we introduce the Cartesian coordinate system (ξ, η) with the origin chosen at the point y_2^0. As follows from Fig. 6.3 and the consideration in Section 6.3,

$$AC = (\Delta s_2) \sin(\pi/2 - \beta)[1 + O(\Delta s_2)]$$

$$= (\Delta s_2) \sin\{\arcsin[\Delta s_2/(2\rho_2)]\} [1 + O(\Delta s_2)] = \frac{(\Delta s_2)^2}{2\rho_2} + O((\Delta s_2)^3),$$

(6.52)

$$BD = \frac{(\Delta s_1)^2}{2\rho_1} + O((\Delta s_1)^3)$$

(cf. Fig. 6.2).

Then in the introduced Cartesian coordinate system (ξ, η) coordinates of the considered points can be expressed as follows:

$$\xi_A = -\Delta s_2, \quad \eta_A = -AC = -\frac{(\Delta s_2)^2}{2\rho_2} + O((\Delta s_2)^3), \quad \xi_1^0 = L_1 \sin \theta_2, \quad \eta_1^0 = L_1 \cos \theta_2,$$

$$\xi_B = \xi_1^0 - \Delta s_1 \cos \alpha + BD \sin \alpha = \xi_1^0 - \Delta s_1 \cos \alpha + \frac{(\Delta s_1)^2}{2\rho_1} \sin \alpha + O((\Delta s_1)^3),$$

(6.53)

$$\eta_B = \eta_1^0 - \Delta s_1 \sin \alpha - BD \cos \alpha = \eta_1^0 - \Delta s_1 \sin \alpha - \frac{(\Delta s_1)^2}{2\rho_1} \cos \alpha + O((\Delta s_1)^3).$$

Therefore,

$$|y_1 - y_2| = \left[(\xi_B - \xi_A)^2 + (\eta_B - \eta_A)^2 \right]^{1/2} = \left\{ \left[\xi_1^0 - \Delta s_1 \cos \alpha + \frac{(\Delta s_1)^2}{2\rho_1} \sin \alpha + \Delta s_2 \right]^2 \right.$$

$$\left. + \left[\eta_1^0 - \Delta s_1 \sin \alpha - \frac{(\Delta s_1)^2}{2\rho_1} \cos \alpha + \frac{(\Delta s_2)^2}{2\rho_2} \right]^2 \right\}^{1/2} + O((\Delta s)^3),$$

$$(\Delta s)^2 = \min \left\{ (\Delta s_1)^2, (\Delta s_2)^2, (\Delta s_1)(\Delta s_2) \right\}.$$

(6.54)

Raising to the square in the braces here and keeping all powers up to the second one only, we can obtain

$$|y_1 - y_2| = \left[L_1^2 - 2\Delta s_1 \left(\xi_1^0 \cos \alpha + \eta_1^0 \sin \alpha \right) + (\Delta s_1)^2 + (\Delta s_2)^2 - 2(\Delta s_1)(\Delta s_2) \cos \alpha \right.$$

(6.55)

$$\left. + \frac{(\Delta s_1)^2}{\rho_1} \left(\xi_1^0 \sin \alpha - \eta_1^0 \cos \alpha \right) + 2\xi_1^0 \Delta s_2 + \eta_1^0 \frac{(\Delta s_2)^2}{\rho_2} \right]^{1/2} + O((\Delta s)^3).$$

Note that $\alpha = \pi - \theta_1 - \theta_2$; hence

$$\xi_1^0 \cos \alpha + \eta_1^0 \sin \alpha = L_1 \sin(\alpha + \theta_2) = L_1 \sin \theta_1,$$

$$\xi_1^0 \sin \alpha - \eta_1^0 \cos \alpha = -L_1 \cos(\alpha + \theta_2) = L_1 \cos \theta_1,$$

(6.56)

$$\cos \alpha = -\cos(\theta_1 + \theta_2).$$

As a result, expression (6.55) can be written as follows:

$$
\begin{aligned}
|y_1 - y_2| = |AB| = L_1 &\left[1 - 2\frac{\Delta s_1}{L_1} \sin\theta_1 + 2\frac{\Delta s_2}{L_1} \sin\theta_2 + \frac{(\Delta s_1)^2}{L_1^2} \left(1 + \frac{L_1}{\rho_1} \cos\theta_1 \right) \right. \\
&\left. + \frac{(\Delta s_2)^2}{L_1^2} \left(1 + \frac{L_1}{\rho_2} \cos\theta_2 \right) + 2\frac{(\Delta s_1)(\Delta s_2)}{L_1^2} \cos(\theta_1 + \theta_2) \right]^{1/2} + O\left((\Delta s)^3\right).
\end{aligned}
\tag{6.57}
$$

Now, Taylor expansion of this root square expression in powers of the small quantities (Δs_1) and (Δs_2) leads to the following representation:

$$
\begin{aligned}
\frac{|y_1 - y_2|}{L_1} = 1 &- \frac{\Delta s_1}{L_1} \sin\theta_1 + \frac{\Delta s_2}{L_1} \sin\theta_2 + \frac{(\Delta s_1)^2}{2L_1^2} \left(1 + \frac{L_1}{\rho_1} \cos\theta_1 - \sin^2\theta_1 \right) \\
&+ \frac{(\Delta s_2)^2}{2L_1^2} \left(1 + \frac{L_1}{\rho_2} \cos\theta_2 - \sin^2\theta_2 \right) + \frac{(\Delta s_1)(\Delta s_2)}{L_1^2} \left[\cos(\theta_1 + \theta_2) + \sin\theta_1 \sin\theta_2 \right] \\
&+ O\left((\Delta s)^3\right).
\end{aligned}
\tag{6.58}
$$

Further, taking together all results (6.51) and (6.58) we can conclude that the phase function $S(y_1, y_2)$ in Eq. (6.50) is

$$
\begin{aligned}
S(y_1, y_2) = L_0 + L_1 + L_2 &+ \frac{(\Delta s_1)^2}{2} \cos^2\theta_1 \left(\frac{1}{L_0} + \frac{1}{L_1} + \frac{2}{\rho_1 \cos\theta_1} \right) \\
&+ \frac{(\Delta s_2)^2}{2} \cos^2\theta_2 \left(\frac{1}{L_2} + \frac{1}{L_1} + \frac{2}{\rho_2 \cos\theta_2} \right) + \frac{(\Delta s_1)(\Delta s_2)}{L_1} \cos\theta_1 \cos\theta_2 + O\left((\Delta s)^3\right).
\end{aligned}
\tag{6.59}
$$

This expression is free of linear terms (Δs_1) and (Δs_2). This proves the theorem.

From results of this theorem, by applying the two-dimensional stationary phase method (see Section 1.4), we can derive the leading asymptotic term for the amplitude of the doubly re-reflected acoustic ray. This is based on the following expressions for the second-order derivatives:

$$
\frac{\partial^2 S}{\partial(\Delta s_1)^2} = \cos^2\theta_1 \left(\frac{1}{L_0} + \frac{1}{L_1} + \frac{2}{\rho_1 \cos\theta_1} \right), \qquad \frac{\partial^2 S}{\partial(\Delta s_1)\partial(\Delta s_2)} = \frac{\cos\theta_1 \cos\theta_2}{L_1},
$$

$$
\frac{\partial^2 S}{\partial(\Delta s_2)^2} = \cos^2\theta_2 \left(\frac{1}{L_2} + \frac{1}{L_1} + \frac{2}{\rho_2 \cos\theta_2} \right),
\tag{6.60}
$$

and, according to Section 1.4, has the following form:

$$
p(x) = \frac{\exp\{i[k(L_0 + L_1 + L_2) + (\pi/4)(\delta_2 - 2)]\}}{\sqrt{L_0 L_1 L_2}\ \sqrt{|\det(D_2)|}}, \qquad k \to \infty.
\tag{6.61}
$$

Here D_2 is the Hessian (a certain factor is omitted in front of all its elements)

$$
D_2 = \begin{pmatrix} \dfrac{1}{L_0} + \dfrac{1}{L_1} + \dfrac{2}{\rho_1 \cos\theta_1} & \dfrac{1}{L_1} \\[2ex] \dfrac{1}{L_1} & \dfrac{1}{L_2} + \dfrac{1}{L_1} + \dfrac{2}{\rho_2 \cos\theta_2} \end{pmatrix},
\tag{6.62}
$$

and $\delta_2 = \operatorname{sign}(D_2) = \nu_+(D_2) - \nu_-(D_2)$ is a difference between the number of positive and negative eigenvalues of the matrix D_2.

Let us consider the generic case when the number of reflections is arbitrary being equal to N. Let the acoustic wave be emitted from the point x_0, then re-reflected N times, from the points y_m^0, $y_m^0 \in l_m^+$, $m = 1, \ldots, N$, sequentially, and finally received at the receiving point x. All these reflecting points are situated so that the incident and reflected angles at the reflection point y_m^0 are equal to each other being θ_m. Let us denote the distance between sequential reflecting points as $L_m = |y_{m+1}^0 - y_m^0|$, $m = 1, \ldots, N-1$. Besides, we denote the distance between the source x_0 and the first point of reflection y_1^0 as $L_0 = |y_1^0 - x_0|$, and the distance between the receiver x and the last reflection point y_N^0 as $L_N = |x - y_N^0|$.

It follows from the previous consideration that the leading asymptotic term for the amplitude of N-times re-reflected acoustic wave can be constructed by a high-frequency estimate of the multiple integral

$$
p(x) \sim e^{-(\pi/4)Ni} \left(\frac{k}{2\pi} \right)^{N/2} \frac{\prod\limits_{m=1}^{N} \cos\theta_m}{\sqrt{\prod\limits_{m=0}^{N} L_m}} \int_{l_1^+} \cdots \int_{l_N^+} e^{ikS} \, dl_1 \ldots dl_N,
$$

(6.63)

$$
S = S(y_1, \ldots, y_N) = |y_1 - x_0| + \sum_{m=1}^{N-1} |y_{m+1} - y_m| + |x - y_N|.
$$

By analogy with the above case $N = 2$, we can conclude that

$$
\frac{\partial^2 S}{\partial(\Delta s_m)^2} = \cos^2\theta_m \left(\frac{1}{L_{m-1}} + \frac{1}{L_m} + \frac{2}{\rho_m \cos\theta_m} \right), \qquad m = 1, \ldots, N,
$$

$$
\frac{\partial^2 S}{\partial(\Delta s_m)\partial(\Delta s_{m+1})} = \frac{\cos\theta_m \cos\theta_{m+1}}{L_m}, \qquad m = 1, \ldots, N-1,
$$

(6.64)

$$
\frac{\partial^2 S}{\partial(\Delta s_m)\partial(\Delta s_n)} = 0, \qquad n \neq m, \; n \neq m+1.
$$

Hence, the multiple stationary phase method (Section 1.4) leads to the explicit formula for the (geometric optical) leading asymptotic term:

$$
p(x) = \frac{\exp\left\{ i\left[k \sum\limits_{m=0}^{N} L_m + (\pi/4)(\delta_N - N) \right] \right\}}{\sqrt{\left(\prod\limits_{m=0}^{N} L_m \right) |\det(D_N)|}}, \qquad k \to \infty,
$$

(6.65)

where

$$
D_N = (d_{nm}) \quad (n, m = 1, \ldots, N),
$$

$$
d_{mm} = \frac{1}{L_{m-1}} + \frac{1}{L_m} + \frac{2}{\rho_m \cos\theta_m} \quad (m = 1, \ldots, N),
$$

$$
d_{m+1,m} = d_{m,m+1} = \frac{1}{L_m} \quad (m = 1, \ldots, N-1),
$$

(6.66)

$$
d_{nm} = 0 \quad (n \neq \{m, m+1, m-1\}), \qquad \delta_N = \text{sign}(D_N).
$$

Helpful remarks

1°. As in the case of single reflection, singular cases can occur in (6.66), too. This happens every time when the denominator is equal to zero, $\det(D_N) = 0$. In such cases the leading asymptotic term has a different qualitative behavior, and its dependence on the frequency parameter k is absolutely different.

2°. Recall that in the present section we studied only the case of acoustically hard boundary. It can easily be proved that in the case when all reflecting lines are acoustically soft, the generic formula (6.65) should be multiplied by $(-1)^N$. Heuristically, this directly follows from the evident observation that at each reflection the phase changes its value by 180° in jump.

3°. It is important to stress the case when $\delta_N = \text{sign}(D_N)$ can be calculated in a simple way. Let us assume that all $\rho_m > 0$, $m = 1, \ldots, N$, i.e., all reflecting curves are convex. In this case the matrix D_N can be represented as the sum of two positive definite matrices (for definition, see Horn and Johnson, 1986; and Section 4.1, which main results can be applied also to the case of finite-dimensional space):

$$D_N = A_N + B_N, \qquad A_N = (a_{nm}), \quad B_N = (b_{nm}),$$

$$a_{mm} = \frac{1}{L_{m-1}} + \frac{1}{L_m} \quad (m = 1, \ldots, N),$$

$$a_{m+1,m} = a_{m,m+1} = \frac{1}{L_m} \quad (m = 1, \ldots, N-1), \tag{6.67}$$

$$a_{nm} = 0 \quad (n \neq \{m, m+1, m-1\}),$$

$$b_{mm} = \frac{2}{\rho_m \cos\theta_m} \quad (m = 1, \ldots, N), \quad b_{nm} = 0 \quad (n \neq m).$$

Positive definiteness of the matrix B_N is obvious, since it has a diagonal structure with diagonal entries all positive. Let us prove that the matrix A_N is positive definite too.

First of all, let us prove by induction that

$$I_N = \det(A_N) = \frac{\sum\limits_{m=0}^{N} L_m}{\prod\limits_{m=0}^{N} L_m}. \tag{6.68}$$

For $N = 1$ we have $I_1 = \det(A_1) = 1/L_0 + 1/L_1 = (L_0 + L_1)/(L_0 L_1)$, so in this case the correctness of Eq. (6.68) is evident. For $N = 2$ we have

$$I_2 = \det(A_2) = \begin{vmatrix} \dfrac{1}{L_0} + \dfrac{1}{L_1} & \dfrac{1}{L_1} \\ \dfrac{1}{L_1} & \dfrac{1}{L_1} + \dfrac{1}{L_2} \end{vmatrix} = \frac{1}{L_0 L_1} + \frac{1}{L_0 L_2} + \frac{1}{L_1 L_2} = \frac{L_0 + L_1 + L_2}{L_0 L_1 L_2}, \tag{6.69}$$

which coincides again with formula (6.68). For arbitrary $N > 2$ we give a proof by induction.

Let us assume that Eq. (6.68) is valid for all $M < N$, and we consider I_N, $N > 2$. Then, expanding this determinant by elements of the first column, we obtain

$$I_N = \left(\frac{1}{L_{N-1}} + \frac{1}{L_N}\right) I_{N-1} - \frac{1}{L_{N-1}^2} I_{N-2} = \left(\frac{1}{L_{N-1}} + \frac{1}{L_N}\right) \frac{\sum\limits_{m=0}^{N-1} L_m}{\prod\limits_{m=0}^{N-1} L_m} - \frac{1}{L_{N-1}^2} \frac{\sum\limits_{m=0}^{N-2} L_m}{\prod\limits_{m=0}^{N-2} L_m}$$

$$= \frac{(1/L_{N-1} + 1/L_N)\sum\limits_{m=0}^{N-1} L_m - (1/L_{N-1})\sum\limits_{m=0}^{N-2} L_m}{\prod\limits_{m=0}^{N-1} L_m} = \frac{1 + (1/L_N)\sum\limits_{m=0}^{N-1} L_m}{\prod\limits_{m=0}^{N-1} L_m} = \frac{\sum\limits_{m=0}^{N} L_m}{\prod\limits_{m=0}^{N} L_m}. \tag{6.70}$$

Now, when relation (6.68) is strictly proved, we will show that the matrix A_N is positive definite, by using the well known Sylvester criterion (see Horn and Johnson, 1986). It claims that any symmetric matrix is positive definite if all its principal determinants (i.e., the ones arranged by cancelling equal numbers in the last columns and rows) are positive. For the matrix A_N under consideration its principal submatrices coincide with the same A_M of smaller dimensions: $M = 1, \ldots, N$. It now becomes clear that, since all principal determinants of the matrix A_N are positive, this matrix is positive definite by virtue of the Sylvester criterion.

Further, since the sum of positive definite matrices is always again positive definite, we can conclude that $D_N = A_N + B_N$ is positive definite, so $\nu_-(D_N) = 0$, $\nu_+(D_N) + N$ and therefore in Eq. (6.65) $\delta_N = \text{sign}(D_N) = \nu_+(D_N) - \nu_-(D_N) = N$.

4°. It should be noted that the matrix A_N is related to the problem with all $\rho_m = \infty$, $m = 1, \ldots, N$, i.e., to the case when each reflector is locally a plane *mirror*. With such a system of plane reflectors, by substituting equality (6.68) into Eq. (6.65) with $D_N = A_N$, $\delta_N = N$, we arrive at the following expression for the amplitude of N times re-reflected acoustic wave:

$$p(x) = \exp\left(ik \sum_{m=0}^{N} L_m\right) \bigg/ \sqrt{\sum_{m=0}^{N} L_m}, \qquad (6.71)$$

which is in complete agreement with clear *optical* ideas that presence of plane reflectors do not modify the value of the incident amplitude.

6.5. Application: Examples of High-Frequency Multiple Diffraction

As an example of the application of the theory let us consider symmetric back far-field reflection from a couple of equal acoustically soft circles of radius R with a distance between their centers of $3R$ (see Fig. 6.4). There are two rays of the usual reflection with $N = 1$ at point A and B, and two rays with $N = 2$ ($CDEF$ and backwards). Due to $\rho_n = R > 0$, $\sigma_N = N$, $N = 1, 2$, the full reflection coefficient predicted by the ray theory is

$$p = -\sqrt{\pi k R} + \sqrt{2\pi k} - \frac{\exp[(5 - 2\sqrt{2})kRi]}{\sqrt{L_1\left[\left(1/L_1 + 2\sqrt{2}/R\right)^2 - 1/L_1^2\right]}}, \qquad L_1 = |DE| = \left(3 - \sqrt{2}\right)R.$$

$$(6.72)$$

Here, and in all the examples considered below, the reflected pressure is taken with such a scale that for back scattering by a disk of radius R the reflection coefficient is equal to $\sqrt{\pi k R}/2$. Figure 6.4 demonstrates the dependence of the reflected pressure upon the parameter R/λ ($\lambda = 2\pi/k$ is the wavelength). The exact solution was calculated using the boundary element technique. For all further examples the obstacle boundary was chosen as acoustically soft because of the convenience for exact calculations.

Symmetric reflection from a semicircle.

The problem of far-field symmetric back reflection from the concave side of a semicircle is very interesting, important, and not trivial, so it is considered separately. Figure 6.5 shows that for every $N > 1$ there are two rays (such as $ABCDE$ and backward), for which $L_0 = L_N = L/2 + \infty$, $L_n = L = 2R \cos\theta_N$, $n = 1, \ldots, N - 1$, $\rho_n = -R$, $\theta_N = \frac{1}{2}\pi(N-1)/N$. For $N = 1$, only the ray of direct *mirror* reflection exists. Hence in formula (6.65), $\det(D_N) = L^{-N} \det(G_N)$, with the entries of the matrix G_N being: $g_{11} = g_{NN} = -3$, $g_{nn} = -2$, $n = 2, \ldots, N - 1$, $g_{n+1,n} = g_{n,n+1} = 1$, $n = 1, \ldots, N - 1$, with the others $g_{nm} = 0$. It can be proved by induction that $\det(G_N) = (-1)^N 4N$. The most difficult task is to determine

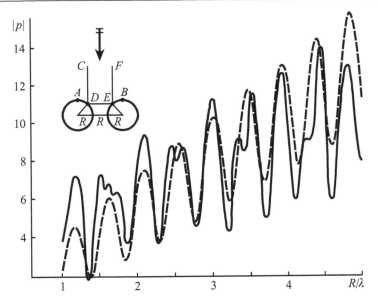

Figure 6.4. Reflection from a couple of acoustically soft disks: solid line represents the exact solution and dashed line represents the ray solution

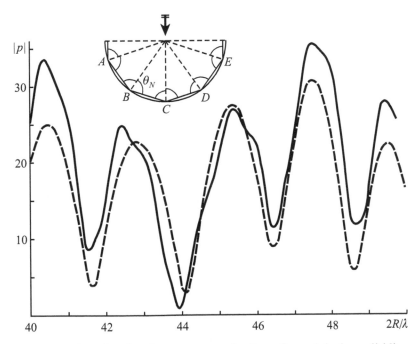

Figure 6.5. Symmetric reflection from an acoustically soft semicircle: solid line represents the exact solution and dashed line represents the ray solution

the value of δ_N. For this purpose localization of eigenvalues for the matrix G_N can be achieved using the well-known Geršgorin disks. It is shown that the full set of eigenvalues lies in the complex plane inside of a union of the disks

$$|z - g_{nn}| < \sum_{m \neq n} |g_n m|, \quad 1 \leq n \leq N. \tag{6.73}$$

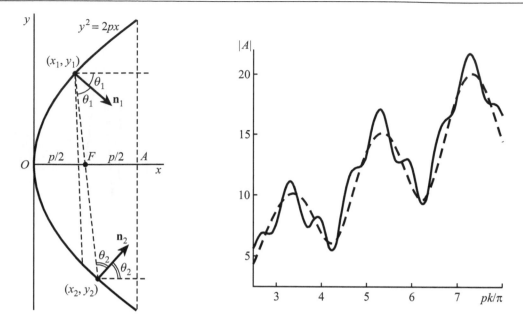

Figure 6.6. Symmetric reflection from an acoustically soft parabolic obstacle

In the problem considered this set coincides with a couple of disks

$$|z + 3| \le 1 \quad \text{and} \quad |z + 2| \le 2,$$ (6.74)

and their union is located in the left complex half-plane. Hence, every real eigenvalue of the Hermitian matrix G_N (as well as D_N) is negative. Thus $\delta_N = v_+ - v_- = -N$. Now it is easy to write an explicit result for the reflection coefficient (for a soft obstacle):

$$p = \sqrt{\frac{\pi k}{8}} \sum_{N=1}^{\infty} i^N \in N \sqrt{\frac{L_N}{N}} \, e^{ikNL_N}, \quad L_N = 2R \sin(\pi/2N), \quad \epsilon_1 = 1, \quad \epsilon_N = 2 \text{ if } N \ge 2.$$ (6.75)

The convergence of the series (6.75) is very slow; this is confirmed by a numerical investigation. To obtain the first three real digits in p, a few hundred terms should be taken in (6.75). Figure 6.5 shows a comparison between the exact solution, and that the semicircle diameter increases from 40λ to 50λ.

The case of a nonisolated stationary point.

The previous consideration was based on the assumption that, when evaluating the integral (6.63), the stationary point of its phase is isolated. Nevertheless, for various important problems this assumption appears to be violated. The clearest example of such a type of problem is symmetric reflection from a parabola (see Fig. 6.6).

It is well known that every ray parallel to its axis is reflected backwards passing through its focus. Hence it might be expected that the main term of the asymptotic expansion for the reflected pressure would contain higher powers of k, and the general result (6.65) is violated here. Ray consideration of one-time reflection by the point O is trivial. Reflections with $N \ge 3$ are absent. Consider the case $N = 2$. For more concreteness let the parabola have finite size: $y^2 = 2px$, $0 \le x \le p$. Then double reflection is possible only by two equal finite parts l of upper and lower halves of the parabola, with both arcs l being separated by a finite distance from the origin O. Thus double re-reflection from the upper to the lower arc l is given by the following expression for the reflection, see (6.50):

$$A = -i \frac{k}{2\pi} \int_l \int_l e^{ik\varphi} \frac{\cos \theta_1 \cos \theta_2}{\sqrt{L_1}} \, ds_1 \, ds_2,$$ (6.76)

where the phase φ is counted from the point A and is equal to $\varphi = (p - x_1) + (p - x_2) + L_1$. Taking into account that $ds_1 = dy_1/\cos\theta_1$, $ds_2 = dy_2/\cos\theta_2$, (6.76) can be rewritten as

$$A = -\frac{ik}{2\pi}\sqrt{2p}\int_l\int_l e^{ik\varphi}\frac{dy_1\,dy_2}{\sqrt{y_1^2 + y_2^2 + 2p^2}}, \qquad (6.77)$$

because the function L_1 has a slow derivative under the square root, and it is defined by a segment crossing the focus F, which leads to: $L_1 = (x_1 + p/2) + (x_2 + p/2) = p + (y_1^2 + y_2^2)/2p$. Here, the main property of the parabolic curve was used which states that the distance between its general point (x, y) and its focus F is equal to the distance between this point and the directness, i.e., equal to $(x + p/2)$. The exact value for phase φ is given by

$$\varphi = \left(p - \frac{y_1^2}{2p}\right) + \left(p - \frac{y_2^2}{2p}\right) + \sqrt{\left(\frac{y_1^2 - y_2^2}{2p}\right) + (y_1 - y_2)^2}. \qquad (6.78)$$

Thus the reflected pressure coefficient can be written as

$$A = -\frac{ik}{\pi}\sqrt{\frac{2}{p}}\int_\epsilon^{p,\frac{p}{2}}\int_{-p,\frac{p}{2}}^{-\epsilon} e^{ik\varphi}\frac{dy_1\,dy_2}{\sqrt{y_1^2 + y_2^2 + 2p^2}}, \qquad \epsilon > 0, \qquad (6.79)$$

$$2p\,\varphi = 4p^2 - y_1^2 - y_2^2 + \sqrt{(y_1 + y_2)^2 + 4p^2}(y_1 - y_2). \qquad (6.80)$$

It can be shown that every pair y_1, y_2 for which $y_1, y_2 = -p^2$ gives a stationary value for φ, this proves the nondisconnection of the stationary point. To evaluate an asymptotic behavior of the integral (6.79), let us introduce the new variables:

$$z_1 = \sqrt{(y_1 + y_2)^2 + 4p^2} - (y_1 - y_2), \qquad z_2 = y_1 + y_2, \qquad (6.81)$$

with the Jacobian $J \equiv -1/2$. Then

$$4p\,\varphi = 12p^2 - z_1^2. \qquad (6.82)$$

Detailed consideration gives the result that the two-dimensional region of integration in (6.79) is such that

$$A = -\frac{ik}{\pi}\sqrt{\frac{p}{2}}\left(\frac{1}{2}\right)2e^{3kpi}\int_0^{p,\frac{p}{2}}dz_2\int_{a(z_2)}^{b(z_2)}\frac{e^{-ikz_1^2/4p}\,dz_1}{\sqrt{(\sqrt{p^2 + z_2^2} - z_1/2)^2 + z_1^2/4}}, \qquad (6.83)$$

$$a(z_2) = z_2 - 2\sqrt{2}\,p + \sqrt{z_2^2 + 4p^2}, \qquad b(z_2) = -z_2 - 2\epsilon + \sqrt{z_2^2 + 4p^2}. \qquad (6.84)$$

The value $z_1 = 0$ with any z_2 is stationary. This confirms once more that the stationary point is not isolate. First of all, the internal integral should be estimated separately. The one-dimensional method of stationary phase can be applied for this. As could be expected *a priori*, its stationary value $z_1 = 0$ exists not for arbitrary z_2, but only if $0 < z_2 < p\sqrt{2}$. This means that the arcs l mentioned above are located on $p/4 < x_1, x_2 < p/2$. Estimation of the integral in (6.83) gives

$$\int e^{-ikz_1^2/4p}\frac{dz_1}{\sqrt{\left(\sqrt{4p^2 + z_2^2} - z_1/2\right)^2 + z_1^2/4}} \sim 2\sqrt{\frac{\pi p}{k}}\frac{e^{-(\pi/4)i}}{\sqrt{4p^2 + z_2^2}}. \qquad (6.85)$$

Therefore,

$$A = -\sqrt{\frac{2k}{\pi}} p\, e^{(\pi/4)i} e^{3kpi} \int_0^{p, \frac{p}{2}} \frac{dz_2}{\sqrt{z_2^2 + 4p^2}} = -\sqrt{\frac{k}{2\pi}} p\, e^{(\pi/4)i} \ln 2 e^{3kpi}. \tag{6.86}$$

In order to write out the final result, the contribution of the ray with $N = 1$ should be recalled. Besides, the expression (6.86) should be taken twice because for every ray considered above, an inverse ray exists too. Taking into account a special scale, the final result can be represented as follows:

$$A = -kp \ln 2 e^{i(3kp+\pi/4)} + i\left(\sqrt{\pi kp}/2\right) e^{2kpi}. \tag{6.87}$$

It should be noted that the first power of k appears for the first time in this paper: compare, for instance, with (6.75). Thus the parabolic reflector is the most powerful among all known reflectors.

Comparison between the exact solution and the solution given by (6.87) is shown in Fig. 6.6(b). The leading term (of the order of $O(kp)$) in (6.87) would give a straight line in Fig. 6.6(b) that passes through the average values of the oscillating graph.

Helpful remarks

It is interesting to note that two very similar geometries represented respectively in Figs. 6.5 and 6.6 yield absolutely different reflected amplitude. In the case of semi-circle the amplitude is quasi-constant, and the calculation of this quantity requires summation of a slowly convergent series. In the case of parabolic reflector the reflected amplitude grows as the first power of the frequency, and this amplitude is expressed by a very simple formula. The marked different qualitative and quantitative reflecting properties of so close in shape reflectors demonstrates that high-frequency analysis should be very refined to estimate correctly the reflected amplitude.

6.6. Application: Physical Diffraction Theory for Nonconvex Obstacles

The ideas discussed in the previous section can be applied to extend Kirchhoff's physical diffraction theory to the case of nonconvex obstacles, where multiple ray re-reflections may take place.

As we remember from Chapter 2, the diffraction problem in the generic case can be completely studied and numerically solved for arbitrarily shaped obstacle by the boundary element methods. Here we can mark the cases when the BIE method at high frequencies, within the framework of the physical diffraction theory, can be reduced to calculation of some integrals. As we will see below, generally this is expressed by some multiple integrals, and in a certain sense such an approach is indeed a generalization of the classical physical diffraction theory (see Section 2.6). For simplicity, let us restrict our consideration by the scalar problem only and only by the case of a back scattering, when the point source and the receiver coincide with each other.

Let us study the diffraction problem for an acoustically hard obstacle, whose boundary contour l is an arbitrary smooth closed contour of finite length. Let us first emit an extremely dense set of irradiated rays from the source x_0 and denote by l^+ the part of the boundary contour l containing the points of first reflections for any emitted ray. The contour l^+ is thus an *illuminated* part of l. As follows from the Kirchhoff's theory, boundary value of acoustic pressure vanishes in the shadow, i.e., outside l^+: $p(y) = 0$, $y \in l \setminus l^+$. Now, according to the

Kirchhoff–Helmholtz integral formula (see Chapter 2), exact value of the diffracted wave amplitude is

$$p^{sc}(x_0) = \int_{l^+} p(y) \frac{\partial \Phi}{\partial n_y} \, dl_y, \qquad (6.88)$$

where the basic question is: how the value of the boundary pressure over illuminated part l^+ can be determined in Eq. (6.88).

The considered illuminated part l^+ may contain both convex and nonconvex zones. Let us denote again as l_1^+ and l_2^+, respectively, the convex and nonconvex parts of the boundary, on the illuminated side: $l^+ = l_1^+ \cup l_2^+$. As indicated above, the solution of the basic BIE on l_1^+ is calculated in the same way as predicted by classical Kirchhoff's theory, because no re-reflected ray can attain these parts:

$$p(y)|_{l_1^+} = 2p_0 = 2 \frac{e^{ik|x_0 - y|}}{\sqrt{|x_0 - y|}}. \qquad (6.89)$$

If the considered contour l would be convex, then this value (6.89) could be directly substituted as a known boundary value of acoustic pressure to the Kirchhoff–Helmholtz integral, to calculate wave amplitude at any observation point. However it is not so easy to calculate *a priori* the pressure on the existing nonconvex parts l_2^+ of the boundary, since there are multiple re-reflections on these parts. Our approach to simplify this calculation is based on the following evident observation. The geometrical diffraction theory in the case of multiple re-reflections operates with asymptotic estimates of some multiple integrals by the stationary phase method. If we calculate these multiple integrals exactly, not by applying any asymptotic estimates, then we arrive at a certain analogue to Kirchhoff's physical diffraction theory for nonconvex obstacles admitting re-reflections of the incident wave by their boundaries. Following this idea, we should first trace the paths of all rays that depart from and come back to the same point x_0, with the calculation of the maximum number N for any ray during its re-reflections over the arc l_2^+.

If $N = 1$, then for every ray, only simple reflections without re-reflections can take place. In this case the Kirchhoff–Helmholtz integral has the same form as in the case of convex obstacle, that is, it is given by a single integral over illuminated side of the boundary, since we have $p(y) = 2p_0$, both on l_1^+ and l_2^+, all over the *light* zone:

$$p^{sc}(x_0) = \int_{l^+} 2p_0 \frac{\partial \Phi}{\partial n_y} \, dl_y = -e^{3\pi i/4} \frac{k}{2\pi} \int_{l^+} \frac{e^{2ik|x_0 - y|}}{|x_0 - y|} \cos(n_y, x_0 - y) \, dl_y. \qquad (6.90)$$

For $N > 1$ we introduce the characteristic function $\chi(y_1, y_2)$, $y_1, y_2 \in l^+$ by the following definition: $\chi(y_1, y_2) = 1$ if the piece y_1–y_2 does not intersect the contour (i.e., these two points are mutually visible), and $\chi(y_1, y_2) = 0$ in the opposite case.

Suppose, for example, $N = 2$. This means that every ray irradiated from the point x_0 and returned back to this point does reflect from the obstacle boundary l no more than twice its during full flight. Hence, before its last hit at the boundary, which indeed defines the boundary value of the function $p(y)$ in Eq. (6.88), each ray did undergo no more than only one reflection. So, when calculating $p(y)$, $y \in l^+$, we may add to the result of a direct hit $2p_0$ also a single Kirchhoff–Helmholtz integral that is responsible for the influence of one-time repeated reflections:

$$p(y) = 2p_0(y) + \int_{l^+} 2p_0(y_1) \frac{\partial \Phi}{\partial n_1} \chi(y, y_1) \, dl_1, \qquad y \in l^+. \qquad (6.91)$$

Here we have taken into account that the point y_1 can influence the boundary value of $p(y)$ at the point y only in the case when the latter is *visible* from y_1. As a result, for $N = 2$ the physical diffraction theory is reduced to calculation of the double integral which is obtained by the substitution of relation (6.91) into Eq. (6.88).

It is clear that for arbitrary $N \geq 3$ the physical diffraction theory is reduced to a certain N-fold integral. The corresponding formulas are omitted here for the sake of brevity. If for a certain contour the quantity N is known approximately from any *a priori* understanding, then it may be taken *on the safe side*, i.e., a little higher by a few units. Correctness of this statement follows from the evident observation: if contribution of any re-reflected wave to the boundary value $p(y)$, $y \in l^+$ is absent, then the corresponding multiple integral will be asymptotically small compared to the result of direct hit. It should be noted however that for $N \geq 3$ the efficiency of the thus generalized Kirchhoff's theory decreases considerably because the number of required arithmetic operations grows as M^N, where M is the number of nodes in the corresponding quadrature formulas when being implemented on a computer. At the same time direct numerical treatment of the BIE by a collocation technique (see Section 1.5) requires around M^3 arithmetic operations since this is typically reduced to an $M \times M$ linear algebraic system (some helpful simple and clear ideas on numerical treatment of linear algebraic systems, with a FORTRAN code, can be found in Forsythe et al., 1977).

Helpful remarks

It should be noted that, as in geometrical diffraction theory, for acoustically soft boundary the final result differs from that for acoustically hard obstacle by a factor $(-1)^N$, which goes in front of the corresponding formula for the receiving acoustic pressure $p(x_0)$. This leads to a very important physical conclusion. As shown in Section 2.6, for convex obstacles, when there are present only single integrals with $N = 1$, classical Kirchhoff's theory predicts the real amplitude of acoustic pressure in the reflected wave to be identical for acoustically hard and soft boundaries, because the complex amplitudes differ only by their (opposite) signs. By contrast, for nonconvex obstacles with $N > 1$ we face with the summation of contributions of rays reflected from the boundary at different times. For all that, the contribution of the rays with even N for both types of boundaries are equal to each other, but with odd N contribute in anti-phase. Therefore, the result of summation of such contributions in the generic case will be different. This means that two geometrically identical objects, one of them being acoustically hard and another acoustically soft, even at extremely short-wave regime may possess absolutely different back-scattered patterns.

Nevertheless, there can be indicated relatively wide class of nonconvex obstacles, for which re-reflections are impossible, and so for them we may always put $N = 1$. For obstacles of this class, reflection from acoustically hard and acoustically soft boundaries will be the same in absolute value. It is proved in the author's work (Druzhinina and Sumbatyan, 1992) that if on a concave arc of the boundary the angle between two arbitrary normals is less than the right angle, then ray re-reflections are surely impossible on this part of the boundary contour. We thus can extract a class of nonconvex domains with *weakly concave* boundaries. This class is covered by obstacles with the boundaries, for which on any concave part the angle between an arbitrary pair of normals is less than 90°. For the obstacles of this class the physical diffraction theory, like for convex domains, is described by a usual single Kirchhoff–Helmholtz integral.

6.7. Short-Wave Integral Operator in Diffraction by a Flaw in Elastic Medium

This problem is of great interest as regards its application to ultrasonic NDT. The general approach developed above is suitable for the elastic case. Let p and s refer to the longitudinal (pressure) and transverse (shear) wave, respectively.

At the beginning, consider p–p reflection. Let an oscillating unit force be located at the point x_0, and its projection into the x_0y direction be Q, $y \in l$. The value of Q depends upon the mutual disposition of x_0 and y. For instance, if the initial unit force is perpendicular to the x_0y segment, then $Q = 0$.

It can be shown that the longitudinal component of Cartesian displacements $u_1^{\text{inc}}, u_2^{\text{inc}}$ caused by the force $\mathbf{Q}$ and related to the tangent and the normal at the point y are defined as

$$u_1^{\text{inc}}(y) = -Q \sin\theta \frac{e^{ik_p|x_0-y|}}{\sqrt{|x_0-y|}}, \tag{6.92}$$

$$u_2^{\text{inc}}(y) = -Q \cos\theta \frac{e^{ik_p|x_0-y|}}{\sqrt{|x_0-y|}}. \tag{6.93}$$

The short-wave representation for the reflected longitudinal component of displacement is given by

$$u_r(x) = \frac{k_p^2}{4k_s^2}\sqrt{\frac{2k_p}{\pi}} e^{-(\pi/4)i} \int_l \left[-\sin 2\theta u_1(y) + \left(\frac{k_s^2}{k_p^2} - 2\sin^2\theta \right) u_2(y) \right] \frac{e^{ik_p|x-y|}}{\sqrt{|x-y|}} ds_y,$$

$$u_\psi(x) = 0, \tag{6.94}$$

where (r, ψ) are polar coordinates related to the point y and (u_1, u_2) is the vector of full displacements over the contour l.

In order to use Kirchhoff theory, those values for $u_1(y), u_2(y)$ should be chosen in (6.94), which are defined by a solution of the diffraction problem for reflection of a plane wave from a half-plane cavity. As is well known, full displacements on the boundary are defined by the incident wave (6.92) and (6.93) and are equal to

$$u_1(y) = \left(1 + V_{pp} - \frac{k_s}{k_p \sin\theta}\sqrt{1 - \frac{k_p^2}{k_s^2}\sin^2\theta}\, V_{ps} \right) u_1^{\text{inc}}(y), \tag{6.95}$$

$$u_2(y) = (1 - V_{pp} - \tan\theta V_{ps}) u_2^{\text{inc}}(y), \tag{6.96}$$

where V_{pp} and V_{ps} are reflection coefficients for p–p and p–s reflection, respectively. After substituting (6.95), (6.96) and (6.92), (6.93) into (6.94), the expression in brackets of (6.94) becomes equal to

$$\frac{k_p^2}{2k_s^2}\left\{ -\sin 2\theta \left[-\sin\theta(1 + V_{pp}) + \frac{k_s}{k_p}\sqrt{1 - \frac{k_p^2}{k_s^2}\sin^2\theta}\, V_{ps} \right] \right.$$
$$\left. + \left(\frac{k_s^2}{k_p^2} - 2\sin^2\theta \right) [-\cos\theta(1 - V_{pp}) + \sin\theta V_{ps}] \right\} = \cos\theta V_{pp}, \tag{6.97}$$

and here we give a proof for this identity.

We start from Snell's law (see Brekhovskikh, 1980)

$$k_s \sin\theta_1 = k_p \sin\theta \quad \sim \quad \frac{k_s}{k_p} = \frac{\sin\theta}{\sin\theta_1}, \tag{6.98}$$

where θ_1 is the reflection angle of the transverse when θ is an angle between the incident longitudinal wave and the normal. We should also recall that (see Brekhovskikh, 1980)

$$V_{pp} = \frac{4\cot\theta\cot\theta_1 - (1 - \cot^2\theta_1)^2}{4\cot\theta\cot\theta_1 + (1 - \cot^2\theta_1)^2}, \quad V_{ps} = \frac{4\cot\theta\,(1 - \cot^2\theta_1)}{4\cot\theta\cot\theta_1 + (1 - \cot^2\theta_1)^2}. \tag{6.99}$$

Then expression in the braces in Eq. (6.97) is

$$
2\sin^2\theta\cos\theta\,(1+V_{pp})-2\frac{k_s}{k_p}\sin\theta\cos\theta\cos\theta_1\,V_{ps}+\left(2\sin^2\theta-\frac{k_s^2}{k_p^2}\right)\cos\theta\,(1-V_{pp})
$$

$$
+\left(\frac{k_s^2}{k_p^2}-2\sin^2\theta\right)\sin\theta V_{ps}=\frac{k_s^2}{k_p^2}\cos\theta\left\{V_{pp}+\frac{k_p^2}{k_s^2}\left(4\sin^2\theta-\frac{k_s^2}{k_p^2}\right)+\frac{\sin\theta}{\cos\theta}\frac{k_p^2}{k_s^2}\right.
$$

$$
\times\left[-2\frac{\sin\theta}{\sin\theta_1}\cos\theta\cos\theta_1+\frac{\sin^2\theta}{\sin^2\theta_1}-2\sin^2\theta\right]V_{ps}\right\}=\frac{k_s^2}{k_p^2}\cos\theta\left\{V_{pp}\right.\qquad(6.100)
$$

$$
+(4\sin^2\theta_1-1)+\tan\theta\frac{\sin^2\theta_1}{\sin^2\theta}\left[-2\frac{\sin\theta}{\sin\theta_1}\cos\theta\cos\theta_1+\frac{\sin^2\theta}{\sin^2\theta_1}-2\sin^2\theta\right]V_{ps}\right\}
$$

$$
=\frac{k_s^2}{k_p^2}\cos\theta\left\{V_{pp}+(4\sin^2\theta_1-1)+[-2\sin\theta_1\cos\theta_1+(1-2\sin^2\theta_1)\tan\theta]V_{ps}\right\}.
$$

Now, by the substitution of relation (6.99) as V_{ps} into (6.100) the braces in (6.100) can be rewritten in the following form:

$$
V_{pp}+\frac{\sin^2\theta_1}{4\cot\theta\cot\theta_1+\left(1-\cot^2\theta_1\right)^2}\left[(3-\cot^2\theta_1)\left(4\cot\theta\cot\theta_1+(1-\cot^2\theta_1)^2\right)\right.
$$

$$
-\left(2\cot\theta_1+(1-\cot^2\theta_1)\tan\theta\right)4\cot\theta(1-\cot^2\theta_1)\Big]\qquad(6.101)
$$

$$
=V_{pp}+\frac{4\cot\theta\cot\theta_1-\left(1-\cot^2\theta_1\right)^2}{4\cot\theta\cot\theta_1+\left(1-\cot^2\theta_1\right)^2}=V_{pp}+V_{pp}=2V_{pp}.
$$

It becomes finally clear that expression (6.100) is equal to $2(k_s^2/k_p^2)V_{pp}\cos\theta$, and so expression (6.97) is

$$
\frac{k_p^2}{2k_s^2}2\frac{k_s^2}{k_p^2}V_{pp}\cos\theta=V_{pp}\cos\theta,\qquad(6.102)
$$

which was to be proved.

Therefore, formula (6.94) for u_r reads

$$
u_r(x)=Qe^{-(\pi/4)i}\sqrt{\frac{k_p}{2\pi}}\frac{\cos\theta}{\sqrt{L_0L}}V_{pp}\int_l e^{ik_r(|x_0-y|+|x-y|)}\,ds.\qquad(6.103)
$$

Asymptotic estimate of the last integral is the same as for (6.97), and gives the final result

$$
u_r(x)=\frac{QV_{pp}}{\sqrt{L_0L}}\frac{\exp\{i[k_p(L_0+L)+(\pi/4)(\delta-1)]\}}{\sqrt{1/L_o+1/L+2/\rho\cos\theta}}.\qquad(6.104)
$$

As can be shown by analogy, for diffraction with an arbitrary number of reflection points p–p–$\cdots$–p–p the general result coincides with a corresponding expressions for the scalar case multiplied by reflection coefficients at every boundary point.

Deduction of a formula for other cases is more complicated. For one-time p–s reflection one can derive instead of (6.103) the following expression for a shear component of displacement:

$$
u_\psi(x)=Qe^{-(\pi/4)i}\frac{k_s}{k_p}\sqrt{\frac{k_s}{2\pi}}\frac{\cos\gamma}{\sqrt{L_0L}}V_{ps}\int_l e^{i(k_p|x_0-y|+k_s|x-y|)}\,ds,\qquad u_r(x)=0,\qquad(6.105)
$$

in polar coordinates, where γ is a reflection angle of the shear wave. This leads to the following result:

$$u_\psi(x) = QV_{ps}\frac{k_s}{k_p}\sqrt{\frac{k_s}{L_0 L}}\frac{\cos\gamma\,\exp\{i[k_p L_0 + k_s L + (\pi/4)(\delta - 1)]\}}{\sqrt{k_p\cos^2\theta/L_0 + k_s\cos^2\gamma/L + k_p\cos\theta/\rho + k_s\cos\gamma/\rho}},$$

$$\delta = \text{sign}\left(\frac{k_p\cos^2\theta}{L_0} + \frac{k_s\cos^2\gamma}{L} + \frac{k_p\cos\theta}{\rho} + \frac{k_s\cos\gamma}{\rho}\right).$$

(6.106)

Further, every type of multiple reflection can be described by a corresponding formula. For example, p–s–p double reflection is given by

$$u_r(x_3) = Q\sqrt{\frac{k_s}{k_p}}V_{ps}(\theta_1)V_{sp}\theta_2\cos\gamma_1\cos\theta_2$$

$$\times\,\frac{\exp i[k_p L_0 + k_s L_1 + k_p L_2 + (\pi/4)(\delta_2 - 2)]}{\sqrt{L_0 L_1 L_2}\sqrt{|\det(D_2)|}}, \qquad \delta_2 = \text{sign}\,D_2,$$

(6.107)

where

$$D_2 = (d_2)_{ij}, \quad i,j = 1,2;$$

$$(d_2)_{11} = \frac{\cos^2\theta_1}{L_0} + \frac{k_s\cos^2\gamma_1}{k_p L_1} + \frac{\cos\theta_1}{\rho_1} + \frac{k_s\cos\gamma_1}{k_p\rho_1},$$

$$(d_2)_{12} = \frac{\cos\gamma_1\cos\gamma_2}{L_1}, \qquad (d_2)_{21} = \frac{\cos\gamma_1\cos\gamma_2}{L_1},$$

$$(d_2)_{22} = \frac{k_s\cos^2\gamma_2}{k_p L_1} + \frac{\cos^2\theta_2}{L_2} + \frac{k_s\cos\gamma_2}{k_p\rho_2} + \frac{\cos\theta_2}{\rho_2}.$$

(6.108)

Here, θ_1 and γ_1 are incident and reflected angles at the first point of reflection and γ_2 and θ_2 are incident and reflected angles at the second point.

Helpful remarks

Obviously, due to presence of reflection coefficient the reflected wave field from the obstacle in a scalar acoustic and elastic media are absolutely different. This is demonstrated by the following example. Let us consider symmetric reflection of a plane longitudinal wave from a concave side of a cosine-type curve: $y = \sqrt{2}\,a[1 - \cos(x/a)]$, $|x| \le (\pi/3)a$. It can be shown that besides the single *direct* reflection at $x = 0$ with $\rho = -a\sqrt{2}$, two mutually inverse rays p–p–p with $N = 2$ exist for which $\theta_1 = \theta_2 = \pi/4$ (at $x = \pm\pi/4$), $\rho_1 = \rho_2 = -a\sqrt{8}$, $\delta_2 = 0$. Other types of reflection are absent. Thus the ray asymptotics has the following form:

$$A = \sqrt{\pi k_p a i}\left(\frac{1}{32^{1/4}} - \frac{V_{pp}^2(\pi/4)}{\sqrt{1 - \pi/4}}e^{ik_p a}\left[(\pi/2) + 2 - 2\sqrt{2}\right]\right).$$

(6.109)

It was taken into account here that $V_{pp}(\theta = 0) = -1$. Comparison of this ray solution with an exact one and with a scalar case is shown in Fig. 6.7 for an elastic aluminum medium. Note that $V_{pp}(\pi/4) = -0.567$ here. Scalar asymptotics would differ from (6.109) by the absence of the reflection coefficient V_{pp}. As can be seen, this leads to a totally different numerical result compared with the elastic case.

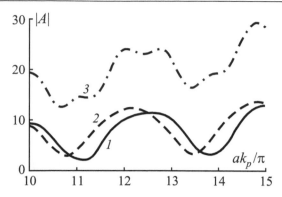

Figure 6.7. Symmetric reflection from a cosine-type crack in elastic medium: 1, exact solution; 2, ray solution; 3, prediction by scalar theory

6.8. High-Frequency Asymptotics of Integral Operator in a Three-Dimensional Diffraction Theory

Here we extend some results of the previous sections, obtained in the 2D case, to the three-dimensional theory. We consider diffraction of a high-frequency (i.e., short-wave) acoustic wave, generated by a point source, from arbitrary acoustically hard smooth obstacle Γ, which provides the boundary condition $\left(\partial p/\partial n\right)\big|_\Gamma = 0$. As in the previous sections, our approach is based on an asymptotic estimate of Kirchhoff's diffraction integral by a multiple stationary phase method. This allows us to obtain the amplitude of the reflected high-frequency wave in an explicit form. A similar result can be obtained starting from Keller's geometric diffraction approach (Keller, 1962), and this can be found in more detail, for example, in McNamara et al. (1990).

Let a point source x_0 generate a spherical acoustic wave, which is incident onto a surface Γ. During interaction of the incident wave with the surface Γ on its convex parts there are only points of simple specular (mirror) reflection. Reflection process from concave parts of the surface is more complex; this is connected with possible re-reflections of acoustic rays (compare with Section 6.4), and is not considered here. The pressure in reflected wave, in the high-frequency limit, is defined by the direction of incident wave and by a small neighborhood of the point of specular reflection. Therefore, with the frequency increasing, the amplitude of the reflected wave can be obtained by a multiple stationary phase method.

If the ray x_0-y-x is reflected from the surface only once, then the Kirchhoff–Helmholtz integral formula reads

$$p^{\text{sc}}(x) = \iint_{\Gamma^+} 2p^{\text{inc}}(y)\frac{\partial\Phi(kr)}{\partial n_y}\,ds_y, \quad \Phi(kr) = \frac{e^{ikr}}{4\pi r}, \quad r = |x - y|, \quad x \in \mathbb{R}^3 \setminus D, \quad (6.110)$$

where we have taken into account that the leading asymptotic term of the pressure $p(y)$, $y \in \Gamma^+$ is given by Kirchhoff's physical diffraction theory as $p(y) = 2p^{\text{inc}}(y)$, $y \in \Gamma^+$ (cf. Section 2.6). As usual, here $p^{\text{inc}}(y)$ is the value of pressure in the incident wave on the boundary surface Γ, $\Phi(kr)$ is the Green's function, n_y is the outer normal to the surface Γ at the point y of a small neighborhood of the specular reflection point y_0. Note that asymptotically at $k \to \infty$ we may put

$$p^{\text{inc}}(y) = \frac{e^{ikr_0}}{r_0}, \quad r_0 = |x_0 - y|, \quad \frac{\partial\Phi(kr)}{\partial n_y} = ik\cos\gamma\,\frac{e^{ikr}}{4\pi r}\left[1 + O(k^{-1})\right], \quad (6.111)$$

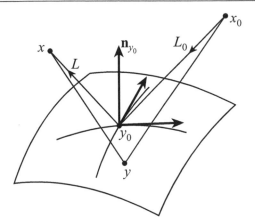

Figure 6.8. Reflection of acoustic wave from a smooth surface

where γ is the angle between the normal n_y and the direction of propagation of the incident ray $r_0 = x_0 - y$: $\gamma = \widehat{n_y \, r_0}$.

Since the incident and reflected rays lie on the same plane with the normal to the surface at the reflecting point n_{y_0}, formulas (6.110)–(6.111) lead to the following basic integral representation (after taking nonoscillating terms outside of the integral, in the considered short-wave approximation:

$$p^{\text{sc}}(x) = \frac{ik \cos \gamma}{4\pi L_0 L} \iint_{\Gamma^+} e^{ikS(y)} \, ds_y, \quad S(y) = r_0 + r = |x_0 - y| + |x - y|,$$

$$L_0 = |x_0 - y_0|, \ L = |x - y_0|.$$

(6.112)

We will obtain the leading (geometrical diffraction) asymptotic term by applying a multiple stationary phase method (see Fedorjuk, 1962) to the integral (6.112).

The estimate procedure should take into account all points y from a small neighborhood of the point y_0. Let us link this small neighborhood to a right Cartesian coordinate system, which is defined by the normal n_{y_0} and two lines of principal curvatures of the surface at the point y_0. Then an arbitrary point y from the neighborhood of y_0 has the following coordinates: $y = \left\{ \Delta s_1, \ \Delta s_2, \ -\frac{1}{2} \left[k_1 (\Delta s_1)^2 + k_2 (\Delta s_2)^2 \right] \right\}$, where Δs_1 and Δs_2 are arc differentials along the lines of principal curvatures, $k_1 = R_1^{-1}$ and $k_2 = R_2^{-1}$ are the principal curvatures, R_1 and R_2 are the principal curvature radii of the surface S at the point y_0, and $\left[k_1 (\Delta s_1)^2 + k_2 (\Delta s_2)^2 \right]$ is the second quadratic form of the surface at the point y_0 related to the principal curvature lines (see, for example, Pogorelov, 1961).

Let us apply cosine theorem to the triangles $x_0 y_0 y$ and $x y_0 y$ (see Fig. 6.4):

$$|x_0 - y|^2 = L_0^2 + |\Delta s|^2 - 2L_0 |\Delta s| \, \cos(x_0 y_0, y_0 y),$$

$$|x - y|^2 = L_0^2 + |\Delta s|^2 - 2L |\Delta s| \, \cos(x y_0, y_0 y),$$

(6.113)

where vector Δs is: $\Delta s = \left\{ \Delta s_1, \ \Delta s_2, \ -\frac{1}{2} \left[k_1 (\Delta s_1)^2 + k_2 (\Delta s_2)^2 \right] \right\}$.

Let us denote the unit vector along direction $(y_0 x_0)$ as $\{\cos \alpha, \ \cos \beta, \ \cos \gamma\}$, then $\{-\cos \alpha, \ -\cos \beta, \ \cos \gamma\}$ is the unit vector along direction $(y_0 x)$. Now, by taking the scalar products of these vectors with the vector Δs, we can obtain

$$|\Delta s| \cos(x_0 y_0, y_0 y) = (\Delta s_1) \cos \alpha + (\Delta s_2) \cos \beta$$
$$- \frac{1}{2} \left[k_1 (\Delta s_1)^2 + k_2 (\Delta s_2)^2 \right] \cos \gamma,$$

$$|\Delta s| \cos(x y_0, y_0 y) = -(\Delta s_1) \cos \alpha - (\Delta s_2) \cos \beta$$
$$- \frac{1}{2} \left[k_1 (\Delta s_1)^2 + k_2 (\Delta s_2)^2 \right] \cos \gamma,$$

(6.114)

hence

$$|x_0 - y| = L_0 - (\Delta s_1) \cos \alpha - (\Delta s_2) \cos \beta + \tfrac{1}{2} \left(L_0^{-1} \sin^2 \alpha + k_1 \cos \gamma \right) (\Delta s_1)^2$$
$$+ \tfrac{1}{2} \left(L_0^{-1} \sin^2 \beta + k_2 \cos \gamma \right) (\Delta s_2)^2 - L_0^{-1} \cos \alpha \cos \beta (\Delta s_1)(\Delta s_2) + O(|\Delta s|^3),$$
$$|x - y| = L + (\Delta s_1) \cos \alpha + (\Delta s_2) \cos \beta + \tfrac{1}{2} \left(L^{-1} \sin^2 \alpha + k_1 \cos \gamma \right) (\Delta s_1)^2$$
$$+ \tfrac{1}{2} \left(L^{-1} \sin^2 \beta + k_2 \cos \gamma \right) (\Delta s_2)^2 - L^{-1} \cos \alpha \cos \beta (\Delta s_1)(\Delta s_2) + O(|\Delta s|^3). \tag{6.115}$$

From these relations we can extract the behavior of the phase function $S(y)$ defined in Eq. (6.112), in the following form:

$$S(y) = L_0 + L + \tfrac{1}{2} d_{11}(\Delta s_1)^2 + d_{12}(\Delta s_1)(\Delta s_2) + \tfrac{1}{2} d_{22}(\Delta s_2)^2 + O(|\Delta s|^3),$$
$$d_{11} = \left(L_0^{-1} + L^{-1} \right) \sin^2 \alpha + 2k_1 \cos \gamma, \quad d_{22} = \left(L_0^{-1} + L^{-1} \right) \sin^2 \beta + 2k_2 \cos \gamma, \tag{6.116}$$
$$d_{12} = - \left(L_0^{-1} + L^{-1} \right) \cos \alpha \cos \beta.$$

Absence of the first powers (Δs_1) and (Δs_2) here confirms that the point y_0 of a (mirror) specular ray reflection provides a stationary value for the phase function $S(y)$. Thus, the multiple stationary phase method (see Section 1.4) implies that the leading asymptotic term of the integral (6.112) is determined by coefficients in front of $(\Delta s_1)^2$, $(\Delta s_1)(\Delta s_2)$, $(\Delta s_2)^2$. This finally yields

$$p(x) = \frac{\exp \left\{ i \left[k(L_0 + L) + (\pi/4)(\delta_2 - 2) \right] \right\}}{L_0 L \sqrt{|\det(D_2)|}}, \tag{6.117}$$

where D_2 is the Hessian of a symmetric structure: $D_2 = \{d_{mj}\}$, $d_{mj} = d_{jm}$ $(m, j = 1, 2)$, and $\delta_2 = \text{sign}(D_2) = \nu_+ - \nu_-$ is the difference between positive and negative eigenvalues of the matrix D_2.

Taking into account that $d_{21} = d_{12}$, the final result can be obtained in the following form:

$$p(x) = \frac{\exp \left\{ i \left[k(L_0 + L) + (\pi/4)(\delta_2 - 2) \right] \right\}}{\left| (L_0 + L)^2 + 2L_0 L(L_0 + L)(k_2 \sin^2 \alpha + k_1 \sin^2 \beta)(\cos \gamma)^{-1} + 4L_0^2 L^2 K \right|^{1/2}}. \tag{6.118}$$

Here $K = k_1 k_2$ is the Gaussian curvature of the surface at the point y_0. Note also that the vector $\{-\cos \alpha, -\cos \beta, -\cos \gamma\}$ determines direction of incidence of the ray $x_0 y_0$ in the chosen coordinate system.

Helpful remarks

$1°$. Certainly, in the cases when expression in the denominator of Eq. (6.118) is trivial, our result (6.118) is not valid. In this case we are faced with an irregular stationary point, and the dependence of the reflected amplitude upon the frequency is more complex.

$2°$. As in the 2D problem, formula (6.118) was derived for the case when a high-frequency wave is incident on a convex surface. If the wave is incident on a concave surface, then the principal curvatures k_1 and k_2 should be taken negative.

$3°$. Let us stress two limiting cases when formula (6.118) can be considerably simplified. First of all, in the case when the reflecting surface is locally plane, we have $k_1 = k_2 = 0$. It can be easily seen that in this case expression (6.118) reduces to

$$p(x) = \frac{e^{ik(L_0 + L)}}{L_0 + L}, \tag{6.119}$$

which conforms to clear heuristic ideas.

Another simple case is related to a far-field scattering. In this case L_0, $L \gg R_1$, R_2, and the last term in the denominator predominates over two other ones. Then formula (6.118) reduces to

$$p(x) = \frac{\sqrt{R_1 R_2} \exp\left\{ i \left[k(L_0 + L) + (\pi/4)(\delta_2 - 2) \right] \right\}}{2 L_0 L}, \qquad (6.120)$$

which coincides with Shenderov (1972). As you can see, the amplitude of the far-field scattered wave is proportional to the square root of the quantity inverse to the Gaussian curvature: $p(x) \sim \sqrt{R_1 R_2} \sim 1/\sqrt{K} = 1/\sqrt{k_1 k_2}$.

Chapter 7

Inverse Problems of the Short-Wave Diffraction

7.1. Some Basic Results in a Local Differential Geometry of Smooth Convex Surfaces

Since almost the whole of this chapter will be devoted to the inverse (reconstruction) problem for smooth convex obstacles, we first give a short survey of some classical results from differential geometry devoted to the concept of the Minkowski *support function*, which is essentially used in our study.

Let us consider a smooth convex closed surface S. We introduce the function $P(\mathbf{q})$ related to the distance $p(\mathbf{q})$ from the origin O to the tangential plane with the normal $\mathbf{q}$, as follows (see Blaschke, 1930; Pogorelov, 1973, 1978; Ramm, 1986):

$$P(\alpha_1, \alpha_2, \alpha_3) = r\, p(\alpha_1/r,\ \alpha_2/r,\ \alpha_3/r), \quad \mathbf{n} = (\alpha_1,\ \alpha_2,\ \alpha_3), \quad r = (\alpha_1^2 + \alpha_2^2 + \alpha_3^2)^{1/2}, \quad (7.1)$$

where $\mathbf{n} = -\mathbf{q}$ is a unit normal vector to the surface. Note that the Cartesian coordinates of the boundary surface can be expressed in terms of the function $P(\alpha_1, \alpha_2, \alpha_3)$ as $x_i = \partial P/\partial \alpha_i$.

First, for some mathematical transformations we consider the function $P(\alpha_1, \alpha_2, \alpha_3)$ as a function of three variables $(\alpha_1, \alpha_2, \alpha_3)$, each of them varying from $-\infty$ to $+\infty$. And then, in final treatment we should take into account that the real surface corresponds to the value $r = 1$.

The introduced function $P(\alpha_1, \alpha_2, \alpha_3)$ is called Minkowski *support function*. Let us derive a partial differential equation that gives the relation between the support function and the corresponding Gaussian curvature.

The function $P(\alpha_1, \alpha_2, \alpha_3)$ is a homogeneous function of the first degree, with respect to its arguments:

$$P(\mu\alpha_1, \mu\alpha_2, \mu\alpha_3) = \mu P(\alpha_1, \alpha_2, \alpha_3), \qquad \mu > 0. \tag{7.2}$$

According to Euler theorem, for the homogeneous function P the following identity holds:

$$\alpha_1 P_1 + \alpha_2 P_2 + \alpha_3 P_3 = P, \qquad P_i = \frac{\partial P}{\partial \alpha_i}. \tag{7.3}$$

The tangential plane to the surface can be written in the form

$$\alpha_1 x_1 + \alpha_2 x_2 + \alpha_3 x_3 = P, \tag{7.4}$$

which is a classical result of analytic geometry. Let us apply to Eq. (7.4) the partial derivatives with respect to α_i, $i = 1, 2, 3$. Then one comes to Cartesian coordinates of the points of tangency in terms of the derivatives P_i:

$$x_i = P_i(\alpha_1, \alpha_2, \alpha_3). \tag{7.5}$$

Let us express principal curvatures of the surface in terms of the support function, following to Rodrigues relations (see, for example, Pogorelov, 1961). On the surface, along the curvature lines, these formulas take the form

$$dx_i + R \, d\xi_i = 0. \tag{7.6}$$

It follows from (7.5) that

$$dx_i = \sum_{k=1}^{3} P_{ik} \, d\alpha_k, \qquad P_{ik} = \frac{\partial^2 P}{\partial \alpha_i \partial \alpha_k}. \tag{7.7}$$

For real surface, when $r = 1$, the quantities α_i become the coordinates of the unit normal ξ_i: $\alpha_i = \xi_i$. Then the substitution of (7.7) into (7.6) yields the following identity:

$$\sum_{k=1}^{3} P_{ik} \, d\xi_k + R \, d\xi_i = 0 \qquad (i = 1, 2, 3). \tag{7.8}$$

When moving along the curvature line all three $d\xi_i$ cannot vanish simultaneously. Therefore, determinant of system (7.8) must be equal to zero: $\det(P_{ik} + \delta_{ik} R) = 0$, $i, k = 1, 2, 3$, where δ_{ik} is Kronecker's delta.

The last determinant is

$$R^3 + (P_{11} + P_{22} + P_{33}) R^2 + (P_{11} P_{22} + P_{11} P_{33} + P_{22} P_{33}$$
$$- P_{12}^2 - P_{13}^2 - P_{23}^2) R + \det(P_{ik}) = 0 \qquad (i, k = 1, 2, 3). \tag{7.9}$$

Let us prove that determinant in (7.9) is equal to zero. To this end, we apply to Eq. (7.3) differentiation with respect to α_i, $i = 1, 2, 3$. As a result we arrive at the homogeneous system $\sum_{k=1}^{3} P_{ik} \alpha_k = 0$, $i = 1, 2, 3$, where all three α_i cannot be equal to zero simultaneously, so determinant of the system is trivial: $\det(P_{ik}) = 0$.

Hence, equation (7.9) is a quadratic equation, whose roots R_1 and R_2 are principal curvature radii satisfying the following identities:

$$P_{11} + P_{22} + P_{33} = -(R_1 + R_2), \tag{7.10}$$

$$P_{11} P_{22} + P_{11} P_{33} + P_{22} P_{33} - P_{12}^2 - P_{13}^2 - P_{23}^2 = R_1 R_2 = K^{-1}(\mathbf{q}), \tag{7.11}$$

according to the Vieta theorem (K is the Gaussian curvature).

It is clear from Eqs. (7.10) and (7.11) that reconstruction of the surface S, through the Minkowski support function P, is possible in two cases: 1) if we know the average curvature, i.e., the function $R_1 + R_2$ is known; 2) if we know the Gaussian curvature K of the surface, i.e., the function $R_1 R_2$ is known. Below in this section we study in more detail the first problem, which is covered by the Christoffel theorem:

THEOREM 7.1 (CHRISTOFFEL). *If there is a one-to-one correspondence between the smooth convex closed surface S and a unit sphere arranged by the endpoints of outer unit normals to this surface, all beginning from the same fixed point of the 3D space, then the function $R_1 + R_2$ defines the surface uniquely.*

Proof. The proof is constructive and follows Blaschke (1930) and Pogorelov (1973). Let us introduce the spherical coordinate system

$$\begin{cases} \alpha_1 = r \sin\theta \cos\varphi, \\ \alpha_2 = r \sin\theta \sin\varphi, \\ \alpha_3 = r \cos\theta. \end{cases} \tag{7.12}$$

Further treatment is based upon some classical results on series in spherical functions (see, for example, Hobson, 1955; Tikhonov and Samarsky, 1977). The spherical functions $U_m(\alpha_1, \alpha_2, \alpha_3)$, $m = 0, 1, 2, \ldots$, are homogeneous polynomials of the power m, being simple solutions of the Laplace equation $(U_m)_{11} + (U_m)_{22} + (U_m)_{33} = 0$. In the spherical coordinate system these functions can be expressed as

$$U_m(\alpha_1, \alpha_2, \alpha_3) = r^m \left\{ P_m^{(j)}(\cos\theta) \right\} \begin{Bmatrix} \sin(j\varphi) \\ \cos(j\varphi) \end{Bmatrix} \qquad (j = 0, 1, \ldots, m), \qquad (7.13)$$

where the $P_m^{(j)}(x) = (1 - x^2)^{j/2} d^j P_m(x)/dx^j$ are associated Legendre functions, and the $P_m(x)$ are Legendre polynomials.

It is proved that the system of spherical functions, considered as functions of (θ, φ), is complete and orthogonal on the spherical surface $0 \le \theta \le \pi, 0 \le \varphi \le 2\pi$. This predetermines the most natural and sufficient way to solve the considered Christoffel problem. Let us expand the support function, defined initially on a unit sphere, to a series in spherical functions:

$$P(\alpha_1, \alpha_2, \alpha_3) = \sum_{m=0}^{\infty} c_m U_m(\alpha_1, \alpha_2, \alpha_3), \qquad r = (\alpha_1^2 + \alpha_2^2 + \alpha_3^2)^{1/2} = 1, \qquad (7.14)$$

where c_m are some unknown coefficients. Then, by taking into account that $P(\alpha_1, \alpha_2, \alpha_3)$ is a homogeneous function of degree 1, and $U_m(\alpha_1, \alpha_2, \alpha_3)$ is a homogeneous function of the power m, we can conclude that representation of the support function in the full 3D space is

$$P(\alpha_1, \alpha_2, \alpha_3) = \sum_{m=0}^{\infty} c_m \frac{U_m(\alpha_1, \alpha_2, \alpha_3)}{r^{m-1}}, \qquad 0 < r = (\alpha_1^2 + \alpha_2^2 + \alpha_3^2)^{1/2} < \infty. \qquad (7.15)$$

From this representation we can calculate $\Delta P = P_{11} + P_{22} + P_{33}$:

$$\Delta P = \sum_{m=0}^{\infty} c_m \left\{ \frac{\Delta U_m}{r^{m-1}} + 2 \sum_{i=1}^{3} \frac{\partial}{\partial \alpha_i} \left(\frac{1}{r^{m-1}} \right) \frac{\partial U_m}{\partial \alpha_i} + U_m \Delta \left(\frac{1}{r^{m-1}} \right) \right\}. \qquad (7.16)$$

In order to calculate this expression, let us recall that $\Delta U_m = 0$, due to definition of spherical functions. Besides, since these functions are homogeneous with the power m, we have

$$\sum_{i=1}^{3} \frac{\partial}{\partial \alpha_i} \left(\frac{1}{r^{m-1}} \right) \frac{\partial U_m}{\partial \alpha_i} = -\frac{m-1}{r^{m+1}} \sum_{i=1}^{3} \alpha_i \frac{\partial U_m}{\partial \alpha_i} = -\frac{m-1}{r^{m+1}} m U_m. \qquad (7.17)$$

At last,

$$\Delta \left(\frac{1}{r^{m-1}} \right) = \frac{(m-1)(m-2)}{r^{m+1}}, \quad \text{so } \Delta P = -\sum_{m=0}^{\infty} c_m \frac{(m-1)(m+2)}{r^{m+1}} U_m(\alpha_1, \alpha_2, \alpha_3). \qquad (7.18)$$

If we apply the last identity for $r = 1$, so as to consider the Christoffel equation (7.10) on the unit sphere, we arrive at the following representation:

$$\Delta P = -\sum_{m=0}^{\infty} c_m (m-1)(m+2) U_m(\alpha_1, \alpha_2, \alpha_3), \qquad r = (\alpha_1^2 + \alpha_2^2 + \alpha_3^2)^{1/2} = 1. \qquad (7.19)$$

Recall that the set of spherical functions form a complete and orthogonal system of functions. Therefore, let us expand the given function $R_1 + R_2$, as a function of the unit normal, i.e., as a function of the spherical coordinates (θ, φ) with $r = 1$, to a series in the spherical functions:

$$R_1 + R_2 = \sum_{m=0}^{\infty} d_m\, U_m(\alpha_1,\, \alpha_2,\, \alpha_3), \qquad (7.20)$$

where we consider coefficients d_m to be known. Then Eq. (7.10) involves

$$(m-1)(m+2)\, c_m = d_m \quad \sim \quad c_m = \frac{d_m}{(m-1)(m+2)}. \qquad (7.21)$$

Therefore, the solution to the Christoffel problem (7.10) is given by the following equality:

$$P = \sum_{m=0}^{\infty} \frac{d_m}{(m-1)(m+2)}\, U_m. \qquad (7.22)$$

You can notice that this formula defines a correct solution only if the coefficient with $m = 1$ is trivial: $d_1 = 0$. Due to orthogonality of the spherical functions in relation (7.20) this is equivalent to

$$\int_{r=1} (R_1 + R_2)\, U_1\, d\Omega = 0, \qquad (7.23)$$

where $d\Omega$ is an elementary space angle. Taking the concrete form of the spherical function U_1 this in particular involves

$$\int_{r=1} (R_1 + R_2)\, \alpha_i\, d\Omega = 0 \quad \sim \quad \int_{r=1} (R_1 + R_2)\, \mathbf{n}\, d\Omega = 0. \qquad (7.24)$$

The last condition is the necessary condition for the Christoffel theorem to be valid. Pogorelov (1973) proves that sufficient conditions are given by the set of the three relations, where the relation representing the necessary condition is only the first condition, among three other ones:

$$1)\ \int_{r=1} (R_1 + R_2)\, \mathbf{n}\, d\Omega = 0, \quad 2)\ (R_1 + R_2) \geq 0, \quad 3)\ (R_1 + R_2) - (R_1 + R_2)_{ss} \geq 0, \quad (7.25)$$

where differentiation is applied with respect to the arc length (s) of any circle of a unit radius on the considered sphere.

Helpful remarks

The method discussed in this section is not applicable for reconstruction of the convex obstacle shape from the known scattering diagram, since this problem is reduced to a Minkowski problem rather than a Christoffel problem. However this study demonstrates that the inverse problem for convex 3D convex surface, at least in the case of linear Christoffel formulation, can be efficiently resolved with the help of treatment in spherical coordinates.

7.2. Reducing Inverse Problem of the Short-Wave Diffraction to Minkowski Problem

As follows from Eq. (6.120), the far-field scattered amplitude in a direct diffraction problem for convex obstacles is expressed in terms of the Gaussian curvature. Therefore, in practical (inverse) reconstruction problems, when the scattered amplitude is usually known for a set of observation angles being received in a far zone, we arrive at the inverse problem which coincides with the classical *Minkowski problem* well known in differential geometry (see, for example, Blaschke, 1930; Pogorelov, 1973, 1978). This problem can be formulated in the following way.

Let S be a sufficiently smooth closed convex surface with the given positive continuous Gaussian curvature known as a function of the outward normal: $K = K(\mathbf{n})$. It is required to reconstruct the shape of the surface itself. Here we study the questions of existence and uniqueness. This problem was solved in part by H. Minkowski himself; then some important results were obtained by A.D. Aleksandrov and A.V. Pogorelov. Our study follows in the main Pogorelov (1973, 1978). All approaches are based on approximation of the given convex surface by a convex polyhedron.

Let us first derive a necessary condition for the Minkowski problem to be solvable. We first start from the consideration of arbitrary convex polyhedron P_N with N faces, with $\mathbf{n}_i$ being the normal to the ith face and S_i its area. Let $\mathbf{e}$ be an arbitrary fixed unit vector in the space, and we displace the polyhedron as a solid figure in direction of this unit vector, by the distance ε. Displacement of the ith plane involves the change of the polyhedron volume by $S_i(\mathbf{n}_i \cdot \mathbf{e})\varepsilon$. Since the total volume of the polyhedron does not change, we can conclude that the following equality must hold:

$$\sum_{i=1}^{N} (\mathbf{n}_i \cdot \mathbf{e})\, S_i = 0. \tag{7.26}$$

Now, due to arbitrariness of the vector $\mathbf{e}$, it follows from Eq. (7.26) the necessary condition of solvability, valid for arbitrary closed polyhedron

$$\sum_{i=1}^{N} \mathbf{n}_i\, S_i = \mathbf{0}. \tag{7.27}$$

Further, let us consider an arbitrary smooth closed convex surface S. We can arrange a sequence of convex polyhedra $\{P_N\}$, $N = 1, 2, \ldots$, which converges to S. Then for any polyhedron P_N equality (7.27) is valid, so in the limit $N \to \infty$ we arrive at the following condition, which must be valid for any closed convex surface S:

$$\int_S \mathbf{n}\, ds(\mathbf{n}) = \mathbf{0}. \tag{7.28}$$

If $K(\mathbf{n})$ designates Gaussian curvature of the surface at the point with normal $\mathbf{n}$, then as known from differential geometry (see, for instance, Pogorelov, 1961), $ds(\mathbf{n}) = d\omega(\mathbf{n})/K(\mathbf{n})$, where $d\omega$ is an elementary solid angle of space in direction $\mathbf{n}$. Substitution of the last relation into Eq. (7.28) leads to the necessary condition, which the given function $K(\mathbf{n})$ of our inverse problem must satisfy, in order to provide solvability of the inverse problem:

$$\int_\Omega \frac{\mathbf{n}\, d\omega(\mathbf{n})}{K(\mathbf{n})} = \mathbf{0}, \tag{7.29}$$

where Ω is the full solid angle of the space.

We may pose quite naturally the question, whether this condition is also sufficient to guarantee the existence of a closed convex surface given by its Gaussian curvature. Unlike the Christoffel problem, the Minkowski problem cannot be solved by a constructive method, which permits any analytical representation for its solution. A real way to establish sufficient conditions is again to operate with approximation of the convex surface by inscribed polyhedra. The following results were stated by Alexandrov (see Alexandrov and Zalgaller, 1967).

Let $\mathbf{n}_i$, $i = 1, 2, \ldots, N$, be a system of noncoplanar unit vectors, and $S_i > 0$, $i = 1, 2, \ldots, N$, are some positive numbers, satisfying condition (7.27). Then there exists a unique closed convex polyhedron P_N, whose faces have outward normals $\mathbf{n}_i$ and respective areas S_i. From this basic result of the theory of convex polyhedra Pogorelov (1973, 1978) proves the theorem, which declares:

Suppose there is a unit sphere with center at the origin of a Cartesian coordinate system and suppose a continuous positive function $K(\mathbf{n}) > 0$ determined for every point on the sphere with radius vector $\mathbf{n}$ is given. Let condition (7.29) be valid, where the integral is taken over this sphere. Then there exists a unique closed convex surface S, which for any point of the surface with the outward normal $\mathbf{n}$ has the Gaussian curvature $K(\mathbf{n})$.

The results of this section allow us, in principle, to uniquely reconstruct the shape of a convex closed obstacle from the data on the wave amplitude scattered by the present obstacle in a far field.

Helpful remarks

$1°$. The quoted classical results of differential geometry guarantee the existence and uniqueness of a solution to the Minkowski problem, but do not offer you any specific efficient way to construct such surface. Some concrete algorithms of the shape reconstruction of convex obstacle can be found below in the forthcoming sections.

$2°$. It should be noted that reconstruction of the convex obstacle shape, which is reduced in the high-frequency range to the Minkowski problem, allows us to come to some interesting and very important conclusions, related to uniqueness of the studied inverse problem.

Let us assume that in the considered short-wave range there is known the back-scattered amplitude of some nonconvex obstacle as a function of the observation angle α: $A(\alpha)$. Obviously, this function is positive and continuous (some results on analytic properties of the scattered wave field can be found, for example, in Colton and Kress, 1983). As follows from Eq. (6.120), where we put $L = L_0$, the considered back-scattered far-field amplitude is

$$A(\alpha) = \frac{\sqrt{R_1 R_2}}{2L_0^2} = \frac{|p^{\text{inc}}|}{2\sqrt{K}\, L_0}, \qquad p^{\text{inc}} = \frac{e^{ikL_0}}{L_0}, \qquad (7.30)$$

which is evidently of the same dimension as the incident wave.

Let us introduce the auxiliary function

$$K(\alpha) = \frac{|p^{\text{inc}}|^2}{4\, A^2(\alpha)\, L_0^2}. \qquad (7.31)$$

The main result stated above in the present section declares that if condition (7.29) holds, then there is a unique convex obstacle whose boundary surface Gaussian curvature is given by Eq. (7.31). This implies a very important conclusion. Namely, let $A(\alpha)$ be a back-scattered far-field diagram for any (not necessarily convex) obstacle for a fixed high frequency. If

$$\int_\Omega A^2(\alpha)\, \mathbf{n}(\alpha)\, d\alpha = \mathbf{0}, \qquad (7.32)$$

then, in addition to the real obstacle generated by this diagram, there is always some convex obstacle with the scattered diagram $A(\alpha)$.

An interesting conclusion follows from the last observation. If we consider any (non-convex) obstacle symmetric with respect to the origin of the coordinate system, then it is obvious, due to natural symmetry, that relation (7.32) is always valid, because the contributions of the integrand with positive and negative values cancel each other out. The same fact takes place if the object is symmetric with respect to all Cartesian coordinate axes. So, in this case we can guarantee (at least at high frequencies) nonuniqueness of the solution of the considered inverse problem on reconstruction of the boundary surface of the obstacle in acoustic medium by its known back far-field scattering pattern.

7.3. Explicit Results for a Differential Operator of the 2D Inverse Problem

Let us now study the two-dimensional problem. By analogy to the general case, the back-scattered amplitude, in ray approximation, is proportional to $|\rho(\mathbf{q})|^{1/2}$, where ρ is a radius of curvature at the point of specular reflection (see Chapter 6). The problem is again nonlinear, and it is wonderful that it can be reduced to a linear partial differential equation with respect to the Minkowski function.

To derive that equation, let us start from the Rodrigues formula again, which can be written along the contour: $dx_i - \rho\, d\alpha_i = 0$, $i = 1, 2$. Since $x_i = \partial P(\alpha_1, \alpha_2)/\partial\alpha_i$ ($x_i = P_i$), we have $dx_i = \sum_{k=1}^{2} \partial^2 P/\partial\alpha_k \partial\alpha_i$, so

$$\sum_{k=1}^{2} \frac{\partial^2 P}{\partial\alpha_k\, \partial\alpha_i}\, d\alpha_k - \rho\, d\alpha_i = 0, \qquad (i = 1, 2). \tag{7.33}$$

The last homogeneous system has a nontrivial solution, so its determinant is equal to zero, with $i, k = 1, 2$:

$$\det\left(\frac{\partial^2 P}{\partial\alpha_k\, \partial\alpha_i} - \delta_{ik}\rho\right) = 0 \sim \rho\left[\rho - \left(\frac{\partial^2 P}{\partial\alpha_1^2} + \frac{\partial^2 P}{\partial\alpha_2^2}\right)\right] + \det\left(\frac{\partial^2 P}{\partial\alpha_k\, \partial\alpha_i}\right) = 0. \tag{7.34}$$

It can be proved, by analogy with the general case, that a homogeneity of P involves the last determinant to vanish. Hence we obtain

$$\frac{\partial^2 P}{\partial\alpha_1^2} + \frac{\partial^2 P}{\partial\alpha_2^2} = \rho \quad \sim \quad \frac{d^2 p(\theta)}{d\theta^2} + p(\theta) = \rho(\theta), \tag{7.35}$$

where $\rho(\theta)$ is a curvature radius of the obstacle boundary curve. The passage from the first to the second equation in formula (7.35) evidently follows the change of variables, from Cartesian to polar coordinates over the unit circle $r = 1$.

It is well known that the solution of the inhomogeneous ordinary differential equation (7.35), for arbitrary function $\rho(\theta)$, is

$$p(\theta) = p^0(\theta) + p^*(\theta), \quad p^0(\theta) = C_1 \sin\theta + C_2 \cos\theta,$$

$$p^*(\theta) = \sin\theta \int_{\theta_0}^{\theta} \rho(\theta)\cos\theta\, d\theta - \cos\theta \int_{\theta_0}^{\theta} \rho(\theta)\sin\theta\, d\theta, \tag{7.36}$$

where $p^0(\theta)$ is the general solution of the homogeneous equation, and $p^*(\theta)$ is a particular solution of the inhomogeneous one. The latter is constructed by a variation method applied to the constants C_1 and C_2.

We have tested experimentally this theoretical method of reconstruction, which is suitable for 2D convex obstacles, on some artificially produced voids made in thick metallic rods. The results on measurements of the back-scattered far-field amplitude are presented in Fig. 7.1. As can be seen from this figure, measurements on the back-scattered diagram there are carried out with the angular step $30°$.

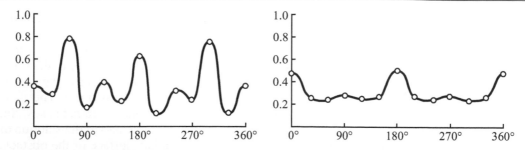

Figure 7.1. Back-scattered diagram for artificially made triangular and oval voids

It should be noted that despite so poor experimental data (collected with the angular step 30°) results of reconstructions show indeed that the proposed method is very stable. The results of reconstructions are demonstrated in Fig. 7.2, where real boundary is marked by a solid and the reconstructed contour by a dashed line.

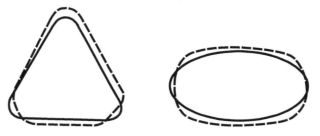

Figure 7.2. Results of reconstruction of triangular and oval voids

Another possible approach here is to directly solve the linear differential equation (7.35) by a finite-difference method, and our detailed analysis confirm efficiency of this method too. It is interesting to demonstrate our scheme on a simple obstacle, such as round disk. If the radius of the disk is equal to R, then in this case $\rho(\theta) \equiv R$, $p(\theta) \equiv R$ if the origin of the coordinate system is taken in the center of the obstacle. Evidently, for these values of $\rho(\theta)$ and $p(\theta)$, Eq. (7.35) holds. Let us study the influence of the origin position inside the obstacle on the reconstruction process. The numerical investigation shows that when we use a symmetrical finite-difference representation for the derivatives: $p_{\theta\theta}^i = (p^{i+1} - 2p^i + p^{i-1})/(\Delta\theta)^2$ (i is the number of a node in the grid, $\Delta\theta$ is the step of the discretization), then we come numerically to the solution $\rho(\theta) \equiv R$ cited above. Moreover, symmetrical location of the origin is obtained for any symmetric obstacle, since we use the symmetric finite-difference form for the derivative. This result can be derived from the observation that any symmetry in the vector $\{p^i\}$ involves the same symmetry in $\rho(\theta)$, by using uniqueness of the inversion considered in Sections 7.1, 7.2. For an arbitrary obstacle, which does not possess any symmetry, the reconstruction algorithm leads to an origin that can change itself with a change of the finite-difference scheme.

Some more details concerning this problem can be found in the author's work Sumbatyan et al. (1993).

Helpful remarks

Note that the studied inverse problem is nonlinear both in the 3D and 2D cases. And it is wonderful that its formulation in terms of the Minkowski support function reduces to the linear ordinary differential equation with constant coefficients (7.35) admitting exact analytical solution in explicit form.

7.4. Exact Explicit Inversion of the Basic Operator in the Case of Axial Symmetry

Let an axially symmetric obstacle bounded by the surface S be spherically irradiated by a high-frequency acoustic wave; the propagation time of the reflected pulse $t(q)$ and the real amplitude of the reflected wave $|A(q)|$ are known for any direction q. We denote the convex parts of the surface S by S_{ex} and the nonconvex parts by S_{in}. We restrict our consideration to an obstacle with a simply connected surface S allowing no more than two points of single mirror reflection at any direction of irradiation in the echo mode. It is shown in Druzhinina and Sumbatyan (1992) that this condition is necessarily satisfied if the angle between the normals erected to the surface S at any two of its points within one nonconvex region is acute. All these facts will be used in the forthcoming study, but here let us restrict the consideration by axially symmetric convex obstacles.

We consider the axially symmetric surface S bounding an obstacle in a Cartesian coordinate system $OXYZ$, where the OZ-axis coincides with the axis symmetry of S, since the surface S is axially symmetric, for its intersection with any axial plane. For the sake of definiteness, we consider the axial plane XOZ. In this plane, the unit vector q has the coordinates $\{\sin\theta, 0, \cos\theta\}$, and $A = A(\theta)$. Here, θ is the angle between the axis of symmetry OZ and the direction q.

First of all, by the known travel time of the reflected pulse $t(\theta)$, we determine the distance $p(\theta)$ to a tangent at the point of specular reflection, which, in the axial plane, is given by the equation

$$x \sin\theta + z \cos\theta - p(\theta) = 0. \tag{7.37}$$

By this known function $p(\theta)$, we construct a convex hull of the surface:

$$x(\theta) = -p'(\theta)\cos\theta + p(\theta)\sin\theta, \qquad z(\theta) = p'(\theta)\sin\theta + p(\theta)\cos\theta, \tag{7.38}$$

which determines the smallest convex contour S_0 enclosing the given one (see Preparata and Shamos, 1985).

This ill-posed procedure will be described in more detail in Section 8.9. Instead here we describe a stable (well-posed) approach of reconstruction on the basis of the known back-scattered amplitude, i.e., by using the function $A(\theta)$. In this case, the back-scattered amplitude in a far-field zone (we omit here a certain insignificant coefficient) is expressed in terms of the Gaussian curvature. It will be shown in the next section that in the case of axial symmetry this is described by the following nonlinear ordinary differential equation of the second order:

$$\left(\frac{d^2 p}{d\theta^2} + p\right)\left(\frac{dp}{d\theta}\cot\theta + p\right) = \gamma^{-1}(\theta), \tag{7.39}$$

where $\gamma(\theta)$ is the known Gaussian curvature.

Note that the first factor of operator (7.39) of the axially symmetric inverse problem is the operator of the inverse problem for a plane case discussed in the previous section, the main equation of which is quoted here again

$$\frac{d^2 p(\theta)}{d\theta^2} + p(\theta) = \rho(\theta), \tag{7.40}$$

with its solution

$$p(\theta) = p^0(\theta) + p^*(\theta), \quad p^0(\theta) = C_1 \sin(\theta) + C_2 \cos(\theta),$$
$$p^*(\theta) = \sin\theta \int_{\theta_0}^{\theta} \rho(\theta)\cos\theta\, d\theta - \cos\theta \int_{\theta_0}^{\theta} \rho(\theta)\sin\theta\, d\theta. \tag{7.41}$$

The solution of the nonlinear equation for an axially symmetric case (7.39) can be obtained in an explicit form by reducing it to a sequential solution of two linear differential equations of the first order with variable coefficients. Here we should note that, by using a differential operator of the first order with variable coefficients, the first differential factor with constant coefficients on the left-hand side of (7.39) can be expressed in terms of the second factor (also with variable coefficients):

$$\frac{d^2p}{d\theta^2} + p = \tan\theta \frac{d}{d\theta}\left(\frac{dp}{d\theta}\cot\theta + p\right) + \left(\frac{dp}{d\theta}\cot\theta + p\right). \tag{7.42}$$

This representation makes it possible by the substitution

$$\frac{dp(\theta)}{d\theta}\cot\theta + p(\theta) = U(\theta) \tag{7.43}$$

to reduce the initial equation to a Bernoulli equation

$$\tan\theta \frac{dU(\theta)}{d\theta} + U(\theta) = U^{-1}(\theta)\gamma^{-1}(\theta). \tag{7.44}$$

Following the general theory of differential equations, by the substitution $U^2(\theta) = f(\theta)$, equation (7.44) can be reduced to a linear equation for the function $f(\theta)$,

$$\frac{df(\theta)}{d\theta} + 2\cot\theta f(\theta) = 2\gamma^{-1}(\theta)\cot\theta. \tag{7.45}$$

We seek the solution to this equation in the form of the product of two functions $f(\theta) = u(\theta)v(\theta)$. We select the function $v(\theta)$ as a particular solution of the homogeneous equation $v(\theta) = \sin^{-2}\theta$. Then, $u(\theta)$ satisfies the equation

$$\frac{du(\theta)}{d\theta} = \gamma_2^{-1}(\theta)\sin 2\theta, \tag{7.46}$$

the general solution of which is

$$u(\theta) = \int_{\theta_0}^{\theta} \gamma^{-1}(\theta)\sin 2\theta \, d\theta + C_1. \tag{7.47}$$

Thus, we determine the function $U(\theta)$, which is the general solution to equation (7.44):

$$U(\theta) = f^{1/2}(\theta) = \frac{1}{\sin\theta}\left[\int_{\theta_0}^{\theta} \gamma_2^{-1}(\theta)\sin 2\theta \, d\theta + C_1\right]^{1/2}. \tag{7.48}$$

Once $U(\theta)$ has been found, the right-hand side of equation (7.43) in $P(\theta)$ becomes known. The solution to the linear differential equation (7.43) is obtained in the same way as the solution to equation (7.45). As a result, the function $P(\theta)$ can be represented as

$$P(\theta) = \cos\theta\left\{\int_{\theta_0}^{\theta}\left[\cos^{-2}\theta\left(\int_{\theta_0}^{\theta}\gamma_2^{-1}\theta\sin 2\theta \, d\theta + C_1\right)^{1/2}\right] + C_2\right\}. \tag{7.49}$$

The results presented here follow the author's work (Boyev and Sumbatyan, 1999).

Helpful remarks

It should be noted that both the plane, Eq. (7.41), and axially symmetric, Eq. (7.49), general solutions to problems contain, in accordance with a classical theory of ordinary differential equations, a pair of arbitrary constants C_1 and C_2. In both the plane and axial symmetric cases the function $p(\theta)$ gives a plane boundary curve. The only difference is that in the case of axial symmetry the boundary curve represents a certain axial cut. In order to determine uniquely the function $p(\theta)$ in Eq. (7.49) (as well as in Eq. (7.41) for the plane problem) for a concrete surface S, we need to assign two boundary (or initial) conditions. Usually such conditions are predetermined by a periodicity of the solution.

7.5. Nonlinear Differential Operator of the Three-Dimensional Inverse Problem

In the generic 3D problem the problem cannot be resolved by any analytical method, since this reduces to the Minkowski problem, which can be represented by a strongly nonlinear equation (7.11). In Section 7.2 we established some general theoretical results on solvability and uniqueness of this problem, and remarked that there is no known algorithm in the literature with a concrete reconstruction method. So this is the principal goal of our investigation in the present section.

Let us pass in Eq. (7.11) from Cartesian coordinates to the spherical ones (r, θ, φ). Since we consider obstacles with a closed boundary surface S, the end of the external normal vector traces over a sphere of radius r. When passing to spherical coordinates, in expressions for the derivatives P_i, P_{ij}, let us recall the following properties of homogeneity (see Section 7.1):

$$\frac{\partial P}{\partial r} = \frac{P}{r}, \qquad \frac{\partial P_i}{\partial r} = 0, \qquad i = 1, 2, 3. \tag{7.50}$$

Eq. (7.50) implies that the functions $P(\mathbf{q})$ and $p(\mathbf{q})$ coincide to each other over the unit sphere $r = 1$. It is rather convenient to solve this equation on the unit sphere with respect to the unknown function $p(\theta, \varphi)$. Let us write out, in terms of the function $p(\theta, \varphi)$ on the unit sphere, the expressions of the first derivatives of P,

$$x_1 = P_1 = p \sin \theta \cos \varphi + \frac{\partial p}{\partial \theta} \cos \theta \cos \varphi - \frac{\partial p}{\partial \varphi} \frac{\sin \varphi}{\sin \theta},$$

$$x_2 = P_2 = p \sin \theta \sin \varphi + \frac{\partial p}{\partial \theta} \cos \theta \sin \varphi + \frac{\partial p}{\partial \varphi} \frac{\cos \varphi}{\sin \theta}, \tag{7.51}$$

$$x_3 = P_3 = p \cos \theta - \frac{\partial p}{\partial \theta} \sin \theta,$$

and the second derivatives of P,

$$P_{11} = \left(\frac{\partial^2 p}{\partial \theta^2} + p \right) \cos^2 \theta \cos^2 \varphi + \left(\frac{\partial p}{\partial \theta} \cot \theta + p \right) \sin^2 \varphi$$
$$- 2 \cot \theta \sin \varphi \cos \varphi \frac{\partial}{\partial \varphi} \left(\frac{\partial p}{\partial \theta} - p \cot \theta \right) + \frac{\sin^2 \varphi}{\sin^2 \theta} \frac{\partial^2 p}{\partial \varphi^2}, \tag{7.52}$$

$$P_{22} = \left(\frac{\partial^2 p}{\partial \theta^2} + p \right) \cos^2 \theta \sin^2 \varphi + \left(\frac{\partial p}{\partial \theta} \cot \theta + p \right) \cos^2 \varphi$$
$$+ 2 \cot \theta \sin \varphi \cos \varphi \frac{\partial}{\partial \varphi} \left(\frac{\partial p}{\partial \theta} - p \cot \theta \right) + \frac{\cos^2 \varphi}{\sin^2 \theta} \frac{\partial^2 p}{\partial \varphi^2}, \tag{7.53}$$

$$P_{33} = \left(\frac{\partial^2 p}{\partial \theta^2} + p \right) \sin^2 \theta, \tag{7.54}$$

$$
\begin{aligned}
P_{12} = &\left(\frac{\partial^2 p}{\partial \theta^2} + p \right) \cos^2 \theta \sin \varphi \cos \varphi - \left(\frac{\partial p}{\partial \theta} \cot \theta + p \right) \sin \varphi \cos \varphi \\
&+ \cot \theta \left(\cos^2 \varphi - \sin^2 \varphi \right) \frac{\partial}{\partial \varphi} \left(\frac{\partial p}{\partial \theta} - p \cot \theta \right) - \frac{\sin \varphi \cos \varphi}{\sin^2 \theta} \frac{\partial^2 p}{\partial \varphi^2},
\end{aligned}
\tag{7.55}
$$

$$P_{13} = -\sin \theta \left[\left(\frac{\partial^2 p}{\partial \theta^2} + p \right) \cos \theta \cos \varphi - \frac{\sin \varphi}{\sin \theta} \frac{\partial}{\partial \varphi} \left(\frac{\partial p}{\partial \theta} - p \cot \theta \right) \right], \tag{7.56}$$

$$P_{23} = -\sin \theta \left[\left(\frac{\partial^2 p}{\partial \theta^2} + p \right) \cos \theta \sin \varphi + \frac{\cos \varphi}{\sin \theta} \frac{\partial}{\partial \varphi} \left(\frac{\partial p}{\partial \theta} - p \cot \theta \right) \right]. \tag{7.57}$$

In geographical coordinates (θ, φ) of the unit external normal, Eq. (7.11) can thus be rewritten for the function $p(\theta, \varphi)$ as follows:

$$L_\theta \bar{L}_\theta(p) \left[\cot \theta \bar{L}_\theta(p) + \frac{1}{\sin^2 \theta} L_\varphi \bar{L}_\varphi(p) \right] - \frac{1}{\sin^2 \theta} \left[\frac{\partial}{\partial \varphi} \bar{L}_\theta(p) \right]^2 = \gamma^{-1}(\theta, \varphi), \tag{7.58}$$

where

$$
\begin{aligned}
L_\theta(p) &= \frac{\partial p}{\partial \theta} + p \cot \theta, &\bar{L}_\theta(p) &= \frac{\partial p}{\partial \theta} - p \cot \theta, \\
L_\varphi(p) &= \frac{\partial p}{\partial \varphi} + p \cot \varphi, &\bar{L}_\varphi(p) &= \frac{\partial p}{\partial \varphi} - p \cot \varphi.
\end{aligned}
\tag{7.59}
$$

Such a representation of the main operator (7.11) of the considered inverse problem in the form (7.58)–(7.59) allows us to extract operators of the respective inverse problems for the two-dimensional case and the case with axial symmetry (see Sections 7.3 and 7.4). As we could see above, in both cases the main differential operator permits explicit inversion.

In the considered generic 3D case we can arrange a mapping of the considered closed surface S to the unit sphere. Such a mapping is achieved if one draws unit vectors, with their origin being put at any fixed center, in direction of the unit normal to the corresponding point of the surface.

Reconstruction of the function $p(\theta, \varphi)$ can be realized by a numerical *step by step* algorithm. In this numerical algorithm, on the unit sphere surface, central differences are used for the first (p'_θ, p'_φ) and the second $(p''_{\theta\theta}, p''_{\theta\varphi}, p''_{\varphi\varphi})$ derivatives over the mesh which consists of parallels $\theta = \theta_i$ ($0 \le \theta_i \le \pi$) and meridians $\varphi = \varphi_j$ ($0 \le \varphi_j < 2\pi$), with a uniform mesh interval, which is assumed to be the same along θ and φ: $h_\theta = \pi/N$, $h_\varphi = 2\pi/(2N)$, $h_\theta = h_\varphi = h$. We have at node (i, j)

$$
\begin{aligned}
\frac{\partial p}{\partial \theta} &= \frac{p^{i+1,j} - p^{i-1,j}}{2h}, &\frac{\partial p}{\partial \varphi} &= \frac{p^{i,j+1} - p^{i,j-1}}{2h}, \\
\frac{\partial^2 p}{\partial \theta^2} &= \frac{p^{i+1,j} - 2p^{i,j} + p^{i-1,j}}{h^2}, &\frac{\partial^2 p}{\partial \varphi^2} &= \frac{p^{i,j+1} - 2p^{i,j} + p^{i,j-1}}{h^2}, \\
\frac{\partial^2 p}{\partial \theta \partial \varphi} &= \frac{p^{i+1,j+1} - p^{i-1,j+1} - p^{i+1,j-1} - p^{i-1,j-1}}{4h^2}.
\end{aligned}
\tag{7.60}
$$

If we substitute relations (7.60) into Eq. (7.58), then we come to a nonlinear algebraic system, which can be written in the operator form

$$Ap = \gamma_2^{-1}, \quad p = \{p_1, p_2, \ldots, p_M\}, \quad \gamma^{-1} = \{\gamma^{-1}(\theta_1, \varphi_1), \ldots, \gamma^{-1}(\theta_M, \varphi_M)\}, \tag{7.61}$$

where M is the total number of the mesh nodes. In order to solve this nonlinear operator equation, we apply iterations of the Newton–Kantorovich method (see Kantorovich and Akilov, 1982), to solve this system

$$p_{n+1} = p_n + z_n,$$ (7.62)

with z_n being defined from the linear algebraic system

$$J z_n = \gamma^{-1} - A p_n.$$ (7.63)

Here $J = A'$ is a Jacobian of system (7.61).

As regards a concrete implementation of this method, note that the Newton–Kantorovich method requires in our problem a linear array of unknown elements p^k. In order to arrange a one-dimensional vector of unknowns p^k on the two-dimensional mesh, it is necessary to enumerate the nodes (θ_i, φ_j) so as to obtain a linear array. We propose the following strategy. For each φ_j sequential enumeration corresponds to increasing of θ_i: $h \le \theta_i \le \pi - h$. Then the process jumps to the next value φ_{j+1}, by increasing of the index j. The last two nodes are two poles. As a result, dimension of the nonlinear algebraic system (7.61) is $M = 2N(N-1) + 2$.

Let us give explicit expressions for the elements of the Jacobian.

$$J_{i,j} = -\left[\left(1 - \frac{2}{h^2 \sin^2 \theta_i} \right) \left(P_{\theta\theta}^i + P^i \right) + \left(1 - \frac{2}{h^2} \right) \left(P_{\theta}^i \cot \theta_i + \frac{P_{\varphi\varphi}^i}{\sin^2 \theta_i} + P^i \right) \right],$$

$$J_{i,j\pm1} = -\left[\pm \frac{\cot \theta_i}{2h} \left(P_{\theta\theta}^i + P^i \right) + \frac{1}{h^2} \left(P_{\theta}^i \cot \theta_i + \frac{P_{\varphi\varphi}^i}{\sin^2 \theta_i} + P^i \right) \right],$$

$$J_{i,j\pm N} = -\left[-\frac{1}{h^2 \sin^2 \theta_i} \left(P_{\theta\theta}^i + P^i \right) \pm \frac{\cos \theta_i}{h \sin^3 \theta_i} \left(P_{\varphi}^i \cot \theta_i - P_{\theta\varphi}^i \right) \right],$$ (7.64)

$$J_{i,j\pm(N+1)} = \frac{2}{h^2 \sin^2 \theta_i} \left(P_{\varphi}^i \cot \theta_i - P_{\theta\varphi}^i \right),$$

$$J_{i,j\pm(N-1)} = -\left[\frac{2}{h^2 \sin^2 \theta_i} \left(P_{\varphi}^i \cot \theta_i - P_{\theta\varphi}^i \right) \right].$$

Figure 7.3 demonstrates the reconstruction of a three-dimensional obstacle, which is a circular cylinder with smoothed edges, when the back-scattered amplitude is known with the step of $\pi/8$ in both ϕ and θ. The third iteration is practically indistinguishable from exact surface. In order to produce data on the back-scattered diagram, we have solved the direct diffraction problem by the BIE method (see Chapter 2).

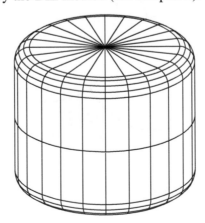

Figure 7.3. Reconstruction of a circular cylinder with smoothed edges

Some more details on the discussed reconstruction method can be found in the author's work (Sumbatyan and Troyan, 1992).

Helpful remarks

It should be noted that the Jacobian is a sparse matrix, with the most part of its nontrivial elements being situated near the main diagonal and its two neighbors. However, generally, the structure of the matrix is not *banded*. Consequently, the method to solve this system should take into account that its matrix is very sparse. A number of efficient methods for solving sparse linear algebraic systems can be found, for instance, in Pissanetski (1988).

7.6. Reconstruction of Nonconvex Obstacles in the High-Frequency Range: 2D Case

We assume that the acoustic medium contains a nonconvex obstacle (Fig. 7.4) with piece-wise smooth positive-curvature boundary L with M corner points. For this contour, we construct a convex hull L_0 (i.e., the smallest convex contour containing L_0). Let $L_m (\bigcup_{m-1}^{M} L_m = L \cap L_0)$ be the common segments of contours L and L_0, and l_m be the segments that belongs only to L $(\bigcup_{m=1}^{M} l_m = L \setminus L_0)$. We assume that, when L is echo sounded at an arbitrary angle α, the number of single scattering points is, at most, two. It has been demonstrated that, at boundary segments l_m, multiple scattering, which induces a back-scattering ray, must be absent if two normals n_{m1} and n_{m2} erected at two arbitrary points of l_m make an acute angle β_m (see Druzhinina and Sumbatyan, 1992). In view of all the limitations imposed on L, each segment l_m may only be constituted by two intersecting convex arcs l_{m1} and l_{m2}, with the angle between their normals at the point of intersection being acute.

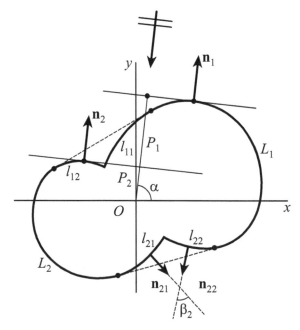

Figure 7.4. Diffraction of a plane acoustic wave by a 2D nonconvex obstacle

When the obstacle is echo-sounded at an arbitrary angle α, there always exists at least one point of specular reflection that lies on one of the boundary segments L_m. In the general case, near l_m, the diffraction pattern is more complicated. However, the above assumptions imply that, for a given angle of incidence, there exists one point of simple singular reflection lying on one of the arcs l_{m1} and l_{m2} of boundary segment l_m.

The reconstruction algorithm rests on the following consideration. We assume that, in circular echo-sounding, the known parameters are the arrival time of the reflected pulse $t(\alpha)$ and the real-valued amplitude of the reflected wave $|A(\alpha)|$, $0 \le \alpha < 2\pi$. The knowledge of $t(\alpha)$ completely defines the convex envelope of the object boundary, because it determines the support function $p(\alpha)$ defining the distance from some selected center O inside the region to the convex boundary $\bigcup_{m=1}^{M} L_m$ at the point with normal $n(\alpha)$ (see Fig. 7.4). The convex hull L_0 of contour L is the envelope of the family of these tangents. Its Cartesian coordinates are defined by

$$x = -p'_\alpha(\alpha) \sin \alpha + p(\alpha) \cos \alpha, \qquad y = p'_\alpha(\alpha) \cos \alpha + p(\alpha) \sin \alpha. \qquad (7.65)$$

Once the convex envelope L_0 (7.65) is constructed using the known function $t(\alpha)$, the arcs l_{m1} and l_{m2} may be determined for boundary segments l_m using the complex amplitude of the reflected wave $A(\alpha)$. In the range $0 \le \alpha < 2\pi$, where there exists only reflection from convex parts of L_m, the amplitude of the reflected wave is

$$A \sim \frac{e^{i(2kR - \pi/4)}}{\sqrt{\pi k}\sqrt{R + \dfrac{R^2}{\rho_1 \cos \vartheta}}}, \qquad (7.66)$$

where R is the distance between the point of observation and the boundary contour L, $\cos \vartheta = 1$, because in the case under consideration, $\vartheta = 0$, k is the wave number, and ρ_1 is the contour curvature radius at a point of specular reflection on a convex boundary segment. In the far field ($R \to \infty$), the last formula may be approximately written as

$$A(\alpha) \sim \sqrt{\rho_1(\alpha)}\, e^{-2ip_1(\alpha)k}. \qquad (7.67)$$

In other intervals of α, waves reflected from one point of $L_m(2)$ and one point on l_m overlap. In this case, the amplitude of the reflected wave reduces to the form

$$A(\alpha) = \sqrt{\rho_1(\alpha)}\, e^{-2ip_1(\alpha)k} + \sqrt{\rho_2(\alpha)}\, e^{-2ip_2(\alpha)k}. \qquad (7.68)$$

In formulas (7.66)–(7.68), $\rho_1(\alpha)$ and $\rho_2(\alpha)$ are the curvature radii at points of segments L_m and l_m respectively; $p_1(\alpha)$ and $p_2(\alpha)$ are the respective support functions. The complex back-scattering amplitude, defined by these expressions, has the form independent of the distance between the obstacle and the point of observation. Then, for the intensity of the reflected wave, we have

$$F(\alpha) = F(\alpha, k) = |A(\alpha)|^2 = \rho_1(\alpha) + \rho_2(\alpha) + 2\sqrt{\rho_1(\alpha)\rho_2(\alpha)}\, \cos[2k(p_1 - p_2)]. \qquad (7.69)$$

We note that formula (7.68) allows for the contributions from the points of ray reflection into the amplitude of the scattering wave. In the lit zone, as $k \to \infty$, the contribution from other points of the smooth boundary segments is a quantity of a higher order of smallness than k^{-1} (see Section 1.4). However the boundary has corner points. In the considered case free from multiple reflections, their contribution into $A(\alpha)$ is of the order of $k^{-1/2}$, and as $k \to \infty$, it becomes a small quantity.

In classical acoustic holography, it is assumed that the main information is contained in the oscillating component of formula (7.69). In this section we suggest an alternative treatment that rests on the following idea. We note that, in formula (7.69), only the last term depends on k. At high frequencies, it is a strongly oscillating function of a. Consequently, the first two terms are slowly oscillating functions of α only if the contour has no segments with sharp variations of curvature. Using filtering (e.g., digital), we retain only the slowly variable component in $F_0(\alpha)$; then

$$F_0(\alpha) = \rho_1(\alpha) + \rho_2(\alpha). \tag{7.70}$$

Now, on the convex part of L_m, the curvature radius $\rho_1(\alpha)$ may be deemed known after the convex envelope of the contour has been constructed. This construction defines the curvature of arcs l_{m1} and l_{m2} on l_m as

$$\rho_2(\alpha) = F_0(\alpha) - \rho_1(\alpha). \tag{7.71}$$

As noted above, for the selected class of boundary contours L, each segment l_m consists of two convex intersecting arcs l_{m1} and l_{m2} having a common tangent with the convex envelope L_0 (see Fig. 7.4). In order to reconstruct each of these arcs, we can use the approach developed in Section 7.3, which reduces the reconstruction of the convex parts of the boundary contour to the solution of an ordinary differential equation of the second kind with constant coefficients, Section 7.3, whose exact analytical solution is given explicitly. However, an alternative efficient approach can be found upon the natural equation of the plane curve (see, for example, Pogorelov, 1961):

$$s'_\alpha(\alpha) = \rho_2(\alpha), \tag{7.72}$$

where s is the arc length counted from the point of tangency. In the Cartesian coordinate system, this arc is described by

$$x(s) = -\int_0^s \sin\alpha(s)\,ds + x_0, \qquad y(s) = \int_0^s \cos\alpha(s)\,ds + y_0, \tag{7.73}$$

where

$$s = s(\alpha) = \int_{\alpha_0}^\alpha [F_0(\alpha) - \rho_1(\alpha)]\,d\alpha, \tag{7.74}$$

and x_0 and y_0 are the Cartesian coordinates of the intersection of the arc and the tangent to the convex envelope L_0. Equations (7.73) take into account the fact that, at any point of the contour, the tangent is directed at $\pi/2$ to the normal. We note also that the angle α_0 is defined by the normal to the point of tangency; this normal is common for adjacent segments L_m and L_{mj}, $j = 1, 2$, of the contour L. Indeed, this point is a point of tangency of a curvilinear arc belonging to the convex boundary segment and the above constructed convex envelope of the object. Since the boundary contour is smooth, the normal $n(\alpha_0)$ is uniquely defined at the point of tangency.

An appropriate choice of the upper bound of α plays an important role in using formulas (7.73) and (7.74). This bound must coincide with the *visible* segment of arc l_{mj}; this is equivalent to the fact that formula (7.74) is valid only for those α which permit specular reflection from arc l_{mj}. Evidently, these intervals of α are associated with the presence of two points of specular reflection (one on L_m, the other on l_{mj}). Thus, the considered intervals refer to α, for which the back-scattered amplitude $|A(\alpha)|$ is a strongly oscillating function in the high frequency range (see formula (7.69)).

For a better insight into this problem, Figure 7.5 shows the pattern of back-scattering from an acoustically soft obstacle configured as two equal circles of radius R with a center-to-center distance $\sqrt{2}\,R$ ($R = 3\lambda$, and $\lambda = 2\pi/k$ is the wavelength). The back-scattering pattern $A(\alpha)$ was determined by solving the direct diffraction problem by the BIE method (see Chapter 2).

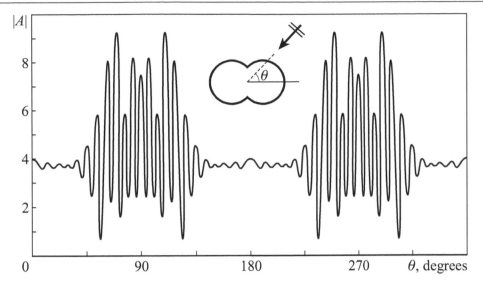

Figure 7.5. Amplitude of scattering by a pair of intersecting circles

The function $|A|$ was defined by solving a direct scattering problem by BIE. Note that the angle between normals to arbitrary pair of points over any internal part S_{in} is less than $\pi/2$. The only case, when a limiting value $\pi/2$ is reached, is related to the points of intersection between the circles. Therefore, as mentioned above, only points of single specular (*mirror*) reflection exist in the region S_{in}, for arbitrary direction of incidence, and multiple ray re-reflections are impossible. It is clear from Fig. 7.5, when passing over the segment $0 \le \theta < 2\pi$, that there are both intervals with slow variance of the amplitude $|A(\theta)|$ and intervals with high oscillations of this function. Approximate values of θ_1 and θ_2 are just chosen as points of *jump* from slow to high oscillations.

In practical implementations of the numerical algorithm, the upper integration limit, α_1, in (7.74) is determined approximately because we cannot exactly determine the boundary between segments of quick and slow variation of the known scattering amplitude. In all considered tests, the arcs of each pair l_{m1}, l_{m2} go somewhat beyond the point of intersection. Therefore, in the reconstruction of the simply connected domain, one must select only those parts of these arcs, which lie from the point of tangency with the convex envelope L_0 to the point of their intersection.

The proposed reconstruction algorithm involves a natural method for extracting the low-frequency component $F_0(\alpha)$ of $F(\alpha) = |A(\alpha)|^2$. It is based on the fact that, upon substituting the experimentally determined $F(\alpha)$ in (7.74), the integration in this formula,

$$s = \int_{\alpha_0}^{\alpha} [F(\alpha) - \rho_1(\alpha)] \, d\alpha, \tag{7.75}$$

provides the filtering of the low-frequency component. For well-defined computation of the integral (7.75), we should define correctly the value of the small integration step. The convex envelope defines the characteristic dimension d of the obstacle. If we assume that ten points per wavelength suffice for well-defined computation of this integral, then the number of nodal points in the integration may be estimated as $10\, d/\lambda$. Formula (7.75) takes into account that the main term of the asymptotic expansion, obtained from the integral of the last oscillating term in (7.69) by the stationary phase method (see Section 1.4) as $k \to \infty$, is of the order of $O(k^{-1})$.

Numerical experiments indicate that this natural method of filtering provides a more accurate reconstruction than many known procedures. We can thus state that the calculation of

Figure 7.6. Reconstruction of an obstacle configured as a pair of intersecting circles

integrals themselves in these problems provide an efficient filtering of the highly oscillating integrands if the step of integration is fixed and does not depend on the wavelength.

A reconstruction of a pair of unit circles (Fig. 7.6 demonstrates the real circular back-scattered amplitude $|A|$) was performed. The accuracy of the reconstruction is not worse than 2–3 % uniformly all over directions of incidence. Note that solid lines here represent the true contour, and dashed lines represent the contour reconstructed by the outlined method.

The results presented here follow the author's work (Boyev et al., 1997).

Helpful remarks

This algorithm was tested by reconstructing the geometries, which do not belong to the considered class. One of them is presented in Fig. 7.7, where the angles between the normals to arcs l_{m1} and l_{m2} may be obtuse and even close to π. The region consists of two tangent circles with $R = 3\lambda$. In this case, segments l_{m1} and l_{m2} admit multiple reflections of rays. However, the accuracy of the reconstruction remains acceptable and, seemingly, the validity range of this method is really wider than the class of contours specified in this paper.

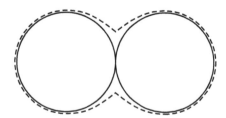

Figure 7.7. Reconstruction of an obstacle in the form of a pair of tangent circles

This conclusion is confirmed also by the reconstruction of a smooth contour in the form of a three-leaf rose,

$$\rho(\varphi) = a(2 + \cos 3\varphi), \qquad 0 \le \varphi < 2\pi, \tag{7.76}$$

shown in Fig. 7.8. This contour does not belong to the considered class because it contains segments with negative curvature. Dashed lines correspond to $a = 2\lambda$ (curve 1), $a = 4\lambda$ (curve 2), $a = 6\lambda$ (curve 3). The characteristic size of the obstacle is 10λ, 20λ, and 30λ, respectively. It is apparent that the reconstruction accuracy increases with frequency.

This contour also does not belong to the class, considered at the present work, because it contains some parts with negative curvature. Dashed lines correspond there to three different values of the parameter a, with λ being the wavelength. For all that, the average size of the obstacle is equal to 10λ, 20λ, 30λ, respectively. As can be seen from Fig. 7.8, the accuracy of the reconstruction improves as frequency increases. Therefore, in practice, the proposed method is applicable to a wider class of boundary surfaces than those assumed.

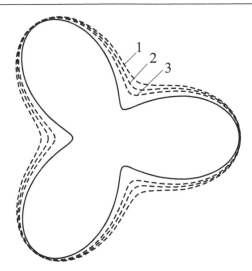

Figure 7.8. Reconstruction of a three-leaf rose: solid line, exact boundary; dashed line, reconstructed boundary; 1, $a = 2\lambda$, 2, $a = 4\lambda$, 3, $a = 6\lambda$

7.7. Reconstruction of Nonconvex Obstacles in the High-Frequency Range: 3D Case

In this section we expand some results of the previous sections to the 3D case. Let there be provided a circular irradiation by high-frequency harmonic waves, for arbitrary angle of incidence, with respect to an obstacle detected in an acoustic medium, under a scanning around the obstacle (S denotes the closed boundary shape of the obstacle). Such an opportunity can be provided in practice, for instance, under immersion scanning of an elastic body placed into a liquid.

With this irradiation the echo-impulse *time of flight* $t(\mathbf{q})$ and the real back-scattered (reflected) amplitude $|A(\mathbf{q})|$ are assumed to be known, for arbitrary direction of incidence $\mathbf{q}$.

Let us start first from the reconstruction of convex part of the boundary surface. Let S_{ex} denote convex (*external*) parts of the surface S, and S_{in} is its nonconvex (*internal*) parts. Let us restrict the consideration by the obstacles which 1) have simply connected boundary surface S with smooth convex parts S_{ex} and smooth junction between convex S_{ex} and nonconvex S_{in} ones; 2) admit no more than two points of single *mirror* (specular) reflection in echo regime, for arbitrary direction $\mathbf{q}$. It is proved in Druzhinina and Sumbatyan (1992) that the latter is automatically satisfied if the angle between two arbitrary external normals over the same nonconvex part S_{in} is acute. Figure 7.9 demonstrates a typical example of obstacle from the considered class of surfaces. An acute angle between two normals is shown on the left side of Fig. 7.10, for axial cut of an obstacle of axial symmetry.

Let us relate the boundary surface S to a Cartesian coordinate system $OX_1X_2X_3$, with the origin O being taken inside the surface. The direction of incidence $\mathbf{q}$ and the unit external normal $\mathbf{n}$ to the surface S may be related to a geographical coordinate system θ, φ (θ is the latitude and φ is the longitude) or, alternatively, may be given by the direction cosines ($\cos\alpha$, $\cos\beta$, $\cos\gamma$).

Under irradiation in echo regime, along arbitrary direction $\mathbf{q} = \{\cos\alpha, \cos\beta, \cos\gamma\}$ there is always a certain point of specular reflection on the convex part S_{ex}. Diffraction by the regions S_{in} is generally more complex. However, assumptions 1) and 2) guarantee that the surface S_{in} has only points of simple *mirror* reflections, and there are no multiple ray re-reflections.

The proposed reconstruction algorithm is the following. First, starting from the known

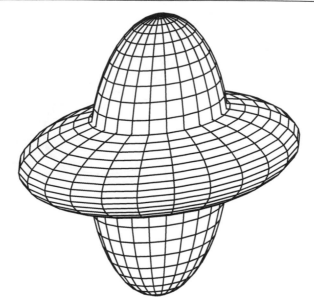

Figure 7.9. A typical shape of the obstacle under consideration

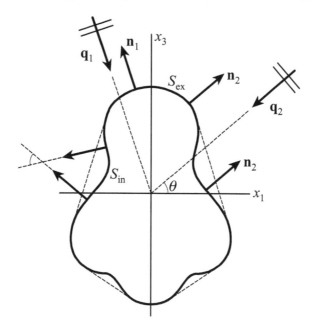

Figure 7.10. Convex hull of the nonconvex obstacle

function $t(\mathbf{q})$ a *convex hull* S_0 of the boundary surface S can be completely defined, i.e., a minimum convex surface containing S inside (see Fig. 7.10). Indeed, $t(\mathbf{q})$ determines the function $p(\mathbf{q})$ that defines the distance from the origin O to the plane tangential to the convex boundary, whose equation is given as follows:

$$Q(x_1, x_2, x_3, \alpha, \beta) = x_1 \cos \alpha + x_2 \cos \beta + x_3 \cos \gamma - p(\alpha, \beta) = 0. \qquad (7.77)$$

The convex hull S_0 is an envelope of a two-parameter family of these tangential planes. We have taken in (7.77), as a pair of independent parameters, the angles α and β, since the third direction angle γ is connected with them, by the trivial identity

$$\cos^2 \alpha + \cos^2 \beta + \cos^2 \gamma = 1. \tag{7.78}$$

Following a classical theory (see, for example, Pogorelov, 1961), the boundary points of the convex envelope have to be defined from the system

$$\begin{cases} Q(x_1, x_2, x_3, \alpha, \beta) = 0, \\ \dfrac{\partial Q}{\partial \alpha}(x_1, x_2, x_3, \alpha, \beta) = 0, \\ \dfrac{\partial Q}{\partial \beta}(x_1, x_2, x_3, \alpha, \beta) = 0. \end{cases} \tag{7.79}$$

The latter is equivalent to

$$\begin{cases} x_1 \cos \alpha + x_2 \cos \beta + x_3 \cos \gamma = p(\alpha, \beta), \\ -x_1 \sin \alpha \cos \gamma + x_3 \sin \alpha \cos \alpha = p'_\alpha \cos \gamma, \\ -x_2 \sin \beta \cos \gamma + x_3 \sin \beta \cos \beta = p'_\beta \cos \gamma, \end{cases} \tag{7.80}$$

since $\gamma'_\alpha \sin \gamma = -\sin \alpha \cos \alpha / \cos \gamma$, $\gamma'_\beta \sin \gamma = -\sin \beta \cos \beta / \cos \gamma$, which directly follows from Eqs. (7.77), (7.78).

The solution of the last linear algebraic system can be obtained, for instance, by using Cramer's rule. Therefore, the Cartesian coordinates of strictly convex parts of the convex hull S_0 are given by the following equations:

$$\begin{aligned} x_1 &= [p(\alpha, \beta) + p'_\beta(\alpha, \beta) \cot \beta] \cos \alpha - p'_\alpha(\alpha, \beta) \sin \alpha, \\ x_2 &= [p(\alpha, \beta) + p'_\alpha(\alpha, \beta) \cot \alpha] \cos \beta - p'_\beta(\alpha, \beta) \sin \beta, \\ x_3 &= [p(\alpha, \beta) + p'_\alpha(\alpha, \beta) \cot \alpha + p'_\beta(\alpha, \beta) \cot \beta] \cos \gamma. \end{aligned} \tag{7.81}$$

Now we proceed to reconstruction of nonconvex regions of the boundary. As soon as the convex hull S_0 is constructed, on the basis of the known *time of flight* function $t(\mathbf{q})$, for reconstruction of internal parts of the boundary surface S the measured values of the echo-impulse amplitude $|A(\mathbf{q})|$ can be used. All directions of incidence $\mathbf{q}$ are divided into two types, which is schematically shown in Fig. 7.10 as unit vectors $\mathbf{q}_1$ and $\mathbf{q}_2$. The first set contains those $\mathbf{q}$, where only one point of specular reflection exists on the convex part S_{ex}, in echo regime of the scanning. For such directions $\mathbf{q}$ the back-scattered far-field amplitude, up to some factor, is (see the previous sections of this chapter)

$$A(\mathbf{q}) = \gamma_1^{-1/2}(\mathbf{q}) \exp[-2ikp_1(\mathbf{q})]. \tag{7.82}$$

For other directions $\mathbf{q}$, for which there are points of specular reflections from the parts S_{in} as well, the echo amplitude is given as follows:

$$A(\mathbf{q}) = \gamma_1^{-1/2}(\mathbf{q}) \exp[-2ikp_1(\mathbf{q})] + \gamma_2^{-1/2}(\mathbf{q}) \exp[-2ikp_2(\mathbf{q})]. \tag{7.83}$$

Here $\gamma_i = (R_1^{(i)} R_2^{(i)})^{-1}$, $i = 1, 2$ are the Gaussian curvatures; $R_1^{(1)}$, $R_2^{(1)}$ and $R_1^{(2)}$, $R_2^{(2)}$ are the principal curvature radii at the points of reflection from S_{ex} and S_{in}, respectively; p_1 and p_2 are the distances from the origin to the tangential planes at the respective reflection points; and k is the wave number.

The main information for reconstruction of the regions S_{in} is contained in the real amplitude of the reflected wave:

$$F(\mathbf{q}) = |A(\mathbf{q})|^2 = \gamma_1^{-1}(\mathbf{q}) + \gamma_2^{-1}(\mathbf{q}) + 2[\gamma_1(\mathbf{q})\gamma_2(\mathbf{q})]^{-1/2} \cos\{2k[p_1(\mathbf{q}) - p_2(\mathbf{q})]\}, \tag{7.84}$$

which is the sum of slowly varying terms γ_1^{-1}, γ_2^{-1} and a highly oscillating function of $(\mathbf{q})$, under high-frequency irradiation by plane acoustic waves. Let us apply any filtration technique, to extract a slowly varying component (at large k) from the function $F(\mathbf{q})$ (see Cappellini et al., 1978):

$$F_0(\mathbf{q}) = \gamma_1^{-1}(\mathbf{q}) + \gamma_2^{-1}(\mathbf{q}). \tag{7.85}$$

So far as the function $\gamma_1(\mathbf{q})$ is known from the convex hull S_0, the Gaussian curvature $\gamma_2(\mathbf{q})$ at the points of the internal parts S_{in} of the surface S can be defined as

$$\gamma_2(\mathbf{q}) = [F_0(\mathbf{q}) - \gamma_1^{-1}(\mathbf{q})]^{-1}. \tag{7.86}$$

Shape reconstruction of the nonconvex parts S_{in} of the surface S can be performed on the basis of the Minkowski function, starting from the known function $\gamma_2(\mathbf{q})$, Eq. (7.86).

Further study can be completed in more detail, for example, in the case of axially symmetric obstacle. Here the problem to reconstruct the function $p(\theta)$ was studied in Section 7.4 and we could see that this admits an explicit analytical solution expressed through two arbitrary constants C_1 and C_2, which can be uniquely determined in the following way.

The surface S_0 determines common points of the convex S_{ex} and internal S_{in} parts of the real boundary surface (see Fig. 7.10, S_{ex} and S_{in} junction). For every region S_{in} such points generate a pair of circles, disposed on parallel planes, with their centers being on the axis of symmetry OZ. As regards the points of these circles of the same axial cross-section, the normals to the surface are parallel to each other and, being related to the unit sphere, represent the same normal with certain coordinates (θ_0, φ_0). For both circles the ends of the unit normals form, in geographical coordinates, a parallel $\theta = \theta_0$. This parallel is disposed in a spherical *belt* $\theta_1 < \theta < \theta_2$ between the parallels $\theta = \theta_1$ and $\theta = \theta_2$, which define a *visible* zone of the region S_{in}.

Now the calculation of $p(\theta)$ can be performed by using the double integral (7.49) over $\theta \in [\theta_1, \theta_2]$. The general solution (7.49) contains a pair of the unknown constants C_1 and C_2. The required initial conditions, to determine these constants, can be formulated on the intervals $[\theta_1, \theta_0]$ and $[\theta_0, \theta_2]$ as follows. Since the convex hull S_0 is already constructed, this means that at the junction points between regions S_{in} and S_{ex}, i.e., at $\theta = \theta_0$, there are known $p(\theta_0) = p_0$ and $dp/d\theta(\theta = \theta_0) = p'_0$. It thus leads to a sequential definition of the constants in (7.49):

$$C_1 = \sin^2 \theta_0 \, (p'_0 \cot \theta_0 + p_0)^2, \qquad C_2 = p_0 / \cos \theta_0. \tag{7.87}$$

Similarly, the unknown constants (7.41) can be found in the two-dimensional problem:

$$C_1 = p_0 \sin \theta_0 + p'_0 \cos \theta_0, \qquad C_2 = -p_0 \cos \theta_0 + p'_0 \sin \theta_0. \tag{7.88}$$

It should be noted that the quantity $(dp/d\theta)$ at $\theta = \theta_0$ is naturally treated, over every segment of integration, as a one-sided limit at $\theta \to \theta_0$, when approaching to the limiting point from the corresponding convex part S_{ex} of the convex hull S_0. Hence, the calculation of the function $p(\theta)$, which describes the internal region S_{in}, is reduced to a numerical treatment of some single integral (two-dimensional case) or to a double integral (the case with axial symmetry), over the intervals $[\theta_1, \theta_0]$ and $[\theta_0, \theta_2]$. For all that, the corresponding parts of the curves (in the two-dimensional problem) and the analogous parts of the curves in the axial cross-section (for the case of axial symmetry) can be reconstructed up to the point of their intersection.

In practice, realization of the proposed method allows one to define the values θ_1 and θ_2 with some error. That is because these values are indicated on the graph of the real amplitude $|A(\theta)|$ of the back-scattered wave as the points, where a slowly varying part of the diagram turns into a highly oscillating one, for corresponding part S_{in}.

This can be clarified again by the two-dimensional example shown in the previous section in Fig. 7.5, which presents a diagram of the real amplitude back-scattered from an acoustically soft obstacle, which is a junction of a pair of equal circles with radius R and distance between their centers $\sqrt{2}\,R$ ($R = 3\lambda$, where $\lambda = 2\pi/k$ is the wavelength).

Let us return back to investigation of the general three-dimensional inverse problem without axial symmetry. We can arrange a mapping of the considered closed surface S to the unit sphere. Such a mapping is achieved if one draws unit vectors, with their origin being put at any fixed center, in direction of the unit normal to the corresponding point of the surface. In this case at any internal part S_{in} of the surface S its *visible* zone generates a certain complex region on this unit sphere. Since there can be only points of single back specular reflections on the internal parts S_{in}, these parts can be a union of some convex pieces only. This guarantees that the region drawn by the ends of external unit normals is simply connected, so it can be defined by the inequalities $\varphi_1 < \varphi < \varphi_2$, $\theta_1(\varphi) < \theta < \theta_2(\varphi)$ (see Fig. 7.11). Particularly, this region may appear to be a spherical *belt* $0 \le \varphi < 2\pi$, $\theta_1(\varphi) < \theta < \theta_2(\varphi)$.

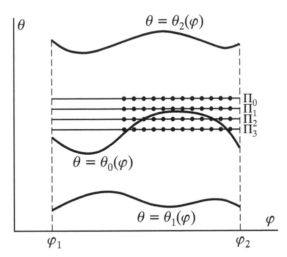

Figure 7.11. A *visible* region of an internal part of the boundary surface

A general sketch of the proposed reconstruction algorithm will be given for the case, when the internal boundary region S_{in} contains a curve of intersection of its composing convex surface parts (see Fig. 7.11). There is a curve $\theta = \theta_0(\varphi)$ that determines, for every φ, a unit normal of the convex hull to lines of junction between the given part S_{in} and the neighbor parts S_{ex}.

Within the framework of the developed approach some numerical experiments were undertaken on reconstruction of a number of concrete nonconvex obstacles. For all examples below we first reconstruct a convex hull of the defect on the basis of measured echo–impulse *time of flight*, as described at the beginning of the paper. Note that both in the two-dimensional case and in the case of axial symmetry the Cartesian coordinates of the convex hull are given as

$$
\begin{aligned}
x(\theta) &= -p'(\theta)\cos\theta + p(\theta)\sin\theta, \\
z(\theta) &= p'(\theta)\sin\theta + p(\theta)\cos\theta,
\end{aligned}
\tag{7.89}
$$

where θ is the angle between external normal and the axis OX_3.

The following two examples were considered, to test efficiency of the proposed algorithm. The first one (see Fig. 7.12) is related to a reconstruction, in the plane X_1OX_3, of an axial cut of the axially symmetric obstacle, bounded by a surface join of two ellipsoids (both with the symmetry axis OX_3):

$$\frac{x_1^2}{4} + \frac{x_2^2}{4} + \frac{x_3^2}{16} = \lambda^2, \qquad \frac{x_1^2}{36} + \frac{x_2^2}{36} + \frac{x_3^2}{4} = \lambda^2. \qquad (7.90)$$

Shape reconstruction of these obstacles was performed on the basis of solution (7.49).

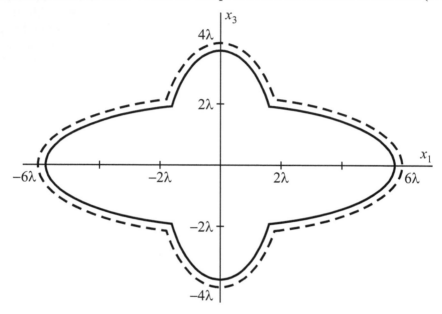

Figure 7.12. Reconstruction of an axially symmetric obstacle: solid line, exact boundary; dashed line, reconstructed boundary

Another example was related to a three-dimensional obstacle without axial symmetry, bounded by a surface join of the two ellipsoids of rotation, around the axes OX_3 and OX_2 (see Fig. 7.9):

$$x_1^2 + x_2^2 + \frac{x_3^2}{9} = 1, \qquad \frac{x_1^2}{4} + \frac{x_2^2}{9} + \frac{x_3^2}{4} = 1. \qquad (7.91)$$

In this case reconstruction of the boundary surface is performed on the basis of the proposed numerical step-by-step method. Note that the differential equation (7.58) should be treated here numerically in a spherical belt $0 \le \varphi < 2\pi$, $\theta_1(\varphi) < \theta < \theta_2(\varphi)$ of the unit sphere.

In both examples just considered, the accuracy of the convex boundary parts reconstruction is around 5–7 %, and that of the reconstruction of the internal parts is no worse than 10 %.

The results presented in this chapter follow the author's work (Vorovich et al., 2001).

Helpful remarks

In order to implement the numerical algorithm developed for the considered inverse problem, we need to have an array of values of the real wave amplitude $|A(\theta)|$, back-scattered from the obstacle along all possible directions. We constructed these values by a numerical treatment of the direct diffraction problem, with the help of the BIE method. Note that a typical size of the obstacles was nearly 8–12 wavelengths, which requires rather detailed mesh with a small mesh step.

Chapter 8

Ill-Posed Equations of Inverse Diffraction Problems for Arbitrary Boundary

8.1. Ill-Posed Problems for Operator Equations of the First Kind: General Properties

The theory of ill-posed (i.e., incorrectly posed) operator equations of the first kind was founded by A.N. Tikhonov (see, for example, Tikhonov and Arsenin, 1977) and had become a further natural development of some ideas of Hadamard and other classicists.

Hadamard first formulated the concept of a well-posed (or correctly posed) problem for the operator equation

$$Au = f, \qquad u \in U, \qquad f \in F, \tag{8.1}$$

where A is a linear operator acting from a Banach space U to a Banach space F. This includes the three points:

1) Equation (8.1) has a solution for any $f \in F$.
2) The solution is unique.
3) The solution is stable with respect to small perturbations of the right-hand side. In other words: small perturbation of f (in the F-metrics) involves small perturbation of u (in the U-metrics). Later on, it was recognized that for a correct treatment the last condition is of most importance. Thus, boundary value problems are well posed for elliptic systems, and initial (Cauchy) problems for hyperbolic systems, but not vice versa.

Obviously, the existence of a bounded operator A^{-1} is sufficient for condition 3) to be valid. However, as shown in Section 1.6, in the case when A is compact the inverse operator A^{-1} does not exist. In the meantime, operator equations of the first kind (8.1) with a compact operator A are widespread in applied mathematics. A typical example is given by the Fredholm integral equation of the first kind

$$(Au)(x) = \int_a^b K(x, \xi) u(\xi)\, d\xi = f(x), \qquad a \le x \le b, \tag{8.2}$$

with a continuous (in both its arguments) kernel $K(x, \xi)$, $a \le x, \xi \le b$. In Section 1.5 it was proved that such an operator Au is compact in $C(a, b)$, i.e., when $U = F = C(a, b)$. It is also well known (see, for example, Kantorovich and Akilov, 1982) that operator (8.2) is compact on the Hilbert space $U = F = L_2(a, b)$.

In order to understand more clearly why the solution of Eq. (8.2) is unstable with respect to small perturbations of the right-hand side, let us suppose that $Au_0 = f_0$, $u_0 \in C(a, b)$, $f_0 \in C(a, b)$, and we take $u_1 = u_0 + \cos \lambda t$. It is obvious that in the case $a = 0$, $b = 2\pi$, $\|u_1 - u_0\| = 1$. If we denote $f_1 = Au_1$, then

$$\|f_1 - f_0\| = \|A(u_1 - u_0)\| = \max_{x \in (0, 2\pi)} \left| \int_0^{2\pi} K(x, t) \cos \lambda t\, dt \right|$$

$$= \tfrac{1}{2} \max_{x \in (0, 2\pi)} \left| \int_0^{2\pi} K(x, t) \left(e^{i\lambda t} + e^{-i\lambda t} \right) dt \right|. \tag{8.3}$$

Let the kernel $K(x,t)$ be differentiable with respect to t. Then, according to results discussed in Section 1.4, $\|f_1 - f_0\| = O(1/\lambda)$, $\lambda \to \infty$. Therefore, for sufficiently large λ small perturbation of the right-hand side corresponds to a finite deviation of the solution.

This example and some general theoretical ideas indicate the possible approach where we can guarantee stability with respect to small variations of f. Actually, the considered set of functions $\{u_0 + \sin(\lambda t)\}$ do not satisfy Arzela–Ascoli criterion (see Section 1.5) of compactness in $C(a,b)$ since these functions are not equicontinuous when $\lambda \to \infty$. Classical results of functional analysis (see, for instance, Kantorovich and Akilov, 1982) show that if we seek a solution to equation (8.1) on any compact set $V \subset U$, then the solution depends continuously upon f, which is provided by the following theorem.

THEOREM. *If a continuous operator $A : U \to F$ determines a one-to-one mapping of a compact set $V \subset U$ onto a set $AV = G \subset F$, then the inverse operator A^{-1} is also continuous on G.*

This theorem gives you a quite natural way to construct a stable solution, which is called the method of *solution by inspection*. It involves the minimization of the discrepancy over some compact set $V \subset U$: u_0 is taken as an approximate solution to equation (8.1) if

$$\min_{u \subset V} \|Au - f\| = \|Au_0 - f\|. \tag{8.4}$$

It is clear that if $u_0 \in V$, then the solution is stable with respect to small change of f, since A^{-1} is continuous on $G = AV$.

These ideas allow us to operate with the concept of *correctness in the sense of Tikhonov*, instead of correctness in the sense of Hadamard.

DEFINITION. *Problem (8.1) is called well posed in the sense of Tikhonov if:*
1) For some f_0 there exists a solution u_0 to this equation belonging to a compact set V: $u_0 \in V$, $Au_0 = f_0$.
2) This solution u_0 is unique.

It is clear that in this case operator A^{-1} is continuous on the set $G = AV$. So, if the right-hand side f of equation (8.1) is given with some small perturbation (which typically occurs when it is known with some small error) and $f \in G$, then the element $u = A^{-1}f$ can be considered as a correct approximation to the exact solution $u_0 = A^{-1}f_0$, due to continuity of the operator A. The considered compact set $V \subset U$ is called a *class of correctness* for the operator A.

There can be listed evident cases when the above ideas can be applied efficiently.

1°. If we seek a solution inside a finite-dimensional subset V of the Banach space U. Actually, any finite-dimensional subset is compact in the Banach space. For example, let us consider the integral equation (8.1) on an interval symmetric with respect to the origin and of the length of periodicity of trigonometric functions

$$(Au)(x) = \int_{-\pi}^{\pi} K(x,\xi)\, u(\xi)\, d\xi = f(x), \qquad |x| \le \pi. \tag{8.5}$$

If the kernel is, for example, even with respect to both its arguments x, ξ, and also the right-hand side $f(x)$ is even with respect to x, then the solution of equation (8.5) is also even, and we may seek an approximate solution in the form

$$u_n(x) = \sum_{m=0}^{n} a_m \cos(mx), \tag{8.6}$$

where a few terms with $n = 2$–5 are usually sufficient to approximate well the exact solution $u_0(x)$ (if the latter exists and is unique). In practice, you should minimize the discrepancy $\rho_n = \|Au_n - f\|$. This can be achieved either by the inspection method or by a strict solution of the corresponding variational problem. In the latter case it is more convenient to consider equation (8.6) in the Hilbert space $L_2(-\pi, \pi)$. Then we arrive at a problem of minimization of the quadratic functional with respect to the unknown coefficients a_m, $m = 0, 1, \ldots, n$

$$\min_{\{a_m\}} \rho_n, \qquad \rho_n = \|Au_n - f\|^2 = (Au_n - f,\ Au_n - f)$$

$$= \sum_{m,j=0}^{n} a_m a_j (A\varphi_m, A\varphi_j) - 2\sum_{m=0}^{n} a_m (A\varphi_m, f) + (f, f), \quad \varphi_m(x) = \cos(mx), \tag{8.7}$$

whose solution can be found by using the well-known criterion of minimum: $\partial\rho_n/\partial a_m = 0$, $m = 1, 2, \ldots, n$. This technique leads to the following linear algebraic system, to define coefficients a_m, $m = 0, 1, \ldots, n$:

$$\sum_{j=0}^{n} (A\varphi_j, A\varphi_m)\, a_j = (f, A\varphi_m), \qquad m = 0, 1, \ldots, n. \tag{8.8}$$

It is interesting to note that this system coincides with the one which could be constructed by a certain type of Galerkin method (see Mikhlin, 1964) if you take a small number of basis functions $\varphi_m(x)$.

Another appropriate structure of the approximate solution is

$$u_n(x) = \sum_{m=0}^{n} a_m x^m, \tag{8.9}$$

which in many cases leads to a more stable method than the one with trigonometric basis. The structure (8.9) represents a linear combination of monotonic functions, and the following paragraph explains why operation with monotonic functions usually leads to stable calculations.

2°. In many problems we possess some *a priori* information about qualitative structure of the solution. For example, in Section 3.6 we could see that in Arctic seas variation of the wave speed with depth represents typically a positive monotonically increasing function. One of the basic results in functional analysis states that any set of monotonic function is compact in $L_2(a, b)$. The proof can be found, for example, in Tikhonov and Arsenin (1977). So, if we solve the problem on reconstruction of the wave-speed function in a stratified ocean from some measured data on acoustic wave scattering, then we can seek the unknown function by inspection in the class of monotonically increasing (or decreasing) functions. As an appropriate method we can quote here representation of the solution in the form (8.9) with unknown but positive (or negative) coefficients a_m, $m = 1, 2, \ldots, n$. The set of such functions consists of monotonic functions only. Note that stable calculations here are provided with arbitrary, even very large, number n, since we do not need to check here that the constructed set is finite-dimensional (in practice, with some small number n).

Helpful remarks

When following the general strategy described above to construct a correct (i.e., stable) solution of the ill-posed operator problem of the first kind, we may encounter the case when the exact solution u_0 related to the exact right-hand side f_0 belongs to a compact set $V \subset U : u_0 \in V$, $Au_0 = f_0 \in G = AV$, but a perturbed right-hand side f is outside of the set $G = AV : f \notin G$. In such cases equation (8.1) may not have solution in the classical sense. To this end, there is introduced the concept of a quasi-solution (see Tikhonov and Arsenin, 1977).

DEFINITION. *Let V be a compact set of the Banach space U. Then an element $u^* \in V$ that provides a minimum for the functional $\min_{u \in V} \rho(u) = \rho(u^*)$, $\rho(u) = \|Au - f\|$, is called a quasi-solution of equation (8.1).*

It is obvious that since V is a compact set, a quasi-solution surely exists (its uniqueness is under question). Besides, we can indicate the case when the quasi-solution is simultaneously a classical solution, namely when $f \in G = AV$. Some interesting and important results on quasi-solutions can be found in Tikhonov and Arsenin (1977).

8.2. Regularization of Ill-Posed Problems with the Help of Smoothing Functional

As we could see from the previous section, a solution of the operator equation $Au = f$, where a compact linear operator A acts from one Banach space U to another Banach space F, i.e., $A : U \to F$, is surely stable with respect to small perturbations of the right-hand side f if this solution is sought on a certain compact subset $V \subset U$ and if $f \in AV$. However, in some cases the perturbed right-hand side f_δ is outside of the set AV: $f_\delta \notin AV$, being at the same time very close to the exact function f: $\|f - f_\delta\| < \delta$, with some small positive quantity $\delta > 0$. In such cases the results of Section 8.1 cannot guarantee stability of the solution, i.e., that the distance $\|u - u_\delta\|$ is small, where $Au_\delta = f_\delta$. This phenomenon seems to be a little strange. The equality $Au_\delta = f_\delta$ implies that a solution of our operator equation exists in the space U, and in many cases it is even unique. However, small changes in f result in a large change in the solution, which is indeed a specific feature of equations of the first kind.

In order to overcome this trouble and also for some other reasons Tikhonov (see Tikhonov and Arsenin, 1977) introduced the concept of regularization for ill-posed problems. The basic idea may be related to some fundamental ideas of Lavrentiev (1959, 1960).

Let for simplicity operator A be self-adjoint and positive in some Hilbert space, $A : H \to H$. If it is not so, then by multiplying the equation $A_u = f$ by A^* from the left, we can achieve both required properties. In this case, as follows from results of Section 1.5, the operator $A + \alpha I$, where I is a unit operator, is invertible with arbitrary positive $\forall \alpha > 0$, i.e., $(A + \alpha I)^{-1}$ is continuous on H.

Indeed, for positive operator A all its eigenvalues λ_n are nonnegative, and so for the operator $A + \alpha I$, $\alpha > 0$, they all are positive. Therefore, the point $\lambda = 0$ is a regular value of the operator $(A + \alpha I - \lambda I)$. Then, according to the fundamental theorem of the operator theory (see Kantorovich and Akilov, 1982), we can conclude that $(A + \alpha I)$ is invertible.

Heuristically, if $0 < \alpha \ll 1$, then the new operator $A + \alpha I$ is in some sense close to the main operator A. So we may hope that the thus constructed solution $u_\alpha = (A + \alpha I)^{-1} f$ can be accepted for a certain small $\alpha > 0$ as an approximate solution to the main equation $Au = f$, stable with respect to small perturbations of $f \in F$, even when f is an arbitrary element in F.

This idea was further developed by Tikhonov who introduced the concept of regularization by smoothing functional. The latter in the simplest treatment consists in the consideration of the variational problem of minimization of the functional

$$M_\alpha(u, f) = \|Au - f\|^2 + \alpha \|u\|^2, \qquad \alpha > 0, \tag{8.10}$$

which is justified by the following theorem.

THEOREM 1. *If A is a one-to-one compact linear operator from a convex closed set $V \subset H$ to the set $AV = G \subset H$ and $u \in V$ is a (unique) solution of the equation $Au = f$*

for a certain given $f \in G$, then there exists a unique solution $u_\alpha \in V$ of the minimization problem

$$u_\alpha : \inf_{u \in V} M_\alpha(u, f) = M_\alpha(u_\alpha, f). \tag{8.11}$$

Proof. The functional (8.10)

$$M_\alpha(u, f) = (Au - f, Au - f) + \alpha(u, u) \tag{8.12}$$

is a quadratic functional with respect to u. So it is infinitely differentiable and its Frechet differentials of the first (which is in fact its gradient) and the second order is easily calculated. Their structure can be obtained from a small variation of the functional as follows:

$$\delta_u M_\alpha(u, f) = 2(Au - f, A\delta u) + 2\alpha(u, \delta u) = 2(A^*Au - A^*f + \alpha u, \delta u); \tag{8.13}$$
$$\text{hence,} \quad \text{grad}_u M_\alpha(u, f) = 2(A^*Au - A^*f + \alpha u).$$

Further,

$$\delta^2_{uu} M_\alpha(u, f) = 2(A^*A\delta u + \alpha \delta u, \delta u) = 2(A\delta u, A\delta u) + 2\alpha(\delta u, \delta u) \geq 2\alpha \|\delta u\|^2, \tag{8.14}$$

so the functional $M_\alpha(u, f)$ is strongly convex. It follows from some classical results of functional analysis (see, for example, Mikhlin, 1965; Groetsch, 1984; Morozov 1984) that any strongly convex functional attains its (unique) minimum value on any convex closed set $V \in H$. The theorem is proved.

THEOREM 2. *If the solution of the variational problem (8.11), $u_\alpha \in V$, is an internal point of the set V, then it can be found from the uniquely solvable equation*

$$A^*Au + \alpha u = A^*f. \tag{8.15}$$

Proof. It is known (see Tikhonov and Arsenin, 1977; Groetsch, 1984; Mikhlin, 1965) that under the conditions of this theorem the problem of minimization of the strongly convex functional by reducing to Euler's equation leads to the necessary and sufficient condition: $\text{grad}\, M_\alpha = 0$, which by taking into account Eq. (8.13) is equivalent to Eq. (8.15). The theorem is proved.

THEOREM 3. *Under the conditions of Theorems 1 and 2, if u_* is a (unique) exact solution of the equation $Au = f$, then*

$$\|u_\alpha - u_*\|, \quad \alpha \to 0, \quad \text{if} \quad \frac{\delta^2}{\alpha} \to 0 \quad \text{and} \quad \|f_\delta - f\| \leq \delta. \tag{8.16}$$

The proof to this theorem can be found in Tikhonov and Goncharsky (1987), and for a more general case, in Tikhonov and Arsenin (1977).

This theorem gives an efficient instrument to choose an appropriate value of the regularization parameter α. Indeed, if the error on the right-hand side f is small enough being of the order of δ, then the regularization parameter α cannot be chosen too small. The optimal choice is such that $\alpha \sim \delta^{2-\varepsilon}, \varepsilon > 0$. In some more details this important question on optimal choice of the parameter α will be discussed below.

The basic idea of Tikhonov's regularization justified by Theorems 1–3 is related to the fact that the smoothing functional $M_\alpha(u, f)$ is strongly convex, so that it attains a minimum at the point $u = u_\alpha$. This allowed Tikhonov to propose an alternative concept of regularization by more smooth stabilizers. One of possible approaches is constructed by introducing the so-called *stabilizing functional* $\Omega(u)$, which is positive: $\Omega(u) \geq 0, \forall u \in V$, and continuous over V, so that the exact solution u_* of the equation $Au = f$ belongs to the set $V : u_* \in V$. Tikhonov proves that if the functional $\Omega(u, f)$ is such that the set $\Omega(u) \leq d$ is a compact subset in V for $\forall d > 0$, then the listed conditions are sufficient for the *smoothing functional*

$$M_\alpha(u, f) = \|Au - f\|^2 + \alpha\Omega(u), \qquad \alpha > 0, \qquad (8.17)$$

to have a unique minimum value $\inf_{u \in V} M_\alpha(u, f)$ for arbitrary positive α, and for all that $u_\alpha \in V$. Moreover, under these conditions Theorems 1 and 3 are valid (see Tikhonov and Arsenin, 1977). Therefore, Eq. (8.15) supplies you with a more general instrument to construct a stable approximate solution to the operator equation of the first kind $Au = f$, due to a general form of the functional $\Omega(u)$. Besides, these results remain valid if the initial space U is not Hilbert but only metric.

It should be noted that the analogue to Theorem 2 in this more general case can be obtained by reducing the minimization problem to the Euler equation, which in this problem can be written out in the following form:

$$A^*Au + \alpha\,\Omega'(u) = A^*f, \qquad (8.18)$$

where $\Omega'(u)$ is the Frechet derivative of the functional, which is the same as its gradient.

The approach described at the last part of the previous discussion turns out very fruitful in numerical treatment of the Fredholm integral equations of the first kind

$$Au = f \quad \sim \quad \int_a^b K(x, \xi)u(\xi)\,d\xi = f(x), \qquad x \in (a, b). \qquad (8.19)$$

Let us consider this operator equation in the Banach space of continuous functions: $K(x, \xi) \in C[(a, b) \times (a, b)]$, $f(x) \in C(a, b)$, $A : C(a, b) \to C(a, b)$. In the ambit of Tikhonov's regularization we can put $V = W_2^1(a, b) \subset C(a, b) = U$. In this case the norm in $W_2^1(a, b)$ can be accepted as the stabilizing functional

$$\Omega(u) = \|u\|_{2,1}^2 = \int_a^b \left[q_0\,u^2(x) + q_1 u'^2(x)\right]\,dx, \qquad q_0, q_1 > 0, \qquad (8.20)$$

since the set $\Omega(u) \leq d$ is here a sphere in $W_2^1(a, b)$, which is known to be compact in $C(a, b)$ (see, for instance, Mikhlin, 1964; Tikhonov and Arsenin, 1977). Here q_0, q_1 are some positive constants. It now becomes clear that the functional (8.20) is a generalization of $\Omega(u) = \|u\|_2^2$ in the space $L_2(a, b)$ since in the case $q_1 = 0$ the former is reduced to the latter.

Within such a treatment the Euler equation for the variational minimization of the smoothing functional can be explicitly constructed since

$$\delta\Omega(u) = 2\int_a^b \left(q_0 u\delta u + q_1 u'\delta u'\right)\,dx = 2\int_a^b \left(q_0 u - q_1 u''\right)\,\delta u\,dx, \qquad (8.21)$$

if we fix the values of the unknown function $u(x)$ at the endpoints of the interval (a, b).

After these preliminary transformations, a stable solution to equation (8.19) can be found from Euler's equation:

$$\int_a^b K_1(x, \xi)u(\xi)\,d\xi + \alpha\left[q_0\,u(x) - q_1 u''(x)\right] = f_1(x), \qquad (8.22)$$

where

$$K_1(x, \xi) = \int_a^b K(t, x)K(t, \xi)\,dt, \qquad f_1(x) = \int_a^b K(t, x)f(t)\,dt. \qquad (8.23)$$

It should be noted that here $f_1 = A^* f$, and the kernel in the integral operator in (8.23) is evidently related to the operator $A^* A$.

Now some words about the choice of the small parameter of regularization α. There are known several approaches to this problem (see Tikhonov and Arsenin, 1977). The following two of them seem to be quite natural and very efficient.

1. If the error of the measurement in the known *data-in* function is δ, and this quantity is *a priori* known from some considerations, then this can helpfully serve to estimate the value α, due to the estimate $\|f - f_\delta\| \leq \delta$ (cf. above). Indeed, if we start to solve equation (8.22) for relatively large α, then the solution u_α will be very stable (as a solution of an operator equation with invertible operator), but too far from the exact solution u_*. In other words, the deviation $\|u_\alpha - u_*\|$ is too large, and so is the discrepancy $\|Au_\alpha - f\|$. If we gradually decrease α, then, according to Theorem 3, $u_\alpha \to u_*$ and so, by continuity of the operator A, the discrepancy $\|Au_\alpha - f\| \to 0$. However, it is not reasonable to arrange this discrepancy less than δ. So a good criterion to choose α is to decrease this parameter until $\|Au_\alpha - f\| \sim \delta$.

2. Numerous implementations of model examples and problems showed that the choice for which

$$\alpha : \inf_{\alpha > 0} \left\| \alpha \frac{du_\alpha}{d\alpha} \right\| \tag{8.24}$$

appears to be very efficient. Heuristic justification to this statement can be clarified as follows. As the parameter α decreases, while it is not too small, the solution u_α weakly depends upon a small change of the value of α, so the quantity $du_\alpha/d\alpha$ is almost constant and the value of $\alpha\, du_\alpha/d\alpha$ seems to be relatively large for such large α. With further decrease of α, when approaching its extremely small values, u_α becomes very sensitive to small variation of α, and so $du_\alpha/d\alpha$ becomes very large. The optimal choice of α should thus be made for the smallest value of the product $\alpha\, du_\alpha/d\alpha$.

It should be noted that the quantity, for which we need to estimate the norm (8.24), admits direct estimate. To clarify this idea, we demonstrate this technique on the simplest case of Eq. (8.15). Let us differentiate Eq. (8.15), where the right-hand side f_δ contains some error δ, with respect to α:

$$A^* A \frac{du_\alpha}{d\alpha} + u_\alpha + \alpha \frac{du_\alpha}{d\alpha} = 0, \qquad A^* A \left(\alpha \frac{du_\alpha}{d\alpha} \right) + \alpha \left(\alpha \frac{du_\alpha}{d\alpha} \right) = -\alpha u_\alpha. \tag{8.25}$$

But from Eq. (8.15) we have

$$-\alpha u_\alpha = A^* A u_\alpha - A^* f_\delta, \tag{8.26}$$

so Eq. (8.25) is:

$$A^* A \left(\alpha \frac{du_\alpha}{d\alpha} \right) + \alpha \left(\alpha \frac{du_\alpha}{d\alpha} \right) = A^* A u_\alpha - A^* f_\delta. \tag{8.27}$$

Thus, the quantity $\alpha\, du_\alpha/d\alpha$ can be found from the same integral equation (8.15) with a certain different right-hand side.

Helpful remarks

1°. Approach applied for investigation of Eq. (8.19) remains very efficient even in the extremely ill-posed case when the interval of variation of the variable $x \in (c, d)$ is different from (a, b) (in practice, the former is a subinterval of the latter), the case rather typical in many inverse problems with very limited input data. You can apply the discussed approach also in this case, which will lead to an equation which is very similar to Eq. (8.22):

$$
\int_a^b K_1(x, \xi) u(\xi)\, d\xi + \alpha \left[q_0\, u(x) - q_1 u''(x) \right] = f_1(x), \qquad a \le x \le b,
$$

$$
K_1(x, \xi) = \int_c^d K(t, x) K(t, \xi)\, dt, \qquad f_1(x) = \int_c^d K(t, x) f(t)\, dt.
$$

(8.28)

$2°$. Operation with the Sobolev Hilbert space $W_2^1(a, b)$ seems to be quite natural from the viewpoint of compactness in the Banach space $C(a, b)$. It is absolutely clear that any set of continuous functions whose derivatives are uniformly bounded by a certain constant satisfy the Arzela–Ascoli theorem (see Section 1.5).

8.3. Iterative Methods for Operator Equations of the First Kind

As follows from results of Chapter 1 (see Sections 1.6 and 1.7), generally there are no results on solvability in the classical sense of operator equations of the first kind, in contrast to the second-kind equations, where the classical Fredholm theory guarantees that in the case of regular values of the parameter λ the operator $(I - \lambda G)$ is invertible and so the equation $u - \lambda G u = f$ is solvable in this case, at least for compact operator G. Some results on solvability of the first-kind integral equations are discussed in Section 1.7, but only for convolution kernels (i.e., when the kernel depends only on the difference of its arguments) and only about a generalized rather than classical solution. Uniqueness of such equations is also a poorly studied question. The only strong (and quite clear) result here is that if A is compact in any Banach functional space, which is not finite-dimensional, then the inverse operator A^{-1} does not exist (i.e., it is not continuous).

From this point of view, it is absolutely unclear how equation (8.1) can be solved in practice by any traditional numerical method, and a powerful *indirect* approach is based upon the Tikhonov's regularization scheme discussed in the previous section. It is rather unexpected that some classical iterative techniques well known in the literature on numerical methods automatically provide convergence of iterations to an exact solution, in the case when the latter exists, perhaps not uniquely.

A natural idea to construct such an approach is based on the following consideration. Let us assume that we know a certain approximation u_n to the exact solution u_* of the equation $Au = f$, where $A : H \to H$ is a linear continuous operator in the Hilbert space H. Let us calculate the discrepancy $\varphi(u) = \|Au - f\|^2$ at the element u_n:

$$
\varphi(u_n) = \|Au_n - f\|^2 = (Au_n - f, Au_n - f),
$$

(8.29)

and try to find a direction where this quadratic functional decreases with the maximum possible rate. This is known to be determined by the gradient to this functional (see Kantorovich and Akilov, 1982; and also Section 8.2 here). The latter can be defined from its elementary variation $\delta\varphi$:

$$
\delta\varphi(u) = \delta(Au - f, Au - f) = 2(Au - f, A\delta u) = 2(A^*(Au - f), \delta u);
$$
$$
\text{hence,} \quad \text{grad}\, \varphi(u) = 2A^*(Au - f),
$$

(8.30)

where A^* designates the adjoint of the operator A. This implies that the next step of iteration can be found along the direction of the calculated gradient:

$$
u_{n+1} = u_n - \gamma\, \text{grad}\, \varphi(u_n) = u_n - \gamma A^*(Au_n - f) \sim \xi_{n+1} = \xi_n - \gamma A^* A \xi_n, \quad \xi_n = u_n - u_*, \quad (8.31)
$$

where u_* is the exact solution to the studied equation:

$$Au_* = f. \tag{8.32}$$

It is obvious that ξ_n designates the deviation of the approximate solution from the exact one. Let us calculate its norm

$$(\xi_{n+1}, \xi_{n+1}) = (\xi_n, \xi_n) - 2\gamma(\xi_n, A^*A\xi_n) + \gamma^2(A^*A\xi_n, A^*A\xi_n)$$
$$\sim \|\xi_{n+1}\|^2 = \|\xi_n\|^2 - 2\gamma\|A\xi_n\|^2 + \gamma^2\|A^*A\xi_n\|^2, \tag{8.33}$$

and try to achieve its minimum value, as much as possible. This is reduced to a simple minimization of the quadratic functional, which involves $\gamma = \|A\xi_n\|^2/\|A^*A\xi_n\|^2$. Therefore, the iteration scheme (8.31) has the optimal form

$$u_{n+1} = u_n - \frac{\|A\xi_n\|^2}{\|A^*A\xi_n\|^2} A^*(Au_n - f), \tag{8.34}$$

or equivalently,

$$\xi_{n+1} = \xi_n - \frac{\|A\xi_n\|^2}{\|A^*A\xi_n\|^2} A^*A\xi_n, \qquad \|\xi_{n+1}\|^2 = \|\xi_n\|^2 - \frac{\|A\xi_n\|^4}{\|A^*A\xi_n\|^2}, \tag{8.35}$$

which is called the *Steepest Descent Method (SDM)*.

The following theorem was first proved by Fridman (1962).

THEOREM 1. *Iterative process (8.34) $\sim$ (8.35) converges monotonically to the exact solution u_* of the equation $Au = f$.*

Proof. Monotonic behavior of the convergence follows from Eq. (8.35), which shows that the distance between the approximate and exact solutions decrease at each step of the iterative process.

In order to give the proof of convergence of the process, let us recall some classical results about geometrical properties of Hilbert spaces (see, for example, Kantorovich and Akilov, 1982). We denote by $\{e_j\}$, $j = 1, 2, \ldots$, an ortho-normalized basis of the space H, constructed from eigenfunctions of the self-adjoint and positive operator A^*A: $A^*Ae_i = \beta_i e_i$, where β_i are its eigenvalues, which we arrange in decreasing order. Then any element $x \in H$ can be expanded into a series in elements of this basis:

$$x = \sum_{j=1}^{\infty} x_j e_j \sim \|x\|^2 = \sum_{j=1}^{\infty} x_j^2, \qquad A^*Ax = \sum_{j=1}^{\infty} \beta_j x_j e_j. \tag{8.36}$$

Let us rewrite Eq. (8.35) in the following form:

$$\xi_{n+1} = \xi_n - \frac{1}{\lambda_n} A^*A\xi_n, \qquad \lambda_n = \frac{\|A^*A\xi_n\|^2}{\|A\xi_n\|^2}, \tag{8.37}$$

and note that $0 \leq m \leq \lambda_n \leq \|A^*\| = \|A\| = M$.

For further consideration it is important whether $m > 0$ or $m = 0$. We give the proof if the former condition is valid. The latter case is proved in a similar way and can be found in Fridman (1962).

So, let us assume that $0 < m \leq \lambda_n \leq M$. First of all, we note that Eq. (8.37) implies

$$(\xi_{n+1}, \xi_{n+1}) = (\xi_n, \xi_n) - \frac{(A\xi_n, A\xi_n)}{\lambda_n}. \tag{8.38}$$

It follows from the last relation that a) $\|\xi_n\|^2$ monotonically decreases with $n \to \infty$ (which was already outlined), and b) that

$$\frac{\|A\xi_n\|^2}{\lambda_n} \to 0, \qquad n \to \infty. \tag{8.39}$$

Since $\lambda_n \leq M$, this implies that $\|A\xi_n\| \to 0$, $n \to \infty$.

Further, from the expansion $\xi_n = \sum_{j=1}^{\infty} \xi_n^j e_j$, $\forall n = 1, 2, \ldots$ it follows that

$$\xi_{n+1} = \xi_n - \frac{1}{\lambda_n} \sum_{j=1}^{\infty} \xi_n^j \beta_j e_j = \sum_{j=1}^{\infty} \xi_n^j \left(1 - \frac{\beta_j}{\lambda_n} \right) e_j$$

$$\sim \xi_{n+1}^j = \xi_n^j \left(1 - \frac{\beta_j}{\lambda_n} \right) = \xi_0^j \prod_{i=0}^{n} \left(1 - \frac{\beta_j}{\lambda_i} \right). \tag{8.40}$$

It follows from the last relation that

$$\|\xi_n\|^2 = \sum_{j=1}^{\infty} (\xi_n^j)^2 = \sum_{j=1}^{\infty} (\xi_0^j)^2 \prod_{i=0}^{n-1} \left(1 - \frac{\beta_j}{\lambda_i} \right)^2. \tag{8.41}$$

Since the convergence of the series $\sum_{j=1}^{\infty} \beta_j^2$ implies that $\beta_j \to 0$ with $j \to \infty$, so there exists the number N such that $\beta_j \geq m$ for $j = 1, 2, \ldots, N$, and $\beta_j \leq m$ for $\forall j \geq N+1$. Then we decompose the sum in Eq. (8.41) to the two ones:

$$\|\xi_n\|^2 = S_1 + S_2, \quad S_1 = \sum_{j=1}^{N} (\xi_0^j)^2 \prod_{i=0}^{n-1} \left(1 - \frac{\beta_j}{\lambda_i} \right)^2, \quad S_2 = \sum_{j=N+1}^{\infty} (\xi_0^j)^2 \prod_{i=0}^{n-1} \left(1 - \frac{\beta_j}{\lambda_i} \right)^2, \tag{8.42}$$

and consider these sums separately. We thus have

$$S_2 \leq \sum_{j=N+1}^{\infty} (\xi_0^j)^2, \tag{8.43}$$

since for these values of j

$$0 \leq \frac{\beta_j}{\lambda_i} \leq 1, \quad \text{and consequently} \quad 0 \leq 1 - \frac{\beta_j}{\lambda_i} \leq 1. \tag{8.44}$$

The quantity (8.43) can be made less than arbitrary small value $\varepsilon/2$, for sufficiently large N: $S_2 < \varepsilon/2$.

When estimating the first sum S_1, we return to the initial representation (8.41):

$$S_1 = \sum_{j=1}^{N} (\xi_n^j)^2 = \sum_{j=1}^{N} \frac{1}{\beta_j} \beta_j (\xi_n^j)^2 \leq \frac{1}{m} \sum_{j=1}^{N} \beta_j (\xi_n^j)^2$$

$$= \frac{1}{m} (A^* A \xi_n, \xi_n) = \frac{1}{m} (A\xi_n, A\xi_n) < \frac{\varepsilon}{2}, \tag{8.45}$$

where the last expression can be made less than the small quantity $\varepsilon/2$, because $\|A\varepsilon_n\| \to 0$, $n \to \infty$. The theorem is proved.

In the case when the operator A is self-adjoint and positive, i.e., for $\forall u, v \in H$ $(Au, v) = (u, Av)$ and $(Au, u) > 0$, the result of the theorem remains valid when the iterative formula is applied in the less complex form:

$$u_{n+1} = u_n = \frac{\|A\xi_n\|^2}{(A\xi_n, A^2\xi_n)} Au_n, \qquad \xi_n = u_n - u_*, \qquad A\xi_n = Au_n - f, \qquad (8.46)$$

which is equivalent to a formulation of Kantorovich (see Kantorovich and Akilov, 1982).

The most difficult problem here is to give the estimate for the rate of the convergence, i.e., to estimate what is the real behavior of $\|\xi_n\| = \|u_n - u_*\|$ as $n \to \infty$. Generally, the only known estimate here is that

$$\varphi(u_n) = \|Au_n - f\|^2 = O\left(\frac{1}{n}\right), \qquad n \to \infty, \qquad (8.47)$$

and no known estimate for $\|u_n - u_*\|$.

In the case of nonlinear operator equations of the first kind, the theory of regularization is not so well developed, but some interesting results can be obtained also in this case. Here our discussion follows Tikhonov and Goncharsky (1987).

First of all, it is clear that if we construct an approximating sequence, then for ill-posed problems the number of iteration steps should be coupled with the *in-put* error, i.e., the value $n(\delta)$ must be related with the error δ. Thus $n(\delta)$ would act as a regularization parameter.

It is clear that the solution of an operator equation like (8.1) also in the nonlinear case can be reduced to the minimization of a discrepancy functional. That is why we consider here the iterative methods simultaneously to solve the first-kind operator equations and to minimize functionals. The most essential difficulties arise because classical iteration algorithms cannot be directly applied to ill-posed problems.

Let $f(u)$ be a rigorously convex differentiable functional defined on a closed convex set U of a Hilbert space H. Then we apply the basic ideas of the method of gradient projection used above, i.e.,

$$u_{n+1} = P_U\big(u_n - \gamma_n f'(u_n)\big), \qquad (8.48)$$

where P_U is the projection operator on $U \subset H$, γ_n is a constant depending on the properties of f. In this formulation classical algorithms like (8.48) typically provide only a weak convergence of the approximating sequence u_n. This is true for any classical method.

Note that the initial nonlinear problem can be re-formulated in the form of a variational inequality. It is known that the minimization of a convex functional on a closed convex set $U \subset H$ of a Hilbert space can be reduced to the solution of the following variational inequality with respect to u:

$$\big(f'(u), (u - z)\big) \leq 0, \qquad \forall z \in U. \qquad (8.49)$$

Here $f'(u)$ means the usual gradient if we assume that the functional is smooth. If $f(u)$ is a convex functional, then $f'(u)$ is a monotonic operator, i.e.,

$$\big(f'(u_2) - f'(u_1), (u_2 - u_1)\big) \geq 0, \qquad \forall u_1, u_2 \in U. \qquad (8.50)$$

Let us formulate the problem of solving the inequality

$$\big(F(u), (u - z)\big) \leq 0, \qquad \forall z \in U, \qquad (8.51)$$

where $F(u)$ is a monotonic operator. In the case $U = H$, the formulation includes the solutions of a nonlinear operator equation with a monotonic operator F. A standard iterative process cannot be applied, as we already noted, to solve this problem.

Let $M(u)$ be a strictly monotonic operator on $U \subset H$, i.e.,

$$\big(M(u_1) - M(u_2), (u_1 - u_2)\big) \geq C_M \|u_1 - u_2\|^2, \qquad C_M > 0. \tag{8.52}$$

and we consider, instead of (8.51), the auxiliary inequality

$$\big(F(u) + \alpha M(u), (u - z)\big) \leq 0, \qquad \forall z \in Q, \qquad \alpha > 0. \tag{8.53}$$

Since the operator $F(x) + \alpha M(x)$ is strictly monotonic (for any $\alpha > 0$), the inequality in (8.53) can be solved by any ordinary classical iterative method. On the other hand, the inequality (8.53) approximates (8.51) when $\alpha \to 0$ (which directly follows from Brauder–Tikhonov lemma). So, let us apply a standard iterative method to solve problem (8.53) for a fixed $\alpha > 0$:

$$u_{n+1} = R(\alpha, u_n, M). \tag{8.54}$$

The iteration algorithm

$$u_{n+1} = R(\alpha_n, u_n, M) \tag{8.55}$$

can solve (8.51) for a certain choice of the sequence $\alpha_n \to 0$.

In order to clarify this statement, let us apply the following theorem, which illustrates the iterative scheme.

THEOREM 2. *Let $f(u)$ be a doubly differentiable functional such that the domain of definition $D(f') = H$, $M(u) = u$, and $\|f''(u)\| \leq N$, $\forall u \in U$. Let $\gamma_n > 0$ and $\alpha_n > 0$ satisfy the following conditions:*

$$(a) \ \lim_{n \to \infty} \alpha_n = 0, \quad (b) \ \alpha_n \leq \alpha_{n-1}, \quad (c) \ \sum_{n=1}^{\infty} \gamma_n \alpha_n = +\infty, \quad (d) \ \varlimsup_{n \to \infty} (1 + \alpha_n)\gamma_n < 2/N. \tag{8.56}$$

Then the iterative process

$$u_{n+1} = P_U\big(u_n - \gamma_n(f'(u_n) + \alpha_n u_n)\big) \tag{8.57}$$

converges from any starting point $u_0 \in H$ to a solution of the variational problem (8.51), which has the minimal norm in H.

The proof can be found in Tikhonov and Goncharsky (1987).

Until now we considered only ill-posed problems with exactly known initial data. Iterative methods such as (8.57) can also be used to solve problems with perturbed input data. This is because the constructed algorithm, with a proper choice of the number of applied iterations linked with the error δ of the input data, yields some regularizing algorithm for (8.51). The following theorem demonstrates how we can determine the applied number of iterations n as $n(\delta)$, to be used in the iterative regularization (see Tikhonov and Goncharsky, 1987)

THEOREM 3. *Let us assume that the gradient of the perturbed right-hand side $\tilde{f}'$ belongs to the same class as the exact f' of Theorem 2, i.e., $\|f''(u)\| \leq N$, $\forall u \in U$, and $\|f'(u) - \tilde{f}'(u)\| \leq \delta\|u\|$. Then the process defined in (8.57) generates a regularizing algorithm for (8.51) if, under the conditions of Theorem 2, $n(\delta)$ is such that*

$$\lim_{n \to \infty} \delta/\alpha_{n(\delta)} = 0. \tag{8.58}$$

The theorems we have represented illustrate the power of iterative regularization. However, the discussed ideas can also be applied to directly solve linear ill-posed operator equations such as (8.1), which can be constructed without using (8.55) or (8.57). For example, a solution of (8.1) with approximate data f_δ and an exact operator A can be found by using the following iterative process (compare with results of Theorem 1):

$$u_{n+1} = u_n + \mu(A^* f_\delta - A^* A u_n), \qquad u_0 = 0, \qquad 0 < \mu < 2/\|A^* A\|. \qquad (8.59)$$

Let $b > 1$ be a fixed constant. Then the regularization parameter can be chosen according to the discrepancy principle in the following way. For $n(\delta)$ we take the first n such that

$$\|A u_n - f_\delta\| \le b\delta. \qquad (8.60)$$

The proof to the following theorem can be found in Tikhonov and Goncharsky (1987).

THEOREM 4. *Let A be a continuous linear operator from a Hilbert space U to a Hilbert space F, with the iteration sequence u_n defined by (8.59). If $n(\delta)$ is chosen as given in (8.60), then $\|u_{n(\delta)} - u_*\| \to 0$ with $\delta \to 0$, where u_* is the normal solution to Eq. (8.1) for $\delta = 0$, i.e., the solution with a minimal norm in U.*

Helpful remarks

$1°$. The discussed iterative method given by Theorem 1 represents an approach, which is an alternative to Tikhonov's smoothing functional construction. A role analogous to the small parameter of regularization $0 < \alpha \ll 1$ here is played by the number of iteration $n \gg 1$.

In order to justify this statement, let us suppose that u_0 is a solution to the main equation: $A u_0 = f$, and f_δ is an approximate (i.e., perturbed) right-hand side with $\|f - f_\delta\| < \delta$, where $\delta > 0$ is a small quantity. Then the explicit form of the SDM, Eq. (8.34) $\sim$ (8.35), which expresses the $(n + 1)$st iteration through the previous nth iteration by a finite combination of the direct operator A, shows that for a finite number of the iteration n the difference $\|u_n^\delta - u_n\|$ can be made arbitrarily small when $\|f - f\|_\delta \to 0$, i.e., the iterative process with finite number n guarantees stability with perturbations of the right-hand side f.

$2°$. Between Tikhonov's and iterative schemes there is a significant difference. If we need to solve the first-kind operator equation (8.1) with exact (i.e., unperturbed) right-hand side f, then the iterative SDM can give us an approximation to the exact solution with arbitrary accuracy, by taking more and more iterations. In contrast with this, Tikhonov's scheme based on the help of a small *smoothing* parameter cannot provide the approximation to the exact solution with arbitrarily small error, since the regularized equation $(A^* A + \alpha I)u = f$ cannot be solved for too small α.

8.4. Comparison of Various Methods for Reconstruction of the Scatterer Geometry

Over the last years, increasing interest has been shown in inverse scattering problems. A good survey of the state of the art can be found in Colton and Kress (1992), where mainly rigorous and abstract mathematical theory is developed. In practical aspect, this interest can be explained by applications to radio-location, ocean acoustics, medical and technical ultrasonics, seismic problems, etc. The current section presents an analysis of the main approaches to the problem of reconstruction of obstacle shape. For all that, we consider only the papers where: 1) the authors do not use any simplified theories (like Kirchhoff's hypothesis) that were a subject of the previous Chapter 7; 2) some examples of reconstruction are presented; 3) proposed algorithms are sufficiently wide to be applicable to more complex problems. It should also be noted that we are only interested in those algorithms which can directly be applied in practice.

We start from the remark that the studied problems are ill-posed and nonlinear simultaneously.

Apparently, the approach proposed by Imbriale and Mittra (1970) was the first one where the authors did not accept any simplified assumptions. If we solve the 2D Helmholtz equation in the exterior of a certain acoustically soft obstacle, then the authors' principal idea is as follows. Let the scattered pattern $F(\varphi)$ be known in the far zone. Then the Fourier coefficients a_m in the series expansion of this function

$$F(\varphi) = \sum_{m=-\infty}^{\infty} a_m e^{im\varphi} \qquad (8.61)$$

uniquely determine the scattered wave field in the polar coordinate system in the following form:

$$p^{sc}(\rho, \varphi) = \sum_{m=-\infty}^{\infty} a_m \, i^m H_m^{(1)}(k\rho) \, e^{im\varphi}, \qquad (8.62)$$

where the series converges at least outside a minimum disk C_0 containing the unknown obstacle. Since the full pressure is trivial on the boundary contour l (due to the boundary conditions), so with decreasing ρ the first point for which $p^{inc} + p^{sc} = 0$ is a point belonging to the boundary of the object. If we use different origins of coordinate systems, then in principle we can restore a set of boundary points. This method can be easily spread to the case of acoustically hard body, since it is obvious that for smooth boundary curves $\partial p/\partial n = \partial p/\partial \rho$.

Despite its simplicity and attractiveness, this method possesses at least the following two disadvantages:

1) This cannot be applied directly to reconstruction of nonconvex objects. In order to overcome this restriction, the authors apply a method of sequential continuation of the scattered wave field. Such an approach, which is contiguous to analytic continuation in the theory of complex-valued analytic functions, consists in multiple re-expansion of the Fourier series in wider and wider domains. Just as the problem of analytic continuation, this problem is one of the most difficult ill-posed problems.

2) Convergence of the series (8.62) in the exterior of the disk C_0 does not mean that it is so easy to calculate its sum in practice. In order to realize in practice the discussed method, we need to calculate the sum (8.62) for points very close to the boundary of its convergence, and it is not clear what is the rate of its convergence in such cases. This question is closely connected with the classical Rayleigh hypothesis if it is possible to continue the series (8.62) inside C_0, and if so, for how long of a distance. For instance, there is a known result (see also Section 8.7) that for elliptic obstacle in the coordinate system, related with its axes, the boundary of the convergence of the series (8.62) passes its foci, situated on the principal axis. Therefore, for elongated ellipses the point at which $p^{inc} + p^{sc} = 0$ is situated very close to the boundary of convergence. However, this feature is a common disadvantage of almost all existing methods of reconstruction: they lose their efficiency for narrow obstacles.

In Colton and Monk (1985) the authors propose an interesting approach, which is based on the concept of a Herglotz function. A Herglotz function is an entire function (in the sense of complex-valued analysis) that satisfies a Helmholtz equation. Then the authors introduce a certain nonnegative functional that vanishes at the actual solution of the problem. Since the considered inverse problem is ill-posed, the proposed numerical algorithm consists in the minimization of the functional on some compact set. In order to apply the proposed algorithm, one needs to know the scattered far-field amplitude on some interval of the wave number, $k \in (k_1, k_2)$, which is too problematic in practice. In the next works of the authors (see Colton and Monk, 1986) this restriction was overcome.

In Kirsch et al. (1988) there is proposed a method where the scattered wave field is sought as a single-layer potential (cf. Section 2.1) over some auxiliary contour (compare with the auxiliary sources method in Section 8.7):

$$p^{sc}(x) = \int_C H_0^{(1)}(k|x - y|)\psi(y)\, ds_y, \tag{8.63}$$

over a certain contour C situated inside the unknown obstacle, whose boundary is under reconstruction. Equation (8.63) implies the following expression for the far-field scattered amplitude:

$$F(\varphi) = \int_C e^{-ik(y_1 \cos \varphi + y_2 \sin \varphi)}\psi(y)\, ds_y, \qquad 0 \le \varphi \le 2\pi. \tag{8.64}$$

If the scattered diagram in the far zone $F(\varphi)$ is known, then Eq. (8.64) is a linear integral equation of the first kind (which is ill-posed) with respect to the function $\psi(y), \quad y \in C$. The second step of the proposed method consists in the determination of the sought boundary contour l for which $p^{inc} + p^{sc} = 0$, where p^{sc} is defined by Eq. (8.63). It is clear that efficiency of such an approach strongly depends upon the choice of the contour C.

Completing the survey of the first three methods let us notice that they all possess the essential restriction: since they use in one or another way analytical representation of the wave field these approaches cannot be directly spread to the case when we know not the full far-field wave amplitude $F(\varphi)$ but only its modulus $|F(\varphi)|$. In the meantime, the problems with given $F(\varphi)$ are of most importance in practice.

So, let us proceed to discussion of methods free of the marked disadvantage.

The method proposed by Angell with co-authors (1989a,b) was tested on some examples. The wave field outside the sought domain is approximated by a finite number of the *elementary* waves

$$p^{sc}(\rho, \varphi) = \sum_{m=-M}^{M} c_m\, i^m H_m^{(1)}(k\rho) \begin{Bmatrix} \cos(m\varphi) \\ \sin(m\varphi) \end{Bmatrix}. \tag{8.65}$$

The chosen approximate representation of the solution *a priori* satisfies the Helmholtz equation and radiation condition. The only remaining conditions to be satisfied are the boundary condition over the contour l: $(p^{inc} + p^{sc})|_l = 0$, and the requirement that the scattered wave field must coincide with the known function $F(\varphi)$:

$$F(\varphi) = \sum_{m=-M}^{M} c_m\, i^{-m} \begin{Bmatrix} \cos(m\varphi) \\ \sin(m\varphi) \end{Bmatrix}, \qquad 0 \le \varphi \le 2\pi. \tag{8.66}$$

Further, by introducing the two discrepancy functionals

$$\Psi_1 = \int_0^{2\pi} \left| F(\varphi) - \sum_{m=-M}^{M} c_m\, i^{-m} \begin{Bmatrix} \cos(m\varphi) \\ \sin(m\varphi) \end{Bmatrix} \right|^2 d\varphi,$$

$$\Psi_2 = \int_l \left| \sum_{m=-M}^{M} c_m\, i^m H_m^{(1)}(k\rho) \begin{Bmatrix} \cos(m\varphi) \\ \sin(m\varphi) \end{Bmatrix} + p^{inc} \right|^2 d\varphi, \tag{8.67}$$

the problem is reduced to a minimization of the functional $\Psi = \Psi_1 + \sigma\Psi_2$ with a certain positive parameter $\sigma > 0$. Since the studied problem is ill-posed, the minimum of Ψ is being sought on a compact set.

By the first view this approach seems to be very similar to the method of Imbriale and Mittra (1970). However, here the question of convergence of any series is of no significance. The difference is quite the same as the difference between the question about convergence of the power expansion of any function $f(x) = \sum_{m=0}^{\infty} a_m x^m$ on the interval $(0, L)$ and the question about approximation of this function on the same interval by a polynomial: $f(x) \approx \sum_{m=0}^{M} a_m x^m$.

It should be noted that the discussed method has the evident merit, because this does not require under the minimization of the functional Ψ multiple repeated solving of the direct scattering problem. It is interesting to note that this merit of the method is at the same time its disadvantage for the following reason. Let for a problem, where the unknown object is irradiated by an incident wave, there be chosen M terms in Eq. (8.65) and equal number of terms in expansion of the contour $\rho(\varphi)$. Then the dimension of the corresponding linear algebraic system, which is solved at each iteration step by the Levenberg–Marquard (see, for example, Gill et al., 1981) method, is $2M$. If there are N incident waves in the problem then the total number of coefficients c_m is equal to MN, and dimension of the corresponding system is $(N = 1) \times M$. For example, for $M = 5$ with one incident wave there is a need to solve a 10×10 system, while for 24 incident waves (a circular scanning with a 15° step), a 125×125 system (compare with the method discussed below in Section 8.4). At the same time, a direct problem would need in this situation to solve a 5×5 matrix system with 24 right-hand sides, which is much easier.

We should agree that in an arbitrary nonlinear problem the most natural way is to apply the Newton method. This approach was realized by Roger (1981), and chronologically that was the second published work, after Imbriale and Mittra (1970), where the author quoted examples on reconstruction of the unknown boundary. In some more detail, the method proposed by Roger is as follows. For a known far-field scattered diagram $F(\varphi)$, in the case of acoustically soft obstacle, the author solves the system of two integral equations

$$F(\varphi) = \int_0^{2\pi} g(\theta) e^{-ik\rho(\theta)\cos(\varphi-\theta)}\, d\theta, \qquad 0 \le \varphi \le 2\pi,$$

$$\int_0^{2\pi} g(\theta) H_0^{(1)}[kr(\varphi,\theta)]\, d\theta = e^{ik\rho(\varphi)\cos\varphi}, \qquad 0 \le \varphi \le 2\pi, \qquad (8.68)$$

$$r(\varphi,\theta) = \left[\rho^2(\varphi) + \rho^2(\theta) - 2\rho(\varphi)\rho(\theta)\cos(\varphi - \theta)\right]^{1/2},$$

with respect to the two unknown functions—the function $g(\varphi)$, related to the normal component of the velocity over the boundary contour, and the function $\rho(\varphi)$, which describes the boundary of the unknown star-like domain. Here the incident wave is assumed to be plane.

For numerical implementation there is used a step-by-step method where at each step a linearized system is being solved. These linear systems, similarly to the full problem, are also ill-posed, and so are studied with the help of Tikhonov's regularization. Such an approach is very fast in the sense of computation time. However, this algorithm possesses an intrinsic disadvantage—the question on convergence of the Newton method in ill-posed problems is poorly studied in the literature. In particular, it is not so evident what is the rate of convergence of this method. In this connection we only note here that all known estimates about convergence of the Newton method applied to nonlinear systems operate with the norm of the operator inverse to the Jacobian of the considered operator equation. However, in the ill-posed problems Jacobian is irreversible.

In the work of Tobocman (1989) the Newton method is extended to the case of acoustically hard boundary. It is also admitted irradiation of the obstacle by several sources with different frequencies. The basic distinction from the method of Roger is that for solving linear systems at each iteration step the author first applies, instead of Tikhonov's regularization, a singular decomposition of the corresponding matrices and then neglects its small eigenvalues.

Kristensson and Vogel (1986) apply a quasi-Newton method to minimization of a functional that corresponds to Tikhonov's smoothing functional under the process of discrepancy minimization for the initial nonlinear system. For all that, as an operator of the corresponding direct problem, the authors use the so-called T-matrix approach (see Boström, 1986).

It should be noted that a good survey of various methods to solve inverse problems in diffraction theory can be found in Burov (1986). Besides, some interesting approaches to inverse problems of diffraction were also proposed in Colton (1984),

Helpful remarks

1°. Let us emphasize the two restrictions inherent in all above discussed approaches:

1) Their possibilities, as a rule, are demonstrated in practice only for the case of a 2D problem. The passage from the model 2D to a real 3D case increases the number of unknowns at least by an order of magnitude, which leads to a significant increase in the computational time.

2) They are efficient only for long or moderate waves. The passage to the real frequency range, where representative size of obstacles is very often by an order larger when compared with the wavelength, implies significant increase in the number of nodes on the grid. Therefore, with the frequency increasing, the dimension of the corresponding finite-dimensional linear (or nonlinear) algebraic systems becomes too large even for modern computers.

If we take into account both remarks 1) and 2), we can conclude that for real practical requirements the total number of the grid nodes increases at least by a factor of $M = 50$ to 100. Since all the above discussed methods reduce the studied inverse diffraction problem to some (linear or nonlinear) system of algebraic equations, this involves an increase in the computation time by a factor of $M^3 \sim 10^5$ to 10^6.

2°. Another important disadvantage inherent in all discussed methods (perhaps, excluding the work of Imbriale and Mittra, 1970) is the fact that they restore the boundary of only star-like domains. Such curves can be described by a one-to-one function $\rho = \rho(\varphi)$ in the polar coordinate system. Operation with arbitrary domains is impossible with such approaches, since it is not clear *a priori* what are the intervals of the polar angle variance for which the function $\rho = \rho(\varphi)$ is not single-valued. In order to overcome this difficulty absolutely different approaches should apparently be developed.

8.5. General Inverse Diffraction Problem: Combination of Iterations and Smoothing

As we could see from the previous Chapter 7, the inverse problem on the reconstruction of the shape of unknown obstacles from the known scattering diagram within the framework of geometrical diffraction theory can be reduced to a Minkowski problem, which is quite classical in differential geometry. At least for the class of convex obstacles this can be numerically studied as a well-posed boundary value problem since this is reduced to a second-order partial differential equation with respect to the support function, the problem known to be very stable with respect to small perturbations of the given data.

In contrast with this, the considered inverse problem in exact formulation is reduced (see below) to a system of nonlinear integral equations, with the principal of them being a

Fredholm equation of the first kind with very smooth (and hence "very bad") kernel. The first three sections of the present chapter dealt with such a type of ill-posed problems, and here we will demonstrate in which way some techniques discussed above can be applied to the studied inverse problem on reconstruction of obstacles' shape in exact formulation.

Let us note once again that inverse problems of mathematical physics in general were studied by many authors (a good state of the art, with applications to inverse diffraction problems, is presented in the monograph of Colton and Kress, 1992).

Let a certain specified wave p_0 be incident on a body situated in an acoustic medium. We will restrict ourselves by the two-dimensional case for the sake of simplicity. As usual, the incident wave is assumed to be plane and for simplicity propagating along the horizontal axis, so that $p_0 = e^{ikx}$ (k is the wave number). We limit ourselves to the class of star-like regions. Then a parametric representation in polar coordinates, $\rho = \rho(\phi)$, $0 \le \phi < 2\pi$, is possible for the contour l, bounds the body under consideration. To be more specific, we assume that contour l is acoustically soft. In this case, the boundary condition has the form $p|_l = 0$. The problem is posed as follows. The field scattered from the body at all angles φ, i.e., the function $F(\varphi) = \int_l \exp[-ik\rho(\theta)\cos(\varphi - \theta)]\partial p/\partial n|_l\, dl$, $0 \le \varphi < 2\pi$ is known (see Chapter 2 for an expression of the scattered wave field).

It is required to determine the function $\rho(\varphi)$. We will assume that contour l is smooth. Then we can apply the equality $dl = (\rho^2 + \rho'^2)^{1/2}\, d\theta$ to introduce the function $g(\theta) = [\rho^2(\theta) + \rho'^2]^{1/2} dp(\theta)/\partial n|_l$, in terms of which the scattered wave field is expressed in the form

$$F(\varphi) = \int_0^{2\pi} \exp\left[-ik\rho(\theta)\cos(\varphi - \theta)\right] g(\theta)\, d\theta, \qquad 0 \le \varphi < 2\pi. \tag{8.69}$$

There are two unknown functions in this problem: $g(\theta)$ and $\rho(\theta)$. It is therefore necessary to attach another relation to (8.69). For example, it might be the boundary integral equation on contour l (see Chapter 2):

$$\int_0^{2\pi} g(\theta)H_0[kr(\varphi, \theta)]\, d\theta = p_0|_l = \exp\left[ik\rho(\varphi)\cos\varphi\right], \qquad 0 \le \varphi < 2\pi,$$

$$r(\varphi, \theta) = [\rho^2(\varphi) + \rho^2(\theta) - 2\rho(\varphi)\rho(\theta)\cos(\varphi - \theta)]^{1/2}. \tag{8.70}$$

Thus, it is possible to reduce the problem to a system of two nonlinear integral equations (8.69), (8.70) for the two unknown functions $g(\theta)$ and $\rho(\theta)$.

Note that the problem posed is the simplest inverse diffraction problem. In more complex cases, it is possible to specify not $F(\phi)$ but only its absolute value $|F(\phi)|$, and not for all $0 \le \phi < 2\pi$. Further, the direction of the incident field may be unfixed.

It should be noted once again that the problem under consideration is ill-posed in the sense of Tikhonov. This is because the operator of the direct problem $G : \rho(\phi) \to F(\phi)$ is compact in the spaces naturally associated with the problem. Then it becomes obvious, as we already remarked above, that the inverse operator $G^{-1} : F(\phi) \to \rho(\phi)$ cannot be continuous. In addition, the problem is nonlinear. All these features complicate its solution since the most efficient numerical methods for ill-posed problems are developed for linear problems.

One of the most universal numerical methods is the gradient descent method (see the previous sections). It has been shown in Section 8.3 that in the linear case (i.e., when the discrepancy functional is quadratic) this method yields a regularizing algorithm. Although there is no rigorous proof for the general case, it turns out that this property of the method also can be extended to nonlinear problems. This follows from the fact that any smooth functional is quadratic in the neighborhood of a local minimum.

The numerical solution algorithm is constructed as follows. First Eqs. (8.69) and (8.70) are written in a finite dimensional form at the nodes ϕ_i, θ_j ($\phi_i = \theta_j$) using a simplest rectangle quadrature formula. The collocation nodes form a uniform grid on the segment $(0, 2\pi)$. The diagonal terms in (8.70) (at $i = j$) must be evaluated as the integral over a small interval of a principal singular part of the kernel (the singularity is logarithmic). With a sufficiently dense grid of nodes, the nonlinear finite-dimensional system obtained in this way

$$f_i(g_1, \ldots, g_j, \rho_1, \ldots, \rho_j) = 0, \qquad i = 1, \ldots, 2J, \tag{8.71}$$

has operator properties similar to those of the original fully continuous operator. For example, some of the eigenvalues of its Jacobian are very close to zero. In constructing the numerical algorithm, therefore, it is necessary to remember that the problem (8.71) is in practice ill-posed.

We write the discrepancy functional $\psi(g, \rho) = \sum_{i=1}^{J} |f_i|^2$. At the zero-th step, we find $\min \psi(g, \rho)$ on all possible circles. Suppose that the values g^0, ρ^0 correspond to a circle found in this way. The next step is to apply an iterative gradient-descent procedure. Here (see, for example, Gill et al., 1981)

$$\operatorname{grad} \psi = A^* f \tag{8.72}$$

where A is the Jacobian of system (8.71), f is the vector of its left-hand side, and the asterisk is the symbol of the conjugate. The Jacobian A can be calculated in explicit form in the present problem. Thus, the direction of descent is given by the equality

$$q = -A^* f, \qquad q = (\Delta g, \Delta \rho)^T. \tag{8.73}$$

It is found that this method makes it possible to solve certain simple problems. For example, it is possible to obtain an ellipse of small eccentricity from a circle. However, the method converges very slowly for more complex problems, hence the convergence should be acceleration by any reasonable procedure.

To provide such an acceleration, we note that any direction $g = -B \operatorname{grad} \psi$ will be a direction of descent if B is a positive definite matrix. A suitable matrix B must be chosen to accelerate the procedure. For this purpose, the functional ψ is approximated quadratically at the point $z = (g, \rho)$ (see Gill et al., 1981): $\psi(z+q) = \psi(z) + (A^* f, q) + \frac{1}{2}(q, Gq)$. Minimization of $\psi(z + q)$ generates the direction (8.73) in linear approximation. Minimization of $\psi(z + q)$ in the quadratic approximation results in the relation $Gq = -A^* f$. It is shown in Gill et al. (1981) that the Hessian $G \approx A^* A$ in the neighborhood of its minimum, so this relation is approximately $A^* Aq = -A^* f$. With (8.72), we might have taken $(A^* A)^{-1}$ as the matrix B. However, since the problem is ill-conditioned, the operator $A^* A$ has no inverse (in the sense specified above), and for this reason we use an idea implicitly related to Tikhonov's regularization. To the operator $A^* A$ we add a small positive operator, for example, αI, $\alpha > 0$. We then obtain

$$(A^* A + \alpha I)q = -A^* f. \tag{8.74}$$

This relation determines a certain direction of descent. Thus, $B = (A^* A + \alpha I)^{-1}$ is correctly defined. It is obvious that (8.74) becomes (8.73) for $\alpha \gg 1$.

It turns out that implementation of the above algorithm faces a difficulty, because it is not so easy to find a universal criterion for selection of the parameter α, so that this will be suitable for various classes of contours l. By our experience, the following approach turns out very efficient. Several values α_m ($m = 1, \ldots, M$) that form a geometric progression (usually $M = 5$ to 8) are chosen in the range $\alpha_{\min} = 10^{-4} < \alpha < 10^1$, and the values α_{opt} at which the functional ψ decreases most sharply on the corresponding direction q_{opt} is identified among them. Naturally, the optimum α_{opt} determined in this way will be different at different steps of the iterative procedure. We find that α_{opt} takes large values in the first few steps and then gradually decreases with iterations. The limit $\alpha > \alpha_{\min}$, which is accepted here, preserves well-posedness for the evaluation of the matrix $(A^*A + \alpha I)^{-1}$. This universal method of selection of the parameter α was used in all of the following examples of reconstruction of the boundary contour l.

It should be noted that system (8.71) is overdetermined. Actually, it contains $3J$ real unknowns: J values of $\{\operatorname{Re} g_j, \operatorname{Im} g_j, \rho_j\}$. At the same time, the system has $4J$ real equations. The proposed approach permits solution of overdetermined systems. For example, the numbers of unknowns and equations would be the same if only the real amplitude $|F(\phi)|$ of the wave was known.

Some examples of reconstruction for certain types of boundary contour l by the method described in this section can be found in the author's work (Vorovich and Sumbatyan, 1990).

Let a body with an unknown boundary l be irradiated successively by plane waves at various angles $\gamma_n, n = 1, \ldots, N$: $p_0^n = \exp[-ik\rho(\varphi)\cos(\varphi - \gamma_n)]$. Suppose that the scattering-wave amplitude is known for several angles φ_i^n, $i = 1, \ldots, I$ for each nth incident wave. Then the problem in finite-dimensional form reduces to the nonlinear system

$$f_i^n = \left| h \sum_{j=1}^{J} \exp[-ik\rho_i \cos(\varphi_i^n - \theta_j)] g_j^n \right|^2 - F_i^n = 0, \qquad (8.75)$$

$$h \sum_{j=1}^{J} H_0[kr(\rho_i, \rho_j)] g_i^n = \exp[-ik\rho_i \cos(\varphi_i - \gamma_n)] \qquad (h = 2\pi/J), \qquad (8.76)$$

where we should remember that the term $i = j$ in the second equality requires a careful treatment.

System (8.75)–(8.76) can be solved by the method described above. Then the linear system (8.74) for determination of the descent direction is a system of $(2N + 1)J$ real equations for the same number of real unknowns $\operatorname{Re} g_j^n$, $\operatorname{Im} g_j^n$, ρ_j. This approach requires a great deal of machine time when J and N are large. To decrease the dimension we make use of two properties of system (8.76), which we treat as a system for g_j^n: it is linear with respect to g_j^n; only the right-hand side changes in this system when n changes. We therefore propose to reduce the problem to minimization of the functional

$$\psi = \sum_{n=1}^{N} \sum_{i=1}^{I} |f_i^n|^2 \qquad (8.77)$$

with condition (8.76), i.e., to a problem of conditional optimization with constraints in the form of equalities (see Gill et al., 1981). As above, we solve this problem by quadratic approximation of functional (8.77) with respect to the increment vector: $q = (\Delta g_j^n, \Delta \rho_j)^T = (x_V, x_U)^T$. Differentiation of (8.76) gives the relation

$$V x_V + U x_U = 0, \qquad V = \begin{pmatrix} v & \cdots & 0 \\ \cdots & v & \cdots \\ 0 & \cdots & v \end{pmatrix}. \qquad (8.78)$$

Here the matrix V consists of N identical matrices $v = (v_{ij})$, $v_{ij} = hH_0[kr(\rho_1, \rho_j)]$ situated on the diagonal of the matrix V. The matrix U can also be written in explicit but more complex form.

We note that the matrix V corresponds to the direct diffraction problem, has a dominant diagonal by virtue of the logarithmic singularity of the kernel, and is therefore properly invertible.

The conditional minimization of the quadratic functional with constraints in the form (8.78) gives the equation (see Gill et al., 1981):

$$(Z^* A^* A Z + \alpha I)\bar{q} = -Z^* A^* f, \qquad (8.79)$$

where the direction of descent is determined from the formula $q = Z\bar{q}$. Here the matrix Z has the form

$$Z = \begin{pmatrix} -V^{-1}U \\ I \end{pmatrix}, \qquad (8.80)$$

the vector $f = (f_i^n)$, and A is the Jacobian of system (8.75). The optimum value of the parameter α is chosen at each step in the same way as in the case of unconditional minimization. The dimension of system (8.79) is equal to J.

Figure 8.1 demonstrates some examples on reconstruction in the case when there is known the back-scattered amplitude for $N = 24$ irradiation directions uniformly distributed in the range of angles $(0, 2\pi)$. For all that, $I = 1$, $\phi^n = \gamma_n$, and $J = 24$. Solid lines designate actual boundary contour, and dashed ones are calculated on results of the proposed reconstruction method.

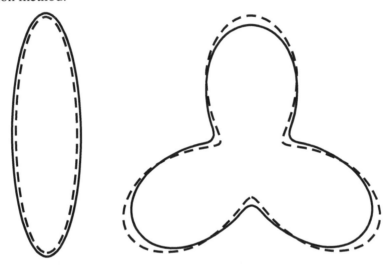

Figure 8.1. Reconstruction of the elliptic and three-leaf obstacles in acoustic medium

A similar approach can be developed in the case of reconstruction of a void of unknown shape in elastic medium. Let in elastic plane (2D case) there be situated a void with unknown boundary l, which is free of load. The given incident wave, which to be more specific we assume to be plane, falls to this void from right to left, parallel to the x_1 axis: $\sigma_{11}^0 = \sigma_0 \exp(-k_p x_1)$, where k_p, k_s are wave numbers related to longitudinal and transverse waves, respectively. For simplicity, we assume that the contour l is star-like, so that it can be represented in the polar coordinate system by a single-valued function $\rho = \rho(\varphi)$, $0 \le \varphi < 2\pi$. The most important problem here is to reconstruct the shape of the void, i.e., the function $\rho(\varphi)$ from the amplitude of the far-field scattered diagram. The direct problem here is to construct this diagram if the shape of the obstacle is known. It can be shown that this diagram is represented by the following integral:

$$\sqrt{R}\,\sigma_{rr}(R, \varphi) \sim F(\varphi), \qquad R \to \infty,$$

$$F(\varphi) = \int_0^{2\pi} e^{-ik_p \rho(\theta)\cos(\theta-\varphi)} \big\{ u_1(\theta)[2\mu\,\gamma(\theta, \varphi)\cos\varphi + \lambda\, n_1(\theta)]$$

$$+ u_2(\theta)[2\mu\,\gamma(\theta, \varphi)\sin\varphi + \lambda\, n_2(\theta)] \big\}\, d\theta, \qquad \gamma(\theta, \varphi) = n_1(\theta)\cos\varphi + n_2(\theta)\sin\varphi,$$

$$n_1(\theta) = \rho(\theta)\cos\theta + \rho'(\theta)\sin\theta, \qquad n_2(\theta) = \rho(\theta)\sin\theta - \rho'(\theta)\cos\theta. \qquad (8.81)$$

In the considered inverse problem function $F(\varphi)$ is supposed to be known. The unknown quantities in Eq. (8.81) are (as in the above scalar acoustic problem) $\rho(\varphi)$, which describes the boundary contour l, and the functions u_1, u_2 which designate the components of the displacement vector over the boundary l in the Cartesian coordinate system x_1, x_2. In order to write out a complete system of equations, we need to add to equations (8.81), for example, a system of boundary integral equations (BIE) on the contour l. In this section we use, both in scalar and elastic problems, the direct BIE method (cf. Section 2.2), where the mentioned system is

$$u_1(x) - 2 \int_l [P_1^{(1)}(y,x)u_1(y) + P_2^{(1)}(y,x)u_2(y)]dl_y = 2u_1^0(x),$$

$$u_2(x) - 2 \int_l [P_1^{(2)}(y,x)u_1(y) + P_2^{(2)}(y,x)u_2(y)]dl_y = 2u_2^0(x),$$

(8.82)

with

$$P^{(k)} = 2\mu \frac{\partial}{\partial n_y}[U^{(k)}(y,x)] + \lambda n_y \cdot \text{div}_y[U^{(k)}(y,x)] + \mu\{n_y \times \text{rot}_y[U^{(k)}(y,x)]\},$$

$$U_j^{(k)}(y,x) = \frac{i}{4\mu k_s^2}\left\{k_s^2 \delta_{kj} H_0^{(1)}(k_p r) - \frac{\partial^2}{\partial y_k \partial y_j}[H_0^{(1)}(k_p r) - H_0^{(1)}(k_s r)]\right\},$$

(8.83)

$$r = \sqrt{\rho^2(\varphi) + \rho^2(\theta) - 2\rho(\varphi)\rho(\theta)\cos(\theta - \varphi)}, \quad x = x(\varphi), \quad y = y(\theta), \quad x \in l.$$

Here, as usual, $H_0^{(1)}$ is the Hankel function, and (u_1^0, u_2^0) are the components of the displacement vector in the incident wave:

$$u_1^0 = \frac{\sigma_0}{(\lambda + 2\mu)\,ik_p}\,e^{-ik_p x_1}, \qquad u_2^0 = 0.$$

(8.84)

Therefore, the problem is reduced to a system of the three nonlinear integral equations (8.81)–(8.82) with respect to the three unknown functions ρ, u_1, u_2.

Since relation (8.81) for the scattered wave field is an integral equation of the first kind with a smooth kernel (which yields a compact integral operator of the first kind), the described problem, similarly to the scalar case, is really an ill-posed problem. That is why this requires a regularization. Here we use the same approach as that developed for the above scalar case. The algorithm consists again of minimization of the discrepancy functional for equations (8.81), (8.82) in the functional space $L_2(0, 2\pi)$. The minimization is carried out again by an iterative steepest descent method, with the help of its quadratic approximation at each step. Recall that this method does coincide neither with the Lewenberg–Marquardt method nor with Tikhonov regularization scheme (see the previous sections), nor with any other known technique. When performing numerical realization, we may write equations (8.81), (8.82) in a finite-dimensional form, which contains the values of the unknown functions $\rho_i, u_{1i}, u_{2i}, \varphi_i, \theta_j$ ($\varphi_i = \theta_i$), $i, j = 1, 2, \ldots, N$, at the nodes of a finite mesh. The principal difficulty is related to the calculation of the Jacobian of the obtained nonlinear finite-dimensional system. Expression for the Jacobian can be written in an explicit form, which reduces the time of computations by around an order compared with its calculation by using a finite-difference scheme.

Figures 8.2–8.5 demonstrate examples on reconstruction of various flaws in elastic medium. Here we used $c_p/c_s = 5.85/3.23$, which corresponds more or less to a steel medium. The results are shown for the total number of nodes $N = 72$ and for the number of iterations near 8–10.

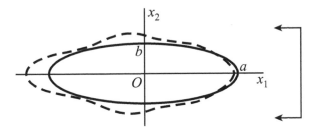

Figure 8.2. Reconstruction of elliptic flaw with semi-axes $ak_p = 1.6$, $a/b = 3$ in elastic medium

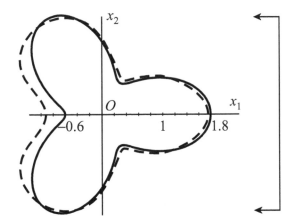

Figure 8.3. Reconstruction of the flaw $k_p\rho(\varphi) = 0.6\,(2 + \cos 3\varphi)$ in elastic medium

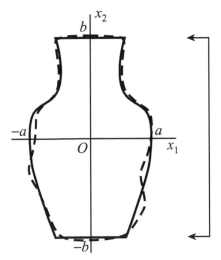

Figure 8.4. Reconstruction of the vase-type flaw with $ak_p = 0.9$, $bk_p = 1.5$ in elastic medium

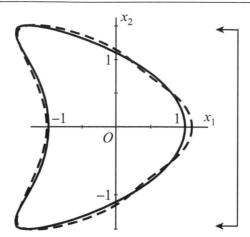

Figure 8.5. Reconstruction of the mask-type flaw in elastic medium: $k_p x_1(t) = \cos t + 0.65 (\cos 2t - 1)$; $k_p x_2(t) = 1.5 \sin t$; $0 \le t \le 2\pi$

Helpful remarks

$1°$. In the case of acoustically hard contour, when the boundary condition has the form $\partial p / \partial n|_l = 0$, the problem reduces to a system of two nonlinear integro-differential equations. However, this is not an obstacle to the application of the method discussed below also to this case.

$2°$. The methods proposed here differ from other known methods: the Gauss–Newton and Lowenberg–Marquardt methods (see Gill et al., 1981), the smoothing functional method (Tikhonov and Arsenin, 1977; and Section 8.2), and others. In the Lowenberg–Marquardt procedure, for example, a unit step along the direction of descent is always fixed. And in the smoothing functional method, α cannot depend on the iteration number (if an iterative procedure is used to minimize the smoothing functional).

$3°$. The method discussed in the present section at first sight seems to coincide with the method proposed by Roger (1981) (see our brief survey in Section 8.4), but this is not so. Roger uses the Newton–Kantorovich method to solve a nonlinear functional equation which, strictly speaking, does not converge in the case of the first-kind operator equation with compact operators (both linear and nonlinear). In contrast with this, our approach is based on the steepest descent method whose convergence is strictly proved (see Section 8.3), at least for quadratic discrepancy functionals, corresponding to linear operator equations. Locally each functional is quadratic, so although we cannot strictly prove the convergence of our method in the nonlinear case, there is a good chance, starting from the proved convergence for quadratic functional, that the proposed algorithm converges also in the more general nonlinear case considered here.

8.6. A Correct Treatment of Ill-Posed Boundary Equations in Acoustics of Closed Regions

Let D be a simply connected two-dimensional finite domain with boundary l occupied by a linear isotropic acoustic medium. Then the wave pressure $p(x_1, x_2)$ satisfies the Helmholtz equation

$$\Delta p + k^2 p = 0, \tag{8.85}$$

where $k = \omega / c$ is the wave number, ω is the circular frequency, and c is the wave speed.

The wave process in D is determined by the boundary condition set on l, which may be a Dirichlet condition (p is given on l), a Neumann condition (the normal derivative $\partial p/\partial n$ is given on l), or a mixed condition (p is given on a part of l; $\partial p/\partial n$, on the rest of it).

We apply the conventional procedure based on the application of the Kirchhoff–Helmholtz integral formula to Eq. (8.85). An auxiliary system of elementary solutions to the Helmholtz equation (8.85) is chosen as the set of plane waves $\exp[ik(\eta, x)]$, where $\eta_1^2 + \eta_2^2 = 1$, $x = (x_1, x_2)$, $\eta = (\eta_1, \eta_2)$, $(\eta, x) = \eta_1 x_1 + \eta_2 x_2$, instead of the Green's function traditionally applied in the classical approach. In general, plane waves of any type (both homogeneous and inhomogeneous) are suitable for our analysis, but we restrict ourselves to homogeneous waves with real η_1 and η_2, which propagate in the plane (x_1, x_2) at arbitrary angles without damping.

We integrate Eq. (8.85) times the chosen elementary solution over D, and apply the conventional procedure based on Green's integral formula to obtain the integral relation

$$\int_l \left[\frac{\partial p(x)}{\partial n} - ik(\eta, n)p(x) \right] \exp\left[ik(\eta, x)\,dl_x\right] = 0, \qquad |\eta| = 1, \qquad (8.86)$$

where $n = (n_1, n_2)$ is the outward unit normal to l at $x \in l$. Subject to any boundary condition, Eq. (8.86) is an ill-posed integral equation of the first kind, since its kernel is dominated by an exponential factor and is an infinitely differentiable function. Another characteristic feature of this equation is that the inner variable runs over l, while the outer one runs over the unit circle $\eta = 1$, and these variables are defined in absolutely different geometrical domains.

Let us reduce the basic equation (8.86) to an infinite algebraic system. In designing a numerical algorithm, we restrict ourselves to the class of star-shaped domains D with smooth boundaries. Then, we have the following unique representation in polar coordinates:

$$x_1 = \rho(\psi)\cos\psi, \quad x_2 = \rho(\psi)\sin\psi, \quad 0 \le \psi < 2\pi,$$
$$\eta_1 = \cos\vartheta, \quad \eta_2 = \sin\vartheta, \quad 0 \le \vartheta < 2\pi, \qquad (8.87)$$

where $\rho(\psi) > 0$. To be more specific, we consider the Dirichlet boundary value problem, with $p(x)$ given on the boundary: $p(x)|_l = f(x)$. By using the elementary relations

$$(\eta, n) = [\rho(\psi)\cos(\psi - \vartheta) + \rho'(\psi)\sin(\psi - \vartheta)]/[\rho'^2(\psi) + \rho^2(\psi)]^{1/2},$$
$$(\eta, x) = \rho(\psi)\cos(\psi - \vartheta), \qquad dl_x = [\rho'^2(\psi) + \rho^2(\psi)]^{1/2}\,d\psi, \qquad (8.88)$$

the basic equation (8.86) is transformed into the following one for the unknown function $g(\psi) = (\rho'^2 + \rho^2)^{1/2}(\partial p/\partial n)$:

$$\int_0^{2\pi} g(\psi)\exp[ik\rho(\psi)\cos(\psi - \vartheta)]\,d\psi$$
$$= ik \int_0^{2\pi} f(\psi)[\rho(\psi)\cos(\psi - \vartheta) + \rho'(\psi)\sin(\psi - \vartheta)] \qquad (8.89)$$
$$\times \exp[ik\rho(\psi)\cos(\psi - \vartheta)]\,d\psi, \quad 0 \le \vartheta < 2\pi.$$

Strictly speaking, this equation is ill-posed, because, in contrast to classical boundary equations based on the conventional application of Green's formula, its kernel does not possess even a weak logarithmic singularity that would ensure the stability of the collocation method (Voronin and Tsetsokho, 1981). However, the right-hand side of Eq. (8.89) has a very special form resulting from the application of a completely continuous operator to a regular function and, hence, has the same degree of smoothness as the operator function on its left-hand side. A natural question arises: is it possible to construct a stable numerical method for Eq. (8.89) that would take into account this property of the right-hand side? The answer to this question is rather difficult. Collocation-type methods are unlikely to be stable. For this reason, we develop a special projection method based on the following idea. For a circular contour l, the Fourier series expansion of the unknown function $g(\psi)$ in Eq. (8.89) yields a known exact solution, which can easily be verified directly from (8.89).

Let us represent $g(\psi)$, in the case of arbitrary contour l, also as a Fourier series in $\psi \in [0, 2\pi)$. For simplicity, we assume that the problem is symmetric about the x_1-axis, which does not lead to a loss of generality. Then, $\rho(\psi)$, $f(\psi)$, and $g(\psi)$ are even functions of ψ, consequently,

$$g(\psi) = \sum_{m=0}^{\infty} g_m \cos(m\psi). \tag{8.90}$$

We substitute (8.90) into Eq. (8.89) and calculate the scalar product of the resulting relation with $\cos(n\vartheta)$ ($n = 0, 1, \dots$), in order to obtain the infinite algebraic system

$$\sum_{m=0}^{\infty} b_{nm} g_m = h_n, \quad n = 0, 1, \dots,$$

$$b_{nm} = 2 \int_0^{2\pi} J_n[k\rho(\psi)] \cos(n\psi) \cos(m\psi) \, d\psi, \tag{8.91}$$

$$h_n = k \int_0^{2\pi} f(\psi) \langle \rho(\psi) \cos(n\psi) \{ J_{n-1}[k\rho(\psi)] - J_{n+1}[k\rho(\psi)] \}$$
$$+ \rho'(\psi) \sin(n\psi) \{ J_{n-1}[k\rho(\psi)] + J_{n+1}[k\rho(\psi)] \} \rangle \, d\psi.$$

To derive the last equation, we have used here the following tabulated integrals:

$$\int_0^{2\pi} \exp[ik\rho(\psi) \cos(\vartheta - \psi)] \cos(n\vartheta) \, d\vartheta = 2\pi i \cos(n\psi) J_n[k\rho(\psi)],$$

$$\int_0^{2\pi} \exp[ik\rho(\psi) \cos(\vartheta - \psi)] \cos(\vartheta - \psi) \cos(n\vartheta) \, d\vartheta$$
$$= \pi i^{n+1} \cos(n\psi) \{ J_{n+1}[k\rho(\psi)] - J_{n-1}[k\rho(\psi)] \}, \tag{8.92}$$

$$\int_0^{2\pi} \exp[ik\rho(\psi) \cos(\vartheta - \psi)] \sin(\vartheta - \psi) \cos(n\vartheta) \, d\vartheta$$
$$= \pi i^{n+1} \sin(n\psi) \{ J_{n+1}[k\rho(\psi)] + J_{n-1}[k\rho(\psi)] \},$$

where J_n is the Bessel function of the first kind of order n.

System (8.91) can be rewritten as an equivalent system of the second kind:

$$g_n + \sum_{m=0}^{\infty} a_{nm} g_m = f_n, \quad n = 0, 1, \dots, \quad f_n = h_n d_n^{-1}, \quad d_n = \int_0^{2\pi} J_n[k\rho(\psi)] \, d\psi,$$

$$a_{nm} = b_{nm} d_n^{-1}, \quad n \neq m, \quad a_{nn} = d_n^{-1} \int_0^{2\pi} J_n[k\rho(\psi)] \cos(2n\psi) \, d\psi. \tag{8.93}$$

The properties of the solution to system (8.93) are determined by the asymptotic behavior of a_{nm} and f_n as $n, m \to \infty$. Using the well-known asymptotics of the Bessel function,

$$J_n(z) \sim \frac{1}{n!} \left(\frac{z}{2} \right)^n, \quad n \to \infty. \tag{8.94}$$

After some algebra, we obtain at $n \to \infty$

$$
\begin{aligned}
b_{nm} &= \int_0^{2\pi} J_n[k\rho(\psi)] \cos\left[(n-m)\psi\right] d\psi + \int_0^{2\pi} J_n[k\rho(\psi)] \cos\left[(n+m)\psi\right] d\psi \\
&\sim \frac{k^n}{n!2^n} \int_0^{2\pi} \rho^n(\psi)(e^{-i|n-m|\psi} + e^{-i(n+m)\psi})\, d\psi, \quad d_n \sim \frac{k^n}{n!2^n} \int_0^{2\pi} \rho^n(\psi)\, d\psi.
\end{aligned}
\tag{8.95}
$$

The coefficients h_n can be represented likewise.

The asymptotics of integrals (8.93) and (8.95) can be found by the saddle-point method and depends on the analytic properties of $\rho(\psi)$, which determines the boundary contour l (see Fedorjuk, 1977). Explicit asymptotic representations of these integrals can be derived when $\rho(\psi)$ has the general form

$$
\rho(\psi) = \sum_{n=0}^{\infty} a_n \cos(n\psi), \quad a_0 > 0, \quad a_n \geq 0, \quad n = 1, 2, \ldots, \quad 0 \leq \psi < 2\pi.
\tag{8.96}
$$

In this case, by introducing the new variable $z = \exp(i\psi)$, the coefficients b_{nm} in (8.95) can be represented in the form

$$
b_{nm} \sim -\frac{ik}{n!2^n} \int_{|z|=1} \{e^{nS_1(z)} + e^{nS_2(z)}\}\, dz, \quad \rho(z) = \frac{1}{2} \sum_{n=-\infty}^{\infty} a_n z^n, \quad a_n = a_{-n},
$$

$$
S_j(z) = \ln \rho(z) - \mu_j \ln z, \quad j = 1, 2, \quad \mu_1 = \frac{|n-m|+1}{n}, \quad \mu_2 = \frac{n+m+1}{n}.
\tag{8.97}
$$

The expression for d_n is analogous, with $\mu_j = 0$.

The general properties of phase functions similar to $S_{1,2}$ were examined in detail in Fedorjuk (1977). In particular, it was proved that the saddle point $z = x_0(\mu_j)$ is always positive and is determined by the equation

$$
z\rho'(z) = \mu_j \rho(z).
\tag{8.98}
$$

However, a stronger result can be stated in our problem: $x_0(\mu_j) > 1$. Indeed, it was proved in Fedorjuk (1977), Lemma 5.2, that

$$
\frac{z\rho'(z)}{\rho(z)} = \left[\sum_{n=0}^{\infty} na_n(z^n - z^{-n})\right] \left[\sum_{n=0}^{\infty} a_n(z^n + z^{-n})\right]^{-1}
\tag{8.99}
$$

is a monotonically increasing function. It vanishes at $z = 1$ and tends to plus infinity as $z \to +\infty$. Consequently, for any $\mu_j > 0$, the function in (8.99) equals μ_j at the single point $z = x_0(\mu_j) > 1$. Thus, the integral in (8.97) has the following asymptotic form (where a certain inessential factor is omitted):

$$
\int_{|z|=1} \exp\left[nS_j(z)\right] dz \sim \frac{\exp[nS_j(x_0)]}{n^{1/2}} = \left[\frac{\rho(x_0)}{x_0^{\mu_j}}\right]^n n^{-1/2}, \quad x_0 = x_0(\mu_j) > 1, \quad n \to \infty.
\tag{8.100}
$$

The crucial point in further analysis in the proof of the fact that $S_j(x_0) = S_j[x_0(\mu_j)]$ is a monotonically decreasing function of μ_j. Indeed, it is easy to see that

$$
\frac{dS}{d\mu_j} = \frac{\partial S}{\partial x_0} \frac{dx_0}{d\mu_j} - \ln x_0 = -\ln x_0 < 0, \quad x_0 > 1,
\tag{8.101}
$$

which follows from $\partial S / \partial x_0 = 0$, since x_0 is a saddle point.

This property implies that the coefficients d_n, which correspond to $\mu_j = 0$ in the integral (8.100) and arise in the denominators of a_{nm} and f_n in (8.93), are exponentially large as compared to the numerators of the corresponding fractions, provided that μ_j in these fractions is not infinitely small. More specifically, for $n \to \infty$, the following holds:

$$a_{nm} = b_{nm} d_n^{-1} \sim e^{n\{S[x_0(\mu_1)] - S[x_0(0)]\}}, \quad \mu_1 = |n - m|/n, \quad x_0(0) = 1, \quad n \neq m,$$

$$a_{nn} \sim e^{n\{S[x_0(2)] - S[x_0(0)]\}}, \quad f_n = h_n d_n^{-1} \sim e^{n\{S[x_0(1)] - S[x_0(0)]\}}. \tag{8.102}$$

Now, it is clear that monotonic decrease in S as a function of μ implies exponential decrease in the coefficients and absolute terms in (8.93) as $n \to \infty$, provided that $\mu_1 = O(1) \sim |n - m| = O(n)$, $n \to \infty$.

Consider the structure of the matrix (8.93) and the structure of its right-hand side for large n in more detail. Obviously, all matrix elements, which are not exponentially small, lie near the principal diagonal. In this sense, the matrix of system (8.93) is similar to a banded matrix. However, its bandwidth increases linearly with the row index n. Infinite algebraic systems with such matrices have not been analyzed previously. In particular, they cannot be referred to any well-known class, such as regular, quasiregular, or Poincaré–Koch normal systems (see, for example, Kantorovich and Krilov, 1958). Nevertheless, the truncation method for system (8.93) converges very rapidly. This is explained by the special form of the right-hand side in Eq. (8.89) and entails exponential decrease in the components of f_n. Indeed, for the truncated system (8.93),

$$g_n = f_n - \sum_{m=0}^{M} a_{nm} g_m, \tag{8.103}$$

we easily prove by induction that $g_n = O(\exp(-\varepsilon n))$, as $n \to \infty$, for $\varepsilon > 0$. The proof is based on the following fact. When n is large, each summand in (8.103) is exponentially small, because so is a_{nm} with a small m (as an element situated far from the diagonal) or g_m with a large m.

Even though the above properties of the infinite system are proved only for $\rho(\psi)$ with positive Fourier coefficients (see (8.96)), some test results obtained by using the method in various examples have shown that the exponential decrease in g_n with increasing n is a common property for smooth contours l with analytic boundary functions $\rho(\psi)$.

In all test examples, a constant boundary function was used: $f(\psi) \equiv 1$. In one example, an ellipse with semi-axes $ka = 3$ and $kb = 1$ was considered. In this case, the Fourier expansion (8.96) of $k\rho(\psi) = ab[a^2 \sin^2 \psi + b^2 \cos^2 \psi]^{-1/2}$ contains both positive and negative coefficients. Nevertheless, the method converges very rapidly, as illustrated by the curves shown in Fig. 8.6.

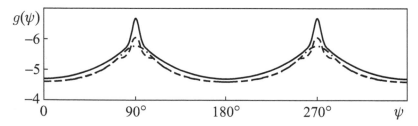

Figure 8.6. Numerical results for an elliptic domain

Figures 8.7 and 8.8 show analogous curves of $g(\psi)$ obtained for contours l shaped as three-leaf roses with different eccentricities, $k\rho(\psi) = 2 + a\cos(3\psi)$ with $a = \frac{1}{2}$ and $a = 1$ corresponding to Figs. 8.7 and 8.8, respectively.

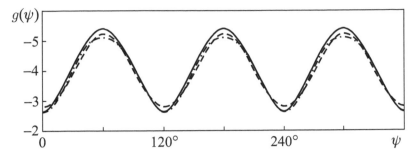

Figure 8.7. Numerical results for a domain in the form of a three-leaf rose: $a = \frac{1}{2}$

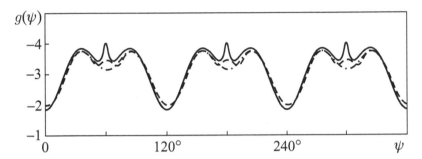

Figure 8.8. Numerical results for a domain in the form of a three-leaf rose: $a = 1$

Solid curves in all figures were calculated by the classical method of boundary integral equations, in which the unknown function $g(\psi)$ is determined by the following Fredholm integral equation of the first kind whose kernel has a logarithmic singularity at the origin:

$$
\begin{aligned}
&\int_0^{2\pi} g(\psi) Y_0(kr)\, d\psi = -2f(\varphi) - k \int_0^{2\pi} f(\psi) \frac{Y_1(kr)}{r} [(\xi_1 - x_1)(\rho' \sin\psi + \rho\cos\psi) \\
&+ (\xi_2 - x_2)(\rho\sin\psi + \rho'\cos\psi)]\, d\psi, \quad \varphi \in [0, 2\pi), \quad r = [(\xi_1 - x_1)^2 + (\xi_2 - x_2)^2]^{1/2}, \\
&x_1 = \rho(\varphi)\cos\varphi, \quad x_2 = \rho(\varphi)\sin\varphi, \quad \xi_1 = \rho(\psi)\cos\psi, \quad \xi_2 = \rho(\psi)\sin\psi,
\end{aligned}
\tag{8.104}
$$

where $Y_1(x)$ is a Bessel function of the second kind. Equation (8.104) was solved numerically by the collocation method with the number of nodes $N = 180$ in $\psi \in [0, 2\pi)$. Dot-and-dash and dashed curves in Figs. 8.6–8.8 corresponded to $M = 10$ and $M = 25$, respectively, where M is the number of equations retained in the truncated system (8.103).

The results displayed in Figs. 8.6–8.8 demonstrate that the algorithm proposed here is highly stable and accurate.

The presented results are discussed in the author's paper (Sumbatyan, 2001).

Helpful remarks

$1°$. Note that the run time of the algorithm is comparable to that of the conventional boundary integral equation method in the isotropic case. Indeed, a greater portion of CPU time in the former case is required to evaluate the integrals in the matrix of the infinite algebraic system. However, test results have shown that dimension of the system in that case can be reduced by an order of magnitude, as compared to the classical method of boundary integral equations, in which the elements of the algebraic system are expressed in explicit form.

2°. The method discussed here seems to be a little alien to the considered direct boundary value problem, when compared for example with the BIE method discussed in Chapter 2. However, the proposed approach looks more efficient in the cases when the Green's function cannot be constructed in a finite form. Indeed, efficiency of the classical BIE approach, based on the Kirchhoff–Helmholtz integral formula, is closely connected with the fact that the Green's function is expressed explicitly in terms of the exponential (3D case) or the Helmholtz (2D case) function. If we consider, for example, the linear acoustic problem in a medium possessing some kind of anisotropy, then the Green's function (i.e., the solution to the basic equation with the delta function $\delta(x - x_0)$ on the right-hand side) is usually expressed by quadrature. Under such conditions the classical BIE method becomes less efficient, but not our approach, since the latter does not require operation with the Green's function.

8.7. Ill-Posed Method of Auxiliary Sources in Diffraction Theory

One of the well-known methods used for solving diffraction problems that has become widely used in recent years is the auxiliary sources (AS) method, in which the unknown current density is disturbed on a certain auxiliary contour (in the two-dimensional case) lying inside the body at which diffraction occurs. This approach is considered to be correct if the auxiliary contour includes all the singularities of the diffracted field, in which case the basic integral equation of the method is uniquely solvable. However, unlike the basic equation of the classical BIE method (Section 2.2), the equation of the AS method is ill-posed in Tikhonov's sense. Thus, even though there is a unique solution, it remains an open question how to select an effective numerical method of solution. Here we investigate the equation of both the direct and the inverse problem of diffraction.

If the solution of the diffraction problem is sought in the form of the single-layer potential with an unknown density $\mu(\theta)$ distributed over an auxiliary contour, then to find the function μ we need to solve the well-known integral equation (see Chapter 2)

$$\int_0^{2\pi} \mu(\theta) H_0^{(1)}[kR(\varphi, \theta)] \, d\theta = f(\varphi), \qquad \varphi \in [0, 2\pi],$$

$$R(\varphi, \theta) = [\rho^2(\varphi) + r^2(\theta) - 2\rho(\varphi)r(\theta)\cos(\varphi - \theta)]^{1/2},$$
(8.105)

where $\rho(\varphi)$ describes the true contour, $r(\theta)$ describes the auxiliary contour, and $f(\varphi)$ is expressed in terms of the incident wave field. For simplicity, both contours are assumed to be star-like. The diagram of the scattered wave field in the far zone is expressed in the form

$$F(\varphi) = \int_0^{2\pi} \mu(\theta) e^{ikr(\theta)\cos(\varphi-\theta)} \, d\theta.$$
(8.106)

The direct diffraction problem requires to solve Eq. (8.105), then the calculation of the scattered diagram (8.106) reduces to a simple integration. It is proved in Apeltsin and Kyurkchan (1990) that (8.105) has a unique solution if and only if the contour $r(\theta)$ encloses all points at which the singularities of the wave field are concentrated, and the numerical solution can be constructed by any classical method. For instance, in the collocation method the integral in (8.105) is replaced by a finite sum; the equation is satisfied at a finite number of nodes φ_n, $n = 1, 2, \ldots, N$, after which the problem reduces to a linear algebraic system of equations, which can be solved by the Gauss method. Strictly speaking, such an approach is not correct, because (8.105) is a first-order equation with a compact operator (in any natural functional space) and, therefore, the problem of its solving is ill-posed. Only if

$r(\varphi) \equiv \rho(\varphi)$, a case corresponding to the BIE method, the kernel of Eq. (8.105) has a weak (logarithmic) singularity. We know (see Voronin and Tsetsokho, 1981) that the collocation method is stable for kernels of this kind. We will therefore call this a well-posed case. Naturally, for general contour, stability of the algorithm in this method will be provided only if the auxiliary contour is very close to the true contour.

The solution of the inverse problem can be found as follows. First, from the given function $F(\varphi)$, we find the potential $\mu(\theta)$ from (8.106). Then, substituting the result for $\mu(\theta)$ into (8.105), we reduce the problem to solving the nonlinear equation (8.105) for the function $\rho(\varphi)$. Unfortunately, there are no results on the solvability of (8.106) in the general case.

After the ill-posed equations (8.105) and (8.106) have been reduced to linear algebraic systems, the properties of stability of the latter depend on the question how rapidly the eigenvalues λ_n of the integral operator (8.105) and (8.106) tend to zero as $n \to \infty$ (cf. Hille and Tamarkin, 1931). The faster they decrease the smaller the modulus of the first eigenvalues of the matrices obtained by the discretization. A study of this spectral problem has been carried out by Hille and Tamarkin (1931), which shows that Eqs. (8.105) and (8.106) belong to different classes with respect to the function $\mu(\theta)$.

The kernel in (8.105) is only analytic in a finite domain of the complex variables φ, θ and, of course, the functions $\rho(\varphi)$ and $r(\theta)$ themselves are regarded as analytic in this case. It follows from Theorem 10.1 of the cited work that $\lambda_n < d^{-n/4}$, $n \to \infty$, $d = (1+b^2)^{1/2} + b$, where b is the dimension of the imaginary semi-axis of the ellipse inside which the kernel of the equation is analytic with respect to the variable φ. In the case under consideration here, analyticity of the kernel of (8.105) is broken at the points where $\rho^2(\varphi) + r^2(\theta) - 2r(\theta)\rho(\varphi)\cos(\varphi - \theta) = 0$. Since $|\cos(\varphi - \theta)| \leq \cosh y$, $\varphi = x + iy$, it is easy to show that the parameter b is determined by the quantity $\min \ln[\rho(\varphi)/r(\theta)]$. Thus $\lambda_n \to 0$ like the terms of a geometric progression, and removing the true contour far from the auxiliary contour makes the quantity $d^{-1/4}$ smaller.

The kernel in (8.106) is an entire function with respect to both its variables φ, θ. It follows from Theorem 11.1 of Hille and Tamarkin (1931) that $\lambda_n \to 0$ here are not slower than $n^{-1/4}$. Hence Eq. (8.106), considered with respect to the function $\mu(\theta)$, is much more ill-conditioned than (8.105).

The results we have obtained here for some simple cases can be made more precise. For example, if $r(\theta) \equiv r$, $\rho(\varphi) \equiv \rho$, $r \leq \rho$, then the eigenfunctions for (8.105) and (8.106) are of the form $\cos(n\theta)$ and $\sin(n\theta)$. Here, respectively,

$$\lambda_n \sim -\frac{2i}{n}\left(\frac{r}{\rho}\right)^n, \qquad \lambda_n \sim \left(\frac{kr}{2}\right)^n \frac{1}{n!}. \tag{8.107}$$

Formula (8.107) confirms the rule obtained for the general case. As one might expect, the closer r to ρ, the better the conditionality of (8.105). In the limiting case corresponding to the BIE method, $\lambda_n = O(1/n)$, leading to a stability of the calculation during the discretization (see also below Section 9.2). The conditionality of (8.106) is independent of the true contour.

From these results, it follows that the basic question arising in the numerical implementation of the AS method concerns the conditionality of the corresponding linear algebraic systems. This is a more relevant factor than the solvability of (8.105) and (8.106). One of criteria for choosing the auxiliary contour proposed in Apeltsin and Kyurkchan (1990) was the absence of strong oscillations of the density $\mu(\theta)$ distributed on that contour. Yet, as specific calculations show, the auxiliary sources for the simplest test example $\rho(\varphi) \equiv 3$, $r(\theta) \equiv 0.5$, $N = 18$ can be uniformly concentrated around a circle $r = 0.5$, with densities of the order of 10^4. The solution by the Gauss method of the (18×18) systems corresponding to the direct and inverse problems then gives results accurate to within 3–4 significant digits, which justifies the use of this approach. Another criterion of correctness of the auxiliary contour choice in Apeltsin and Kyurkchan (1990) was that the discrepancy between the left- and right-hand sides at points situated half-way between selected nodes should be small. It was found that this could only be the case for a contour which encloses all the singularities of the wave field. The specific calculations of the diffraction problem considered in the cited work on an ellipse with semi-axes 3 and 1.2 for small values of N (N = 18, 36, etc.) show that these discrepancies are of the same order in the following cases: (a) $r(\theta)$ corresponds to an ellipse with semi-axis 2.8 and 1 which includes the singularities; (b) the contour $r(\theta) \equiv 1$ does not include the singularities; (c) the contour $r(\theta) \equiv \rho(\theta)$ corresponds to the BIE method and is *a priori* admissible. In all three cases the solutions of the direct and inverse problems are with 3 or 4 significant digits. The scattered field diagram is obtained with the same accuracy.

It follows from the foregoing consideration if the AS method in its discrete modification is not associated with an integral equation (which is what happens when only a small number of nodes is chosen), any interior auxiliary contour can be used, and the conditionality of the resulting system becomes a main parameter which depends on the choice of the contour. It is always true that the closer the auxiliary contour to the true contour, the better the conditionality of the corresponding algebraic system. Since the resulting systems are always ill-conditioned, some regularization method must be used to solve them. The following approach is used here: (a) the method based on orthogonal Hausdorff transformations; (b) the QR-algorithm; (c) the method of steepest descent.

To verify the dependence of the conditionality number on the position of the auxiliary contour, we will find its values in a few cases. Suppose that the true contour is a circle with $k\rho(\varphi) \equiv 3$. Let the interior contour also be a circle, with various possible values of the parameter kr. The performed calculations show for the conditionality number $\beta = M/m$ of the corresponding systems the following results. When kr varies from 1 to 3, β varies from 10^4 to 10^1 in the direct problem and from 10^6 to 10^3 in the inverse problem.

Suppose now that the true contour is an ellipse with semi-axes $ka = 3$ and $kb = 1.2$. The singular points of the wave field are then situated at the foci $kc = \pm 2.75$ (cf. Apeltsin and Kyurkchan, 1990). The first of the three auxiliary curves encloses the singularities, but the second and third do not. The corresponding conditionality number β is such that for the direct problem it is always of the order 10^2, and for the inverse problem varies from 10^7 to 10^{10}.

The following conclusions can be drawn form these results.

1. The conditionality of the inverse problem is always several orders of magnitude worse than for the direct problem.

2. The conditionality of the direct problem depends on the position of the auxiliary contour, improving as it approaches the true contour. We have found that the conditionality depends much more on the position of the auxiliary contour than on whether it encloses the singularities or not.

3. The conditionality of the inverse problem depends much less on the position of the auxiliary contour, always being poor.

It should be noted that the conclusions given here might not be directly applicable to the case where there is a large number of auxiliary sources. The behavior of the computational singularities of the AS method as N increases is basically an open question and requires more thorough study.

As noted above, the solution of the inverse problem can be obtained by the successive solution of (8.106) and (8.105). After substituting the function $\mu(\theta)$ found from the linear Eq. (8.106) into (8.105), this equation becomes nonlinear with respect to the function $\rho(\varphi)$. This approach is genetically close to the method first proposed in Kirsch et al. (1988), in which the resulting nonlinear ill-posed problem is solved by the Tikhonov smoothing functional method. The approach we are using is based on the method of gradient descent for minimizing a functional, which is the square of the norm of the discrepancy. The gradient of this functional in this problem can be easily written out explicitly. As one might expect, the accuracy with which the contour $\rho(\varphi)$ is reconstructed depends closely on the successful choice of the auxiliary contour $r(\theta)$. Two examples on the applications of these results can be found in the author's work (Bratsun and Sumbatyan, 1993). The first example is related to an elliptic domain with semi-axes 2.2 : 1.2, which was restored with auxiliary contour $kr(\theta) \equiv 1.2$. The second *cat-face* contour is described by the parametric equations

$$kx(t) = \cos t + 0.65\,[\cos(2t) - 1], \qquad ky(t) = 1.5 \sin t, \qquad 0 \le t < 2\pi, \qquad (8.108)$$

with $kr(\theta) \equiv 0.8$. Note that the auxiliary contour in the first example does not include the singularities, which are at points ± 1.84. A similar situation also appears to apply in the second case.

Unfortunately, more complicated boundaries $\rho(\varphi)$ cannot be reconstructed by this method. Another drawback is that the method cannot be applied if we study the case where we know only the modulus $|F(\varphi)|$ rather than $F(\varphi)$ itself, then Eq. (8.106) becomes nonlinear with respect to the function $\mu(\theta)$.

Helpful remarks

The method described at the present section can be used as a basis to generically study ill-posed integral equations of the first kind $Au = f$. If, for simplicity, the integral operator A is symmetric and positive, then all its eigenvalues are positive: $\lambda_n > 0$, $\forall n$, and the respective eigenfunctions $\{\varphi\}$ form an ortho-normal system. Then if we seek the solution as a series in the eigenfunctions, $u = \sum u_n \varphi_n$, then $u_n = f_n/\lambda_n$, where $f_n = (f, \varphi_n)$, so the issue of convergence of the series depends on the issue of the asymptotics of f_n and λ_n as $n \to \infty$. In this sense, solvability of the equation is linked with the fact in which degree the kernel $K(x, \xi)$ and the right-hand side $f(x)$ are smooth. This idea was implicitly used in our study in Section 8.6, and now in the present section. We can conclude from the consideration carried out in the last two sections that solvability takes place when the right-hand side is smoother in a sense than the kernel. This idea will also be used in Section 9.2. Some thorough results on asymptotic distribution of eigenvalues λ_n, as noted above, can be found in Hille and Tamarkin (1931).

8.8. A Method of Global Random Search in Inverse Problems

Here we propose an alternative approach to solve ill-posed problems which is related to the operator equation of the first kind

$$Au = f, \qquad (8.109)$$

where A is a compact operator acting in some Hilbert space H. The problem of finding u, when the right-hand side f is known, is therefore an inverse problem. It is known (Tikhonov and Arsenin, 1977) that Eq. (8.109) is considered as Tikhonov's ill-posed problem in the sense that a large change in the solution u may correspond to a small change in the right-hand side. This leads to instability of the basic numerical methods when they are applied to Eq. (8.109) (see the previous sections).

To overcome these difficulties, special numerical methods have been developed. One of possible approaches is based on the ideas of regularizing Eq. (8.109) by means of an appropriate small perturbation of the operator A that was discussed in Section 8.2.

In this section a method which is natural from the physical point of view is proposed. If approximations u_n are selected randomly, it is possible to evaluate their closeness to the exact solution u^*, from the closeness of the left- and right-hand sides, that is, from the smallness of the discrepancy functional,

$$M(u_n) = \|Ax_n - f\|^2. \tag{8.110}$$

But it is known (see Sections 8.1 and 8.2) that the ill-posed equation (8.109) cannot be reduced to minimization of the functional $M(u)$, (8.110). Instead, it is necessary to consider the smoothing functional (see Section 8.2)

$$M_\alpha \alpha(u) = \|Au - f\|^2 + \alpha \Omega(u), \qquad 0 < \alpha \ll 1. \tag{8.111}$$

To minimize functional (8.111) we will use the method of global random search (see Zhiglyavskii, 1985). Unlike direct random search, it has important properties which enable the process of finding a good approximation u_n to be accelerated. This is valid because

1) random sampling of values u_n in the neighborhood of the points X, for which the values of $M_\alpha(X)$ are smaller, happens more frequently than that in the neighborhood of the points Y, where the values of $M_\alpha(Y)$ are larger, and

2) domains, in which random values u_n are chosen, are gradually contracted to the small neighborhoods of the points with small values of $M_\alpha(u)$.

This algorithm has been tested by examples on minimization of finite-dimensional functions of a small number of variables. Here we use it to solve functional equations in the Hilbert space.

We will first consider the integral equation of the first kind

$$Au = \int_0^{2\pi} \cos[\rho \cos(\varphi - \theta)]\, u(\theta)\, d\theta = f(\varphi), \qquad 0 \le \varphi < 2\pi, \qquad \rho = \text{const} > 0, \tag{8.112}$$

as a model example. Its kernel is infinitely differentiable, which implies that this equation is extremely ill-posed.

We will find its solution in the form of a Fourier series with a finite number of terms

$$u(\theta) = \sum_{m=0}^M a_m \cos m\theta, \qquad 0 \le \theta < 2\pi. \tag{8.113}$$

For simplicity, we consider here the case when the right-hand side $f(\varphi)$ (as well as the kernel) is an even function of φ. In this approach the minimization of the functional (8.111), based on the method of global random search described above, implies random sampling of sets of real numbers $(a_0, a_1, \ldots, a_m)_n$. It is proposed to choose a regularizing functional in the form (see above Section 8.2)

$$\Omega(u) = \|u\|_{W_2^2(0, 2\pi)} \tag{8.114}$$

where W_2^2 is Sobolev's space, and the basic space is $H = L_2(0, 2\pi)$.

The results of the application of the proposed algorithm for two right-hand sides, which are $f_1(\varphi) = 2\pi J_0(\rho)$ and $f_2(\varphi) = -2\pi J_2 \cos 2\varphi$, can be found in the author's paper (Sumbatyan, 1992), where we took the values $\rho = 1$, $M = 4$, $\alpha = 10^{-1}$ to 10^{-4}. The number of random samples of the solution defining the number of calculations of the direct operator A was equal to $N = 300$.

The method given in this section is especially effective for problems characterized by a large number of local minima of functional (8.111), as well as by large values of the gradients of the functional. In such problems it is practically impossible to find the global minimum by regular methods. As a detailed analysis indicates, the problems of recognizing an object from the wave field scattered by it are distinguished by these properties.

When investigating the problem on reconstruction of the boundary contour, we will restrict ourselves, for simplicity, to the two-dimensional case and to the simplest model of an acoustic medium described by a single Helmholtz equation. Then it is possible to reduce the problem to a system of two nonlinear integral equations (see the previous sections)

$$\left| \int_l e^{ik(q \cdot x)} g(x) \, ds_x \right| = f(\varphi), \qquad 0 \le \varphi < 2\pi, \tag{8.115}$$

$$\int_l H_0^{(1)}(k|x - y|) g(x) \, ds_x = e^{ik(q \cdot y)}, \qquad y \in l, \qquad q = -\{\cos \varphi, \sin \varphi\}. \tag{8.116}$$

Here k is the wave number and $H_0^{(1)}$ is the Hankel function. In Eqs. (8.115) and (8.116) the unknowns are the function $g(x)$, connected with the normal derivative of the velocity, and the function $x \in l$, which defines the position of the boundary contour l (the latter may be specified, for example, in some parametric form).

The results on reconstruction of two objects, an ellipse with a ratio of its semi-axes equal to 3 : 1, and a semi-circle with a diameter of 10, from the amplitude of circular back scattering, can be found in Sumbatyan (1992). As in the model example, the parametric representation of the contour l was specified in the form of a finite truncation of a Fourier series in the polar coordinate system with $M = 4$.

Helpful remarks

1°. System (8.115), (8.116) holds for the case when the amplitude of the back scattering from the object is known for the whole range of variation of the scanning angle $\varphi \in (0, 2\pi)$. This method of scanning, when the directions of propagation of the incident wave and the reflected wave coincide, corresponds to the echo method, widely used in ultrasonic testing, in radio-location, in underwater acoustics, and many other applications. In this method the same ultrasonic sensor device serves both as the emitter and the receiver. The amplitude of the reflected signal $f(\varphi)$ is therefore known, but the phase, as a rule, is unknown. It is assumed that similar measurements may be carried out in principle for any angle of incidence $\varphi \in (0, 2\pi)$.

2°. The direct problem of diffraction consists in the calculation of the function $f(\varphi)$ along the known contour l. To do this it is necessary, first, to solve the boundary integral equation (8.116) for the function $g(x)$, $x \in l$, and then to calculate the quadrature (8.115). It is obvious that the corresponding operator Au in (8.109) is nonlinear. Calculation of the direct action of a nonlinear operator is therefore reduced to solving one linear equation and one quadrature. This situation is typical for nonlinear inverse problems. Unlike the above case the inverse operator A^{-1} (if it exists) is extremely nonlinear. When using regular methods to invert Eq. (8.109) it would be necessary to calculate the Frechet derivative for the operator A, which is quite difficult to do. The method described here requires only the calculation of the direct operator, which, as noted above, involves linear operations only.

8.9. Ill-Posed Problem on Reconstruction of Convex Hull of the Obstacle in Acoustic Medium

Let us come back to the problem of practical realization of algorithm on reconstruction of convex parts of a smooth obstacle in acoustic medium from the far-field back-scattered diagram, when the known information is contained in the *time-of-flight* data. In this case the problem of reconstruction of the convex hull S_0 can be reduced to calculations by explicit formulas (7.81), on the basis of the support function usually known from experimental measurements. Note that these formulas require operation with numerical differentiation, a process that is well known to be an ill-posed problem (see Tikhonov and Arsenin, 1977).

In practical scanning there is typically a discrete array of directions where they carry out measurements with some error. We thus can conclude that direct usage of formulas is impossible in practice for two reasons. First of all, there is an error in definition of time of flight $t(\mathbf{q})$. Secondly, a finite step of discrete measurements is usually not sufficiently small to provide precise and correct computations of the derivatives in those formulas. Even for convex parts of the boundary surface the obtained function $p(\mathbf{n})$ is only piecewise-smooth, and so its first derivatives p'_α, p'_β used in reconstruction of the convex hull S_0 may differ from actual values not only by their values but even by their sign.

Taking into account the above discussed specifics of the studied problem, let us construct an algorithm, which permits approximation of the function $p(\alpha, \beta)$ by a certain smooth function, for which its values and values of its derivatives are close to really obtained from measurements. In the considered situation when in practical scanning we obtain, instead of a smooth function, its approximate values in some nodes, the bases of our algorithm is an approximation of the sought function by cubic splines (see, for instance, De Boor, 1978).

Let us restrict the study by the 2D case, for simplicity, and assume that we know approximate values p_i, $i = 1, 2, \ldots, N$ of a certain smooth function (in our case, the function $p(\mathbf{n})$) at the points $\mathbf{n}_1, \ldots, \mathbf{n}_N$, and the estimate Δp_i of the mean-square deviation of p_i from the real values $p(\mathbf{n}_i)$. In practice, Δp_i does not exceed 10%. Let us construct the function $f = f_\delta$ that for a given parameter $\delta \in [0, 1]$ determined by the mean-square deviation minimizes the functional (see De Boor, 1978)

$$\pi(f) = \delta \sum_{i=1}^{N} \left[\frac{p_i - f_i}{\Delta p_i}\right]^2 + (1 - \delta) \int_{n_1}^{n_N} \left[f''(\theta)\right]^2 d\theta, \qquad (8.117)$$

over all functions f possessing second-order derivatives.

Minimization of the functional $\pi(f)$ is a compromise between two requirements: to provide a good approximation to the given values p_i, and to provide a sufficiently smooth approximating function.

It is known (see De Boor, 1978) that f_δ is a second-order spline with simple points of interpolation at the points $n_2, \ldots, n_{N-1}$ and satisfying the condition $f''_\delta(n_1) = f''_\delta(n_N) = 0$.

Let us assume, for simplicity, the step Δn between any pair of neighbor normals to be constant, and let us introduce the notations: $a_i = f_\delta$, $c_i = f''_\delta$, $i = 1, 2, \ldots, N$. In practice, for the 2D reconstruction problems discussed below, the dependence of the function $p(\mathbf{n})$ upon the direction of outward normal $\mathbf{n}$ is equivalent to its dependence on the angle of incidence (the same as the angle of observation). Therefore, in practice, we can put $\Delta n = \Delta \theta$, i.e., the angular step when scanning around the obstacle.

Then the condition of continuity of the first-order derivatives f'_δ for a closed contour takes the following form:

$$(c_{i-1} + 4c_i + c_{i+1})(\Delta n)^2 = 6(a_{i-1} - 2a_i + a_{i+1}), \quad i = 2, 3, \ldots, N - 1,$$
$$(c_{N-1} + 4c_N + c_1)(\Delta n)^2 = 6(a_{N-1} - 2a_N + a_1), \qquad (8.118)$$
$$(c_N + 4c_1 + c_2)(\Delta n)^2 = 6(a_N - 2a_1 + a_2).$$

For all that, the functional (8.117) takes the following form:

$$\pi(f) = \delta \sum_{i=1}^{N} \left[\frac{p_i - a_i}{\Delta p_i}\right]^2 + \frac{1 - \delta}{3} \Delta n \sum_{i=1}^{N} \left(c_i^2 + c_i c_{i+1} + c_{i+1}^2\right). \qquad (8.119)$$

Further, by applying in relations (8.118) approximation of the second-order derivatives $c_i = f''_\delta$ by central finite differences, with the help of values of the unknown function at three nodes, we can reduce the condition of minimum of the functional (8.117) to the following form written as a vector equality:

$$-\frac{2\delta}{(\Delta p)^2}(\mathbf{p} - \mathbf{a}) + \frac{1-\delta}{12\Delta n}\mathbf{Xc} = \mathbf{0}, \tag{8.120}$$

$$\mathbf{p} = \{p_1, \ldots, p_N\}, \quad \mathbf{a} = \{a_1, \ldots, a_N\}, \quad \mathbf{c} = \{c_1, \ldots, c_N\},$$

where $\mathbf{X}$ is a certain sparse matrix. Nontrivial elements of this matrix form a diagonal band of five lines, and additionally two triples in the upper right and lower left corners of the matrix. Recall that Δp is a maximum error of measurements, which is in our case around 0.1 of the representative size of the obstacle.

In order to obtain a self-consistent algebraic system, we should add to system (8.120) also relations (8.118) written in the vector form

$$(\Delta n)^2 \mathbf{R}\mathbf{c} = 6\mathbf{Q}\mathbf{a}, \tag{8.121}$$

where $\mathbf{R}$ and $\mathbf{Q}$ are some sparse matrices. By their structure, these are three-diagonal banded matrices, with two additional elements, one in the upper right and another in the lower left corners. Let us substitute (8.121) into (8.120), then we arrive at a system of linear algebraic equations, to define the vector of the second-order derivatives $\mathbf{c}$:

$$\left[\frac{2\delta(\Delta n)^2}{(\Delta p)^2}\mathbf{R} + \frac{1-\delta}{2\Delta n}\mathbf{Q}\mathbf{X}\right]\mathbf{c} = \frac{12\delta}{(\Delta p)^2}\mathbf{Q}\mathbf{p}. \tag{8.122}$$

In investigation of this system it is convenient to introduce the parameter M

$$M = \sum_{i=1}^{N}\left[\frac{p_i - f_i}{\Delta p_i}\right]^2 = \frac{(1-\delta)^2(\Delta p)^2}{(24\,\delta\Delta n)^2}\sum_{i=1}^{N}(\mathbf{X}\mathbf{c})_i^2. \tag{8.123}$$

On the basis of the introduced parameter M we can arrange the estimate of the minimum value of the parameter δ, $0 \leq \delta \leq 1$ so as to provide the value of M less than a certain *a priori* given value L.

As soon as the vector $\mathbf{c}$ is defined from system (8.122), we can determine vector $\mathbf{a}$ of values of the unknown function f_δ at the chosen nodes, from the following relation:

$$\mathbf{a} = \mathbf{p} - \frac{(1-\delta)(\Delta p)^2}{24\,\delta\Delta n}\mathbf{X}\mathbf{c}. \tag{8.124}$$

This method allows us also to determine approximate values of the first three derivatives:

$$f_\delta(n_i) = a_i, \quad f'_\delta(n_i) = \frac{a_{i+1} - a_i}{\Delta n} - \frac{c_i}{2\Delta n_i} - \frac{f'''_\delta(n_i)}{6(\Delta n)^2},$$

$$f''_\delta(n_i) = c_i, \quad f'''_\delta(n_i) = \frac{c_{i+1} - c_i}{\Delta n}. \tag{8.125}$$

The quantity L is proposed in De Boor (1978) to be taken somewhere between $\sqrt{2N}$ and N. Such a choice permits arrangement of the function f_δ to provide a sufficiently accurate approximation $p(\mathbf{n})$ together with its first and second derivatives at the points n_i, and to be simultaneously smooth enough.

The function $p(\mathbf{n})$ obtained by the thus developed method finally gives, after the substitution to formulas (7.81), the sought surface of the convex hull for the considered obstacle.

Below we demonstrate the proposed reconstruction algorithm on some real cylindrical flaws made in aluminum plates. The representative size of all defects was around 10–15 mm. The measured back-scattered diagrams were obtained with the angle of incidence changing in 5° increments. Scattering patterns with the respective results of reconstruction are reflected for some flaws in Figs. 8.9–8.12.

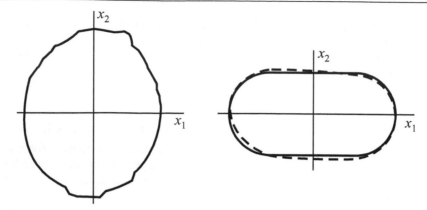

Figure 8.9. Support function for the first obstacle and the corresponding reconstructed convex hull

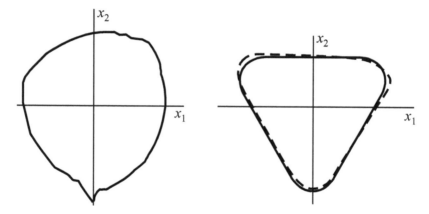

Figure 8.10. Support function for the second obstacle and the corresponding reconstructed convex hull

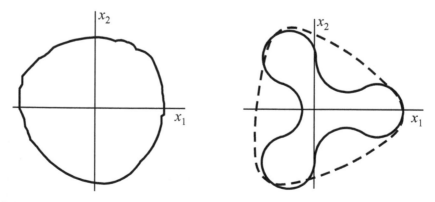

Figure 8.11. Support function for the third obstacle and the corresponding reconstructed convex hull

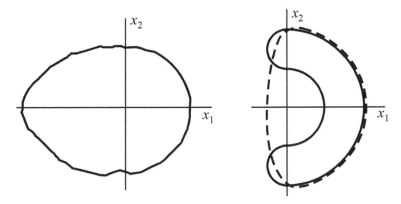

Figure 8.12. Support function for the fourth obstacle and the corresponding reconstructed convex hull

Helpful remarks

It should be noted that in the considered 2D case Cartesian coordinates of the convex hull S_0 are expressed by the following formulas:

$$
\begin{aligned}
x_1 &= p(\theta)\cos\theta - p'_\theta(\theta)\sin\theta, \\
x_2 &= p(\theta)\sin\theta - p'_\theta(\theta)\cos\theta
\end{aligned}
\tag{8.126}
$$

(compare with Eq. (7.81) valid for 3D case), where θ is the scanning angle, equal at the same time to the angle between the normal to the boundary of the obstacle at the reflection point (x_1, x_2) and positive direction of the axis x_1. In our calculations we used the following values: $N = 360°/5° = 72$ and $15 \le L \le 25$. For the results of reconstructions shown in Figs. 8.9–8.12, solid lines designate real boundary contour, and dashed lines the reconstructed ones. Note that we operate both with convex and nonconvex obstacles, and the shown results confirm that the proposed method to construct the convex hull of a smooth obstacle can be used for obstacles of both types.

Chapter 9

Numerical Methods for Irregular Operator Equations

9.1. Steepest Descent Method: Stability and Improvement of the Convergence

As shown in Chapter 2, a diffraction problem in acoustic media may be reduced to integral equations obtained from the Kirchhoff–Helmholtz integral formula. The Dirichlet problem is described by an equation of the first kind and the Neumann problem by an equation of the second kind, if we speak about the direct BIE method:

$$\int_S \psi(y)\, G(y_0, y)\, dS_y = F_1(y_0), \qquad y_0 \in S, \tag{9.1}$$

$$p(y_0) - 2 \int_S p(y) \frac{\partial G(y_0, y)}{\partial n_y}\, dS_y = F_2(y_0), \qquad y_0 \in S. \tag{9.2}$$

Here $\psi(y) = \partial p / \partial n_y$, p is the diffracted pressure on the boundary surface S; F_1 and F_2 are functions expressed in terms of boundary functions. To be more specific, we assume that the surface is closed and sufficiently smooth; the normal n_y is directed outwards the surface S. Equations (9.1) and (9.2) describe the diffraction problem both in the 2D and 3D cases. In such cases we have, respectively

$$G(y_0, y) = \frac{i}{4} H_0^{(1)}(kr), \qquad G(y_0, y) = \frac{1}{4\pi} \frac{e^{ikr}}{r}, \qquad r = |y - y_0|. \tag{9.3}$$

Note that equations (9.1), (9.2) may be written out for full wave field too. This changes only the form of the right-hand sides F_1 and F_2. As we could see from the previous study, equations (9.1), (9.2) are investigated in detail, and a number of analytical and numerical methods have been developed to solve these equations. The exact analytical solutions are known only for canonical regions (circle, sphere, ellipse, segment, etc.). In the high-frequency range ($ka \gg 1$, a is a characteristic size of the S-surface) the short-wave asymptotic methods are efficient for regions of complex shape. As regards direct numerical methods, we note once again that these methods become inefficient for extremely high frequencies. The main disadvantage of numerical methods is their isolation from the physical essence of phenomena. Besides, it is known that numerical methods lose their stability for large values of ka (see also the Preface). Here we propose a method that uses a combination of analytical and numerical techniques. The analytical solution obtained from the Kirchhoff theory is used as an initial-step approximation for a certain iterative method. It is effective for $ka \gg 1$ only, and for lower values of ka this is improved by the iterative method, and we show that each step of the proposed technique improves the chosen analytical solution.

Let us write the considered equations (9.1), (9.2) in the operator form

$$Bx = y, \tag{9.4}$$

and consider the linear operator B acting in some complex Hilbert space H, which may be chosen as $L_2(S)$. We assume the operator B to be self-adjoint and positive. We always can achieve such properties of B representing it in the form A^*A, where A is the operator of the initial integral equation. In this case $y = A^*f$, where f is a right-hand side of the initial integral equation.

Let the spectrum of the operator B lie on the interval $[m, M]$, $0 \le m < M$. For the first-kind equation, $m = 0$ always, which follows from compactness of the integral operator. We proved in Section 8.3 that if equation (9.4) has a solution, x^*, then the iterative steepest descent method (SDM)

$$x_{n+1} = x_n - \frac{\|Lx_n\|^2}{(Lx_n, BLx_n)} L_n x, \qquad Lx_n = Bx_n - y, \tag{9.5}$$

converges monotonically to this solution in the norm of the chosen space H. This means that $\|x_{n+1} - x^*\| \le \|x_n - x^*\|$. For all that, if $m > 0$, then

$$\|x_n - x^*\| \le \frac{\|Lx_0\|}{m} \left(\frac{M - m}{M + m} \right)^n; \tag{9.6}$$

if $m = 0$, then

$$F(x_n) = \left(B(x_n - x^*), x_n - x^* \right) = O\left(\frac{1}{n} \right). \tag{9.7}$$

It is easy to demonstrate that this result is correct if we use some allied methods, instead of SDM, which can be written as (see Section 8.3)

$$x_{n+1} = x_n - \frac{\|Lx_n\|^2}{\|A^*Lx_n\|^2} A^*Lx_n, \qquad (A^*A = B), \tag{9.8}$$

or

$$x_{n+1} = x_n - \frac{(BLx_n, Lx_n)}{\|BLx_n\|^2} Lx_n. \tag{9.9}$$

Recall that for convergence of SDM the existence of the solution x^* is only sufficient. Below we will use this fact in the case when the solution is not unique.

The numerical realization of the considered iterative techniques implies that the integral operator is changed by a finite-dimensional one. In particular, when solving equations (9.1), (9.2) by a collocation method, we arrive at a linear algebraic system $N \times N$. The convergence in (9.5) takes place; however in the case of an equation of the first kind we have $m > 0$ always, except the special cases mentioned below, hence the estimate (9.6) for the rate of convergence is always valid. At the same time, in the case when the integral equation (9.1) is approximated by a finite-dimensional operator equation with the large number of nodes N, the finite dimensional operator is a good approximation to the initial compact one, therefore $m/M \ll 1$. Hence, the matrix of the system is ill-conditioned. It is easily seen that the rate of the convergence here is very slow. In the case of equation (9.2) concrete calculations at $ak \gg 1$ show that the matrix of the corresponding algebraic system is poorly posed too. From this point of view, equations of the second kind differ not very much by their qualitative properties from equations of the first kind, in the case of the high-frequency process.

In connection with this, the problem of the SDM acceleration seems to be very important. Different approaches to acceleration of iterative methods are known in the literature. Here we use an acceleration method, which is based on some asymptotic properties of SDM for large n. It is shown in Samarsky and Nikolaev (1978) that when $n \to \infty$ the iterative process (9.5) has the following properties:

$$\frac{F(x_n)}{F(x_{n-1})} = \frac{\|Ax_n - f\|^2}{\|Ax_{n-1} - f\|^2} = c^2, \tag{9.10}$$

$$x_n - x^* = c^2(x_{n-2} - x^*) \qquad \left(c^2 = \frac{M - m}{M + m}\right). \tag{9.11}$$

It follows from (9.11) that, when SDM approaches its asymptotics, the next iteration may be chosen in the form

$$\tilde{x} = \frac{x_n - c^2 x_{n-2}}{1 - c^2}, \tag{9.12}$$

where c^2 may be calculated from (9.10). The usage of such an approach in concrete numerical examples demonstrates fast acceleration of the iterative process for any matrix equations, including ill-conditioned ones ($m/M \ll 1, c^2 \sim 1$). In some more details, we can observe the following properties of the iterative process. For sufficiently large n the process (9.5) approaches the worst of its estimate in inequality (9.6), and the asymptotic relation (9.10) becomes approximate as an equality. Under such conditions the difference $F(x_n)$ at each step decreases c^2 times; consequently, for ill-conditioned problem with $c^2 \sim 1$ it decreases very slowly. For all that, implementation of one step (9.12) eliminates this *cycling* and further the process (9.5) converges more rapidly. This property is well observed from the table related to the first of the examples considered below.

Recall that the first-kind equation (9.1) is ill-posed in the sense of Tikhonov, hence the iterative process (9.5) defines some family of regularizing operators R_n for equation (9.1).

Indeed, the iterative process at every step defines some nonlinear operator $x_n = R_n y$ and in order to calculate R_n it requires application of a finite number of continuous operators for any finite value of n. This property of R_n and convergence of the SDM process (9.5) proves the fact that R_n is a regularization operator.

It is known (see Chapter 2) that for the closed surface S there exist the values of the parameter $k = k_j$ ($j = 1, 2, \dots$) at which the operators corresponding to equations (9.1), (9.2) are not invertible. Such a situation takes place at frequencies that correspond to resonances of the corresponding interior problem. Some authors used various methods to overcome this difficulty. Here we show that our approach automatically avoids this obstacle. It is proved in Colton and Kress (1983) that for equation (9.1) at the critical frequencies k_j the right-hand side is always orthogonal to solutions of the homogeneous equation. It is also proved that under such conditions equation (9.1) is solvable (not uniquely) for any boundary function in the Dirichlet problem, and the wave field computed on the basis of this solution is defined uniquely. Since a solution to equation (9.1) exists, the described method guarantees the iterative process to convergence to a certain solution (in case the solution is not unique). This means that the proposed method provides a stable solution to the Dirichlet problem at critical frequencies too.

It is also well known that the indirect BIE method permits reducing of the Dirichlet problem to an integral equation of the second kind (see Chapter 2). But in this section we do not use such an opportunity because at the critical frequencies in this case the right-hand side is not orthogonal to solutions of the adjoint homogeneous equation.

It should also be noted that if the size of the complex-valued matrix in discretization of Eqs. (9.1), (9.2) is equal to N, then solution of the matrix equation by any direct method requires around N^3 arithmetic operations. At the same time one step of the process (9.5) requires near N^2 operations. With large ka it is necessary to choose the number N of the nodes rather large for adequate description of the field on the boundary surface. In such situation the SDM is more stable from the computational point of view.

Thus, in our approach we essentially use the following two new points:

1) acceleration of SDM; this allows us to achieve fast practical convergence;

2) the fact that SDM overcomes the classical difficulty connected with presence of critical frequencies in the Dirichlet problem.

The following examples were considered to test efficiency of the proposed method.

First of all, we tested the proposed approach by a certain 2D problem about diffraction of the plane wave by an acoustically hard round disk. We took $ka = 5$, the number of points $N = 72$. For such moderate wavelengths the solution on the Kirchhoff theory as the first step of iterations is ineffective. The proposed method allows us to improve it. The SDM rates of convergence without acceleration and with acceleration by Eq. (9.12) after each 10 iterations are compared in the table below.

n	$F(x_n)$	c^2	$F^*(x_n)$	n	$F(x_n)$	c^2	$F^*(x_n)$
1	2.61×10^0	0.293	2.61×10^0	16	5.27×10^{-3}	0.9271	1.92×10^{-4}
2	7.67×10^{-1}	0.332	7.67×10^{-1}	17	4.89×10^{-3}	0.9271	1.42×10^{-4}
3	2.54×10^{-1}	0.377	2.54×10^{-1}	18	4.53×10^{-3}	0.9271	1.19×10^{-4}
4	9.63×10^{-2}	0.430	9.63×10^{-2}	19	4.20×10^{-3}	0.9272	1.07×10^{-4}
5	4.14×10^{-2}	0.523	4.14×10^{-2}	20	3.89×10^{-3}	0.9272	9.76×10^{-5}
6	2.17×10^{-2}	0.649	2.17×10^{-2}	21	3.61×10^{-3}	0.9272	8.06×10^{-5}
7	1.41×10^{-2}	0.776	1.41×10^{-2}	22	3.35×10^{-3}	0.9273	4.57×10^{-6}
8	1.09×10^{-2}	0.856	1.09×10^{-2}	23	3.11×10^{-3}	0.9273	3.35×10^{-7}
9	9.36×10^{-3}	0.901	9.36×10^{-3}	24	2.88×10^{-3}	0.9273	5.72×10^{-8}
10	8.44×10^{-3}	0.917	8.44×10^{-3}	25	2.67×10^{-3}	0.9273	3.30×10^{-8}
11	2.48×10^{-3}	0.924	1.93×10^{-2}	26	2.48×10^{-3}	0.9274	2.63×10^{-8}
12	7.15×10^{-3}	0.925	6.10×10^{-3}	27	2.29×10^{-3}	0.9274	2.31×10^{-8}
13	6.62×10^{-3}	0.926	2.03×10^{-3}	28	2.13×10^{-3}	0.9274	2.09×10^{-8}
14	6.14×10^{-3}	0.926	7.33×10^{-4}	29	1.97×10^{-3}	0.9274	1.91×10^{-8}
15	5.69×10^{-3}	0.927	3.31×10^{-4}	30	1.83×10^{-3}	0.9274	1.72×10^{-8}

Here c^2 is calculated as in (9.10), n is the number of the iteration, $F(x_n)$ is the value of the discrepancy functional without acceleration, $F(x_n)$ are analogous values of the functional, by applying the proposed acceleration method. We can observe stabilization of the parameter c^2 in the table, which is in agreement with the described asymptotic properties of SDM. We can see that the value of c^2 is around 1 and this is connected with the poor definiteness of the matrix, which results in slow convergence of the process (without acceleration). At the same time, the acceleration makes the process more rapidly convergent: the value of the discrepancy decreases by 8 orders when compared with the first step of the iteration.

Further, we continued to test efficiency of the proposed approach, by consideration of diffraction of the 2D plane acoustic wave by a hard ellipse. Here $ka = 5$, $kb = 0.5$ ($a/b = 10$), $N = 180$. It is known that efficiency of the BIE method decreases considerably for elongated bodies. It is caused by the fact that such bodies require very large number of nodes N. Besides, the numerical investigation shows that the matrix of the corresponding algebraic system becomes extremely ill-conditioned for elongated bodies. The performed computations prove that the proposed method is efficient in this case too.

Then we considered the problem about diffraction of the plane wave by a soft cylinder with $ka = 3.8317$; this corresponds to interior resonance. We took here $N = 72$. The implementation of the SDM with the above described acceleration allows us to obtain a solution efficiently in this case also. The results are presented in more detail, with some diagrams and tables, in the author's work (Druzhinina and Sumbatyan, 1990).

Helpful remarks

We also tested some simple geometries for extremely high frequencies. The problem on comparison between approximate Kirchhoff's solution and results of the direct numerical treatment requires a special investigation. The first problem we studied is related to a circular acoustically hard cylinder at $ka = 20$, $N = 180$. Here we can include also comparison with the solution predicted by the geometrical diffraction theory, which is given as

$$F(\varphi) = \frac{\sqrt{\pi ka}}{2} \sqrt{|\sin \varphi/2|}. \tag{9.13}$$

The results of computation (see Druzhinina and Sumbatyan, 1990) show a perfect accuracy of the asymptotic solutions in the light, and poor accuracy in a shadow zone.

Then we considered an example with higher frequency: $ka = 50$ to study further increase of the frequency parameter. To this end, to adequately describe the problem we needed to take $N = 400$. The 400×400 (complex-valued) algebraic system was solved by a direct Gauss method, which did not give us any reasonable result. The realization of SDM with 20 iterations and with acceleration after each 5 iterations decreases the discrepancy by 4 orders. The comparison with the asymptotic results demonstrates again that asymptotic theories in the shadow zone are of poor accuracy.

9.2. Galerkin Methods for Integral Equations of the First Kind with Weakly Singular Kernels

As shown by the example of the problem considered in Section 8.7, the question of the stable solution of integral equations of the first kind depends upon the rate of asymptotic decrease of the eigenvalues ($\lambda_n \to 0$, $n \to \infty$) of the considered integral operator, as well as upon the asymptotic behavior of the coefficients in expansion of the right-hand side by eigenfunctions. If the kernel is positive and symmetric, then the spectrum $\{\lambda_n\}$ is positive and the eigenfunctions form a complete set of orthogonal functions. As follows from results of Hille and Tamarkin (1931), less regular kernels generate the spectrum $\{\lambda_n\}$, where $\lambda_n \to 0$ more slowly. Therefore, there is a good chance to construct a stable numerical method for the first-kind integral equations, whose kernel possesses a weak singularity.

Let us start from the characteristic equation with the periodic logarithmic kernel

$$-\int_{-\pi}^{\pi} u(\xi) \ln \left| \sin \frac{x-\xi}{2} \right| d\xi = f(x), \quad |x| \le \pi,$$

$$\sim K_0 u = f, \quad K_0(x, \xi) = -\ln \left| \sin \frac{x-\xi}{2} \right|. \tag{9.14}$$

If, for simplicity, $f(x)$ is even, $f(-x) = f(x)$, then it can easily be proved that the solution of equation (9.14) is even too. The spectrum of operator (9.14) can be directly defined on the basis of the following tabulated series (see Gradshteyn and Ryzhik, 1994):

$$\sum_{n=1}^{\infty} \frac{\cos n(x-\xi)}{n} = \ln \left| 2 \sin \frac{x-\xi}{2} \right|, \quad |x| \le \pi. \tag{9.15}$$

This shows that the set of eigenfunctions of the operator K_0 is $\{\varphi_n(x)\} = \{\cos nx\}$, $n = 0, 1, \ldots$, because for $n = 0$ we have (see Gradshteyn and Ryzhik, 1994)

$$
\begin{aligned}
K_0\varphi_0 &= -\int_{-\pi}^{\pi} \ln\left|\sin\frac{x-\xi}{2}\right| d\xi = -\int_{-\pi}^{\pi} \ln\left|\sin\frac{\xi}{2}\right| d\xi = -2\int_0^{\pi} \ln\left(\sin\frac{\xi}{2}\right) d\xi \\
&= -4\int_0^{\pi/2} \ln(\sin\xi)\, d\xi = (-4)\left(-\frac{\pi}{2}\right)\ln 2 = 2\pi\ln 2,
\end{aligned}
\tag{9.16}
$$

where we have used the following property of periodic functions: the value of the integral of any periodic function taken over the interval of the length equal to its period remains without change if we arbitrarily shift the interval of integration keeping it of the same length.

For other $n = 1, 2, \ldots$ it directly follows from Eq. (9.15) that

$$
\begin{aligned}
K_0\varphi_n &= -\int_{-\pi}^{\pi} \cos(n\xi)\left|\sin\frac{x-\xi}{2}\right| d\xi = -\int_{-\pi}^{\pi} \cos n\xi \ln\left|\sin\frac{x-\xi}{2}\right| d\xi \\
&= -\sum_{m=1}^{\infty} \frac{1}{m}\int_{-\pi}^{\pi} \cos(n\xi)\cos[m(x-\xi)]\, d\xi \\
&= -\sum_{m=1}^{\infty} \frac{\cos(nx)}{m}\int_{-\pi}^{\pi} \cos(n\xi)\cos(m\xi)\, d\xi \\
&= \frac{\pi}{n}\cos(nx) = \frac{\pi}{n}\varphi_n(x), \qquad n = 1, 2, \ldots;
\end{aligned}
\tag{9.17}
$$

hence $\{\varphi_n\}$ are indeed eigenfunctions of the operator K_0, with the eigenvalues $\lambda_0 = 2\pi\ln 2$, $\lambda_n = \pi/n$, $n = 1, 2, \ldots$.

Now solvability of equation (9.14) depends only on analytic properties of the right-hand side $f(x)$. If $f(x) \in C_2(-\pi, \pi)$, then

$$
f_n = \int_{-\pi}^{\pi} f(x)\varphi_n(x)\, dx = \int_{-\pi}^{\pi} f(x)\cos\pi nx\, dx = O\left(\frac{1}{n^3}\right),
\tag{9.18}
$$

as follows from the asymptotic estimates of Section 1.4. Then the exact solution to integral equation (9.14) can be explicitly expressed as a series in eigenfunctions

$$
u(x) = \sum_{n=0}^{\infty} u_n\varphi_n(x) = \sum_{n=0}^{\infty} u_n\cos(nx), \qquad u_n = \frac{f_n}{\lambda_n} = O\left(\frac{1}{n^2}\right),
\tag{9.19}
$$

which is uniformly convergent over the integral $x \in [-\pi, \pi]$.

In the case of a full Fredholm integral equation of the first kind, whose periodic kernel contains a weak logarithmic singularity

$$
\int_{-\pi}^{\pi} [K_0(x-\xi) + K_1(x,\xi)]\, u(\xi)\, d\xi = f(x), \quad |x| \leq \pi, \quad K_0(x-\xi) = -\ln\left|\sin\frac{x-\xi}{2}\right|,
\tag{9.20}
$$

where $f(x)$ and $K_1(x,\xi)$ are even with respect to their arguments and $K_1(x,\xi)$ is regular in some sense, we may seek a solution of equation (9.20) again in the form of expansion by eigenfunctions of the kernel $K_0(x-\xi)$:

$$
u(x) = \sum_{m=0}^{\infty} u_m\varphi_m(x), \qquad \varphi_m(x) = \cos(mx).
\tag{9.21}
$$

Then the standard Galerkin method (see, for example, Fletcher, 1984) imply the substitution of the representation (9.21) into Eq. (9.20):

$$\sum_{m=0}^{\infty} u_m \int_{-\pi}^{\pi} [K_0(x-\xi) + K_1(x,\xi)]\varphi_m(\xi)\, d\xi = f(x), \qquad |x| \le \pi,$$

$$\sum_{m=0}^{\infty} u_m \left[\lambda_m \varphi_m(x) + \int_{-\pi}^{\pi} K_1(x,\xi)\varphi_m(\xi)\, d\xi \right] = f(x), \qquad |x| \le \pi, \tag{9.22}$$

and the application of the scalar product of Eq. (9.22) with the functions from the same basis. As a result, we arrive at an infinite system of linear algebraic equations

$$u_n + \sum_{m=0}^{\infty} a_{nm} u_m = f_n, \qquad n = 0, 1, 2, \ldots, \tag{9.23}$$

where

$$a_{nm} = \frac{\delta_n}{\lambda_n} \int_{-\pi}^{\pi} \int_{-\pi}^{\pi} K_1(x,\xi)\varphi_n(x)\varphi_m(\xi)\, d\xi \qquad (n,m = 0, 1, \ldots),$$

$$f_n = \frac{\delta_n}{\lambda_n} \int_{-\pi}^{\pi} f(x)\varphi_n(x)\, dx, \qquad n = 0, 1, \ldots, \qquad \delta_n = \begin{cases} 1/2\pi, & n = 0, \\ 1/\pi, & n = 1, 2, \ldots. \end{cases} \tag{9.24}$$

Here we need to recall some classical result from the theory of regular infinite linear algebraic systems.

DEFINITION 1. *System (9.23) is called regular if $\sum_{m=0}^{\infty} |a_{nm}| < 1$ for $\forall n = 0, 1, \ldots$.*

It is shown in Kantorovich and Krilov (1958) that regular systems have a unique bounded solution if $|f_n| \le K \left(1 - \sum_{m=0}^{\infty} |a_{nm}| \right)$ for all $n = 0, 1, 2, \ldots$, and K is identical for all n.

DEFINITION 2. *System (9.23) is called completely regular if $\sum_{m=0}^{\infty} |a_{nm}| < \theta < 1$ for $\forall n = 0, 1, \ldots$, where θ is the same constant for all n.*

It is proved in Kantorovich and Krilov (1958) that completely regular systems have always a unique bounded solution if f_n are uniformly bounded. They can also be numerically solved by the truncation method, which is treated in the following way. Let us consider the finite-dimensional truncated system

$$v_n + \sum_{m=0}^{N} a_{nm} v_m = f_n, \qquad n = 0, 1, \ldots, N. \tag{9.25}$$

Then for all $n = 0, 1, \ldots$, the solution v_n of this system tends to u_n as $N \to \infty$.

DEFINITION 3. *System (9.23) is called quasi completely regular if for all $n = 0, 1, 2, \ldots$ we have $\sum_{m=0}^{\infty} |a_{nm}| < \infty$ and $\exists M$ such that $\sum_{m=0}^{\infty} |a_{nm}| < \theta < 1$ for all $n \ge M$.*

It is proved in Kantorovich and Krilov (1958) that such a system is uniquely solvable if the finite system (9.25), formed by the first M equations, is uniquely solvable, and if f_n are uniformly bounded. Under this condition such infinite system can be numerically solved by the truncation method.

Let us show that our system, obtained by the application of the Galerkin method to the first-kind Fredholm integral equation with a (weak) logarithmic singularity, is quasi completely regular if the regular part of the kernel is smooth enough. More precisely, this statement is valid if at least $K(x,\xi) \in C_2[(-\pi, \pi) \times (-\pi, \pi)]$ and $f(x) \in C_2(-\pi, \pi)$, since the asymptotic estimates of Section 1.4 give here: $a_{nm} = O(1/n\,m^2)$, $n, m \to \infty$, and $f_n = O(1/n)$, $n \to \infty$. Indeed, with such an estimate $\sum_{m=0}^{\infty} |a_{nm}| = O(1/n^2)$, which can be made less than θ, $0 < \theta < 1$, beginning from a certain $n \geq M$. Besides, the quantities f_n are uniformly bounded.

From the above consideration we can finally conclude that a full Fredholm equation of the first kind (9.20) with a (weak) logarithmic singularity in the kernel can be efficiently solved by the application of the Galerkin method, which reduces the problem to a quasi completely regular infinite system of linear algebraic equations that can be solved by the truncation method.

Let us pass now to equations of the first kind with logarithmic but not periodic kernels. The characteristic equation here is

$$-\int_{-1}^{1} \ln|x - \xi|\, u(\xi)\, d\xi = f(x), \quad |x| \leq 1, \quad G_0 u = f, \quad G_0(x, \xi) = -\ln|x - \xi|, \qquad (9.26)$$

and the eigenfunctions can be constructed on the basis of the following tabulated integral (see Gradshteyn and Ryzhik, 1994):

$$\int_{-1}^{1} \frac{T_n(\xi)}{\sqrt{1 - \xi^2}} \ln|x - \xi|\, d\xi = \begin{cases} \pi \ln 2, & n = 0, \\ \frac{\pi}{n} T_n(x), & n = 1, 2, \ldots, \end{cases} \qquad (9.27)$$

where $T_n(x) = \cos(n \arccos x)$ is the Chebyshev polynomial of order n (see Abramowitz and Stegun, 1965). Hence, following our general line, we can state that the orthogonal basis formed by the eigenfunctions is $\{\varphi_n(x)\} = \{T_n(x)/\sqrt{1 - x^2}\}$, with the respective eigenvalues being equal to $\lambda_n = \pi/n$ $(n = 1, 2, \ldots)$, $\lambda_0 = \pi \ln 2$. Our statement is based on the general property of orthogonality of eigenfunctions generated by a self-adjoint operator. In our case this is given by another tabulated integral, which is the relation of orthogonality (δ_{nm} is Kronecker's delta)

$$\int_{-1}^{1} \frac{T_n(x) T_m(x)}{\sqrt{1 - x^2}} = \varepsilon_n \delta_{nm}, \qquad \varepsilon_0 = \pi/2, \qquad \varepsilon_n = \pi. \qquad (9.28)$$

These relations (9.27) and (9.28) immediately give an exact representation of the solution to equation (9.26) in the form

$$u(x) = \sum_{m=0}^{\infty} u_m \varphi_m(x), \qquad \varphi_m(x) = \frac{T_m(x)}{\sqrt{1 - x^2}}, \qquad (9.29)$$

which is in a complete agreement with the theoretical result established in Section 1.7 about the structure of the solution to Eq. (9.26). Actually, the latter contains a square root singularity at the ends of the interval $(-1, 1)$.

Substitution of the series (9.29) into (9.26), with the help of relation (9.27), leads to an explicit expression for the unknown coefficients u_n:

$$u_n = \frac{f_n}{\varepsilon_n \lambda_n}, \qquad f_n = \int_{-1}^{1} f(x) \frac{T_n(x)}{\sqrt{1 - x^2}}\, dx. \qquad (9.30)$$

Now convergence of the series (9.29) is connected with the question of the asymptotic estimate of f_n as $n \to \infty$. The latter depends on analytic properties of the right-hand side $f(x)$. If this function is at least twice-differentiable, $f(x) \in C_2(-1, 1)$, then

$$f_n = \int_0^{\pi} \tilde{f}(t) \cos(nt)\, dt, \qquad x = \cos t, \qquad \tilde{f}(t) = f(\cos t). \qquad (9.31)$$

If, for simplicity, we consider again only even right-hand sides, then

$$f_n = \frac{1}{2} \int_{-\pi}^{\pi} \tilde{f}(t) \cos(nt) \, dt = O\left(\frac{1}{n^3}\right), \qquad n \to \infty, \tag{9.32}$$

due to the periodicity (cf. Section 1.4). For all that, as follows from Eq. (9.30),

$$u_n = O\left(\frac{1}{n^2}\right), \qquad n \to \infty, \tag{9.33}$$

so for such functions $f(x)$ the series is uniformly convergent on the interval $(-1, 1)$.

In the case of the full equation

$$\int_{-1}^{1} [G_0(x - \xi) + G_1(x, \xi)] \, u(\xi) \, d\xi = f(x), \qquad |x| \le 1, \tag{9.34}$$

where for simplicity we restrict the consideration only by even $f(x)$ and even $G_1(x, \xi)$, with respect to both its arguments, we may seek the unknown function $u(x)$ again in the form of a series in eigenfunctions:

$$u(x) = \sum_{m=0}^{\infty} u_m \varphi_m(x), \qquad \varphi_m(x) = \frac{T_m(x)}{\sqrt{1 - x^2}}. \tag{9.35}$$

Then the substitution of this representation into Eq. (9.34) and the application of the scalar product with $\varphi_n(x)$ lead to a second-kind infinite linear algebraic system

$$u_n + \sum_{m=0}^{\infty} b_{nm} u_m = f_n, \qquad n = 0, 1, 2, \ldots, \tag{9.36}$$

where

$$b_{nm} = \frac{\varepsilon_n}{\lambda_n} \int_{-1}^{1} \int_{-1}^{1} \frac{T_n(x)}{\sqrt{1 - x^2}} \frac{T_m(\xi)}{\sqrt{1 - \xi^2}} G_1(x, \xi) \, d\xi, \qquad f_n = \frac{1}{\lambda_n} \int_{-1}^{1} f(x) \frac{T_n(x)}{\sqrt{1 - x^2}} \, dx. \tag{9.37}$$

By analogy to the case with periodic kernel it is directly proved that system (9.36) is quasi completely regular if $f(x) \in C_2(-1, 1)$ and $G_1(x, \xi) \in C_2[(-1, 1) \times (-1, 1)]$. Therefore, the truncation method may be used to solve numerically this system.

Helpful remarks

It should be noted that the method described here is an alternative to another very efficient collocation technique (see Voronin and Tsetsokho, 1981). Each of these methods has its own intrinsic merits and restrictions. Thus, the collocation technique needs to solve a finite-dimensional algebraic system, in contrast with the method discussed at the present section. However, convergence of this method for integral equations with the logarithmic singularity is a more complex problem. Here we need to solve formally an infinite algebraic system, which can be reduced again to a finite-dimensional system, within the framework of the truncation method. Convergence of such a process is substantiated in a more clear way.

9.3. Integral Equations of the Physical Diffraction Theory in the Case of Nonconvex Obstacles

In Section 2.6 we could see that in the case of short-wave diffraction Kirchhoff's physical diffraction theory can asymptotically predict the value of the unknown pressure over the boundary surface (or contour, in the 2D problem). Here we spread some ideas of the physical diffraction theory, by applying boundary integral equations to the short-wave diffraction by obstacle with an arbitrary smooth boundary. The background of our approach is the specific feature of the incident wave interaction with convex and concave parts of the boundary. Essentially, acoustic rays incident on convex parts of the boundary cannot participate in repeated re-reflections. Vice versa, the rays incident on any concave part can re-reflect only between points of this part, and never fall to any convex part of the boundary. We show all ideas on the 2D case, but the 3D case can be treated just in the same way.

Let a plane incident acoustic wave be incident on an obstacle with a smooth boundary contour l (see Fig. 9.1).

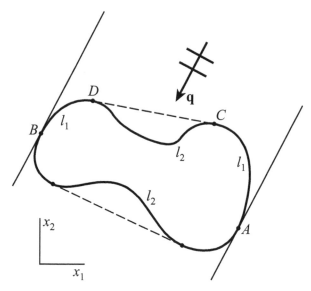

Figure 9.1. Incidence of the plane acoustic wave to a nonconvex obstacle

We formulate the problem simultaneously both for scalar and elastic problems. If the boundary contour is free of load, then (to be more specific) in the scalar case of acoustically soft boundary we have $p|_l = 0$, and in the elastic case: $\mathbf{T}|_l = (T_1, T_2)|_l = 0$, where p is the acoustic pressure in the scalar problem, and $\mathbf{T}$ is the stress vector in the elastic problem. Then, within the framework of the direct method the boundary integral equation in the scalar problem is (see Chapter 2)

$$\frac{i}{4} \int_l H_0^{(1)}(kr)g(y)\, dl_y = p_0(x) = e^{ik(q \cdot x)},$$

$$g(y) = \frac{\partial p}{\partial n_y}\bigg|_l, \quad r = |x - y|, \quad x \in l, \tag{9.38}$$

where k is the wave number, $\mathbf{q}$ is the unit vector which determines direction of the incident wave, and $H_0^{(1)}$ is the Hankel function.

In elastic medium, instead of one equation, we have the system of two equations:

$$u_1(x) - 2 \int_l \left[P_1^{(1)}(x,y)u_1(y) + P_2^{(1)}(x,y)u_2(y) \right] dl_y = 2u_1^\circ(x),$$

$$u_2(x) - 2 \int_l \left[P_1^{(2)}(x,y)u_1(y) + P_2^{(2)}(x,y)u_2(y) \right] dl_y = 2u_2^\circ(x), \quad x \in l,$$

$$\mathbf{P}^{(k)}(x,y) = 2\mu \frac{\partial}{\partial n_y} \left[\mathbf{U}^{(k)}(x,y) \right] + \lambda \, \mathbf{n}_y \, \mathrm{div}_y \left[\mathbf{U}^{(k)}(x,y) \right] \qquad (9.39)$$

$$+ \mu \left\{ \mathbf{n}_y \times \mathrm{rot}_y \left[\mathbf{U}^{(k)}(x,y) \right] \right\},$$

$$U_j^{(k)}(x,y) = \frac{i}{4\mu k_s^2} \left\{ k_s^2 \delta_{kj} H_0^{(1)}(k_s r) - \frac{\partial^2}{\partial y_k \, \partial y_j} \left[H_0^{(1)}(k_p r) - H_0^{(1)}(k_s r) \right] \right\},$$

$$\mathbf{u}^\circ(x) = \mathbf{q} \, e^{ik_p(q \cdot x)}, \qquad r = |x - y|, \qquad k, j = 1, 2,$$

where δ_{kj} is Kronecker's delta, k_p and k_s are the longitudinal and transverse wave numbers, respectively, λ and μ are elastic moduli. Here, to be more specific, we consider incidence of the longitudinal wave.

Let us construct a convex hull of the boundary (see, for example, Preparata and Shamos, 1985; and our Section 8.9), which is the minimum convex contour containing the given contour l. This implies that we need to draw tangent straight lines to the contour l, which separate convex (l_1) and nonconvex (l_2) parts of the boundary. Further, if we draw tangent lines parallel to the vector $\mathbf{q}$, then we separate the *shadow* zone AB of the total boundary. The latter is situated on its back side. It is obvious that the points A and B belong to convex parts of the contour l_1.

Let L be a characteristic size of the obstacle. If in the asymptotic sense, $kL \gg 1$, $k_p L \gg 1$, $k_s L \gg 1$, then the solution of integral equations in the shadow zone is equal to zero. So BIE (9.38) and (9.39) are asymptotically reduced to the arc $ACDB$. In what follows we mean by l_1 and l_2 only those parts of the contour, which lie on the front part $ACDB$ of the boundary contour.

The behavior of the solution of the basic integral equations over the convex parts l_1 differs qualitatively from its behavior on the nonconvex parts l_2. Actually, on l_1 (arcs BD and AC in Fig. 9.1) there is no re-reflections of acoustic rays. By contrast, ray re-reflections play an essential role on l_2. Therefore, at high frequencies the solution of equations (9.38) and (9.39) on the convex parts BD and AC is asymptotically defined as in the case of strictly convex obstacle. This implies for Eq. (9.38) that the leading asymptotic term is

$$g(y) = 2 \frac{\partial p_0}{\partial n_y} = 2ik(\mathbf{q} \cdot \mathbf{n}_y) e^{ik(q \cdot y)}, \qquad y \in l_1, \qquad (9.40)$$

in the scalar problem.

In the case of elastic medium in the neighborhood of every point on the convex parts l_1, the solution of system (9.39) coincides with the known solution of the problem about reflection of the plane wave from a free boundary of the corresponding elastic half-plane.

Let us relate to every point $y \in l_1$ basis unit vectors $\boldsymbol{\tau}_y$, $\mathbf{n}_y$ that form a right Cartesian coordinate system where $\mathbf{n}_y$ is an outward normal. Let $\mathbf{t}_y$ be a unit vector that determines the direction of the transverse wave reflected from the boundary l_1 at the point y. Then, according to Brekhovskikh (1980),

$$u_\tau(y) = \left[-(\mathbf{q} \cdot \boldsymbol{\tau}_y)\left(1 + V_{pp}\right) + \frac{k_s}{k_p}(\mathbf{t}_y \cdot \mathbf{n}_y) V_{ps} \right] e^{ik_p(q \cdot y)},$$

$$\qquad\qquad (9.41)$$

$$u_n(y) = \left[-(\mathbf{q} \cdot \mathbf{n}_y)\left(1 - V_{pp}\right) - \frac{k_s}{k_p}(\mathbf{t}_y \cdot \boldsymbol{\tau}_y) V_{ps} \right] e^{ik_p(q \cdot y)}.$$

Here V_{pp} and V_{ps} are coefficients of reflection from the boundary at the point $y \in l_1$, for longitudinal and transverse wave, respectively.

As a result, components of the displacement vector $\mathbf{u} = \{u_1, u_2\}$ on the convex parts l_1 can be found without solving system (9.39), in the following form:

$$u_1(y) = n_1 u_n(y) + n_2 u_\tau(y), \quad u_2(y) = n_2 u_n(y) - n_1 u_\tau(y), \tag{9.42}$$

where n_1 and n_2 are Cartesian components of the unit outward normal $\mathbf{n}_y = (n_1, n_2)$.

As follows from the previous analysis, now we only need to define the value of the unknown functions in Eqs. (9.38), (9.39) on the nonconvex parts l_2. To this end, we keep on the left-hand sides of the corresponding BIE only integrals over $y \in l_2$. For $y \in l_1$ the value of the unknown functions are already determined by formulas (9.40)–(9.42), and the corresponding integrals over these contours may be carried over the right-hand sides as some already known functions. Therefore, by putting $x \in l_2$, we can reduce in the considered high-frequency case the studied BIEs to some integral equations over l_2. Such a reducing essentially decreases required calculation time, which is of the order of M^3 if M designates the number of the collocation nodes in any quadrature formula. Let us write out, for example, an equation to which we come in the scalar case, after this reducing:

$$\frac{i}{4} \int_{l_2} H_0^{(1)}(k|x - y|) g(y) \, dl_y$$
$$= e^{ik(q \cdot x)} + \frac{k}{2} \int_{l_1} (\mathbf{q} \cdot \mathbf{n}_y) e^{ik(q \cdot y)} H_0^{(1)}(k|x - y|) \, dl_y, \quad x \in l_2. \tag{9.43}$$

Further simplification is possible in the same way. If on the *illuminated* part of the boundary contour there are several arcs of the type l_2, then the key essence of the diffraction process is as follows: acoustic sources, as boundary values of the corresponding unknown functions, lying on a certain such nonconvex part, in the asymptotic sense do not influence the values of wave fields on other such nonconvex parts. By other words, acoustic rays reflected from one nonconvex part cannot hit other nonconvex parts of the boundary. Therefore, if there are J arcs l_{2j}, $j = 1, \ldots, J$, then it suffices to solve equation (9.43) on each contour l_{2j} separately.

At last, additional simplification may be achieved if we neglect the influence of acoustic sources lying on convex parts l_1 to nonconvex ones l_2. Such an approach is quite natural since the rays reflected, for instance, from the arc AC cannot hit the arc CD neither in scalar nor in elastic case. For instance, Eq. (9.43) within the framework of such an approach can be simplified and reduced to a set of more simple independent equations

$$\frac{i}{4} \int_{l_{2j}} H_0^{(1)}(k|x - y|) g(y) \, dl_y = e^{ik(q \cdot x)}, \qquad x \in l_{2j} \quad (j = 1, \ldots, J). \tag{9.44}$$

To demonstrate efficiency of this approach, we consider the calculation of the back-scattered diagram, which is of significant interest for many applications. If the unknown functions are defined from Eqs. (9.38), (9.39), then the back-scattered diagram is determined from the following integral representation:

$$A(\alpha) = -\frac{i}{4} \int_l e^{ik(q \cdot y)} g(y) \, dl_y, \qquad \mathbf{q} = \{-\cos\alpha, -\sin\alpha\}, \tag{9.45}$$

$$A(\alpha) = \frac{1}{4} \left(\frac{k_p}{k_s}\right)^4 \int_l e^{ik_p(q \cdot y)} \left\{\mathbf{u}_y \cdot \left[2(\mathbf{q} \cdot \mathbf{n}_y) \mathbf{q} + \frac{\lambda}{\mu} \mathbf{n}_y\right]\right\} dl_y. \tag{9.46}$$

Below we demonstrate the proposed method for obstacles both of convex and nonconvex shape. In all figures line 1 refers to the exact numerical solution of the full integral equation and line 2 refers to the calculations within the framework of our approximate approach. Figure 9.2 demonstrates the back-scattered diagram $A(\alpha)$, (9.46), for the 2D diffraction problem about elliptic void in an elastic medium. We used the values $k_p a = 10$, $a/b = 3$, $k_s/k_p = c_p/c_s = 5.85/3.23$. For this convex obstacle implementation time required to construct lines 1 and 2 differ by 120 times. The number of nodes here was taken $N = 120$.

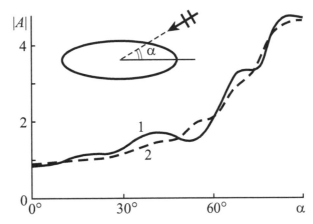

Figure 9.2. Comparison of exact (line 1) and proposed (line 2) methods for the calculation of diffracted waves for an elliptic obstacle

Figure 9.3 shows the back-scattered diagram for an obstacle in the form of a *bagel*, which is represented in the Cartesian coordinate system by the following parametric equations:

$$kx(t) = \sum_{m=0}^{4} a_m \cos(mt), \quad ky(t) = \sum_{m=1}^{4} b_m \sin(mt), \quad 0 \le t < 2\pi,$$

$$a_0 = 2.76; \quad a_1 = 3.38; \quad a_2 = 6.15; \quad a_3 = -1.12; \quad a_4 = -0.36;$$
$$b_1 = 8.48; \quad b_2 = 2.35; \quad b_3 = 1.30; \quad b_4 = -1.74. \tag{9.47}$$

The processor time required for calculating curve 2 is 8 times less than that required for constructing line 1. For all that, $N = 64$.

There is reflected in Fig. 9.4 the back-scattered diagram for a *three-leaf rose* considered as a void in elastic medium. The equation describing boundary contour in the polar coordinate system is here $k_p \rho(\varphi) = 5 (2 + \cos 3\varphi)$. Physical parameters have the same value as in the case shown in Fig. 9.2. We took here $N = 150$. Calculation time between lines 1 and 2 differs by 5 times.

Due to evident symmetry of all considered obstacles we show all results only for a certain part of variation of the angle $\alpha \in (0°, 360°)$.

In order to evaluate more completely the accuracy of the presented results, we note that the characteristic size of the considered obstacles is around 3λ to 4λ, where λ is the wavelength (in elastic case we put $\lambda = \lambda_p$, the length of the longitudinal wave).

Helpful remarks

It should be noted that some alternative approaches to extend the Boundary Integral Equation method to higher frequencies are proposed by Thiele and Newhouse (1975), as well as by Tobocman (1986, 1987).

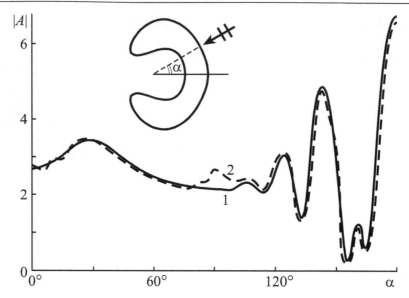

Figure 9.3. Comparison of exact (line 1) and proposed (line 2) methods for the calculation of diffracted waves for a bagel obstacle

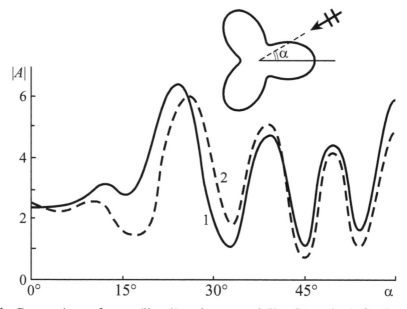

Figure 9.4. Comparison of exact (line 1) and proposed (line 2) methods for the calculation of diffracted waves for a three-leaf rose

9.4. Numerical Methods in Singular Integral Equations with the Cauchy-Type Kernel

This section follows in the main the classical results of Belotserkovsky and Lifanov (1993). Let us start to expound the methods suitable for numerical treatment of singular integrals and integral equations from a quadrature formula for the Cauchy-type singular integrals. Let us introduce the so-called *canonical partition* of the interval (a, b) with its subdivision to n+1 equal subintervals of the length $h = (b - a)/(n + 1)$. This implies two sets of nodes: $\{t_j\}$ with $a = t_0, t_1, t_2, \ldots, t_n, t_{n+1} = b, t_j = a + jh, j = 0, 1, \ldots, n + 1$, and $\{x_i\}$ as the central points of each subinterval (t_i, t_{i+1}), $i = 0, 1, \ldots, n$, so that $x_i = a + (i + 1/2)\, h$, $i = 0, \ldots, n$.

It is proved that if $g(x) \in C_1(a, b)$, then

$$\int_a^b \frac{g(t)\, dt}{x_i - t} = h \sum_{j=1}^n \frac{g(t_j)}{x_i - t_j} + O\left[\frac{\ln n}{n(x_i - a)(b - x_i)}\right], \qquad n \to \infty \sim h \to 0. \quad (9.48)$$

Now let us now study a singular integral equation with a characteristic kernel

$$\int_a^b \frac{g(t)\, dt}{x - t} = f(x), \qquad x \in (a, b), \quad f(x) \in C_1(a, b), \quad (9.49)$$

and we recall some classical results related to this equation (see, for example, Gakhov, 1966; Muskhelishvili, 1965). For equation (9.49) there exist three different classes of its solution: 1) Solution unbounded at both ends of the interval (a, b). Such a solution is not unique, and as in the theory of ordinary differential equations of the first order this is defined to within an arbitrary constant C. 2) Solution bounded at any of the two ends of the integration interval. Such a solution is unique. 3) With some additional condition for the right-hand side (i.e., not for every function $f(x)$) there can exist a solution bounded at both ends of the interval. We will be interested here only in case 1), where the exact analytical solution to equation (9.49) is given explicitly as

$$g(t) = \frac{1}{\pi^2 \sqrt{(t - a)(b - t)}} \left[\pi C - \int_a^b \frac{\sqrt{(x - a)(b - x)}}{t - x} f(x)\, dx\right]. \quad (9.50)$$

Note that arbitrary constant C is related with the integral of the unknown function $g(x)$:

$$C = \int_a^b g(x)\, dx. \quad (9.51)$$

Our further strategy is to construct a direct numerical collocation technique to solve characteristic Eq. (9.49) for arbitrary right-hand side, so that the constructed solution is a correct approximation for exact analytical solution (9.50). Then the method developed for the characteristic case will allow us to spread these results to a full singular integral equation where the analytical solution is not known.

To this end, we apply approximation (9.48) for singular integral:

$$\int_a^b \frac{g(t)\, dt}{x_i - t} \approx h \sum_{j=1}^n \frac{g(t_j)}{x_i - t_j}, \quad (9.52)$$

so that we arrive at some linear algebraic system. Let us prove that the solution of the system

$$\begin{cases} h \displaystyle\sum_{j=1}^n \frac{g(t_j)}{x_i - t_j} = f(x_i), & i = 1, \ldots, n-1, \\[2mm] h \displaystyle\sum_{j=1}^n g(t_j) = C, \end{cases} \quad (9.53)$$

which is indeed an $n \times n$ linear algebraic system with respect to unknown values $g(t_j)$, $j = 1, 2, \ldots, n$, gives a correct approximation to the exact solution (9.51) of equation (9.49), in the sense that if $t = t_l \in (a, b)$ is fixed, then the difference between the solution $g(t_l)$ of system (9.53) and the analytical solution (9.50) tends to zero as $n \to \infty$. The principal determinant of system (9.53) is

$$
D = h^n \Delta, \qquad \Delta = \begin{vmatrix} \dfrac{1}{x_1 - t_1} & \cdots & \dfrac{1}{x_1 - t_n} \\ \cdots & \cdots & \cdots \\ \dfrac{1}{x_{n-1} - t_1} & \cdots & \dfrac{1}{x_{n-1} - t_n} \\ 1 & \cdots & 1 \end{vmatrix}. \tag{9.54}
$$

It is proved in Belotserkovsky and Lifanov (1993) that

$$
\Delta = \frac{\prod\prod_{q<p} (t_q - t_p) \prod\prod_{q<p} (x_p - x_q)}{\prod\prod_{q,p} (x_q - t_p)}, \tag{9.55}
$$

where the lower limit in all products is 1 and the upper is n.

According to Cramer's rule, one needs to calculate the determinant Δ_l where the right-hand side column is substituted for the lth column of Δ:

$$
D_l = h^{n-1} \Delta_l, \qquad \Delta_l = \begin{vmatrix} \dfrac{1}{x_1 - t_1} & \cdots & f(x_1) & \cdots & \dfrac{1}{x_1 - t_n} \\ \dfrac{1}{x_2 - t_1} & \cdots & f(x_2) & \cdots & \dfrac{1}{x_2 - t_n} \\ \cdots & \cdots & \cdots & \cdots & \cdots \\ \dfrac{1}{x_{n-1} - t_1} & \cdots & f(x_{n-1}) & \cdots & \dfrac{1}{x_{n-1} - t_n} \\ 1 & \cdots & C & \cdots & 1 \end{vmatrix}. \tag{9.56}
$$

The last determinant may be calculated with the help of expansion by elements of the lth column:

$$
\begin{aligned}
\Delta_l = \sum_{j=1}^{n-1} f(x_j)(-1)^{j+l} \frac{\prod\limits_{1\leq q<p\leq n;\; q,p\neq l} (t_q - t_p) \prod\limits_{\substack{1\leq q<p\leq n-1;\; q,p\neq j}} (x_p - x_q)}{\prod\limits_{1\leq q\leq n-1} \prod\limits_{\substack{1\leq p\leq n \\ p\neq l}} (x_q - t_p)} \\
+ C(-1)^{n+l} \frac{\prod\limits_{1\leq q<p\leq n;\; q,p\neq l} (t_q - t_p) \prod\limits_{\substack{1\leq q\leq n-1;\; 1\leq p\leq n-1 \\ p\neq l}}^{q<p} (x_p - x_q)}{\prod\limits_{1\leq q\leq n-1} \prod\limits_{1\leq p\leq n} (x_q - t_p)},
\end{aligned} \tag{9.57}
$$

so

$$
g(t_l) = \frac{D_l}{D} = \frac{\prod\limits_{\substack{1\leq q\leq n-1 \\ q\neq l}} (x_q - t_l)}{h \prod\limits_{1\leq q\leq n} (t_q - t_l)} \left[C - \sum_{j=1}^{n-1} \frac{f(x_j)}{t_l - x_j} \frac{\prod\limits_{1\leq q\leq n} (x_j - t_q)}{\prod\limits_{\substack{1\leq q\leq n-1 \\ q\neq j}} (x_j - x_q)} \right]. \tag{9.58}
$$

Let us simplify the products present in the last formula. We have

$$
\frac{\displaystyle\prod_{\substack{1\leq q\leq n-1\\q\neq l}} (x_q - t_l)}{\displaystyle\prod_{1\leq q\leq n} (t_q - t_l)} = (x_l - t_l)\,\frac{\displaystyle\prod_{q=1}^{l-1}\left(x_q - t_l\right)}{\displaystyle\prod_{q=1}^{l-1}\left(t_q - t_l\right)} \times \frac{\displaystyle\prod_{q=l+1}^{n-1}\left(x_q - t_l\right)}{\displaystyle\prod_{q=l+1}^{n}\left(t_q - t_l\right)}
$$

$$
= \frac{h}{2}\,\frac{\displaystyle\prod_{q=1}^{l-1}(q-1/2)h}{\displaystyle\prod_{q=1}^{l-1} qh} \times \frac{\displaystyle\prod_{q=1}^{n-l-1}(q+1/2)h}{\displaystyle\prod_{q=1}^{n-l} qh} = \frac{h/2}{(n-l+1/2)h} \tag{9.59}
$$

$$
\times \prod_{q=1}^{l-1}\left(1 - \frac{1}{2q}\right) \times \prod_{q=1}^{n-l}\left(1 + \frac{1}{2q}\right) \sim \frac{h/2}{(n-l+1/2)h} \times \frac{l^{-1/2}}{\Gamma(1/2)} \times \frac{(n-l)^{1/2}}{\Gamma(3/2)}
$$

$$
\sim \frac{h}{b-t_l} \times \frac{\sqrt{b-t_l}}{\pi\sqrt{t_l - a}} = \frac{h}{\pi\sqrt{(b-t_l)(t_l - a)}}, \qquad n \to \infty.
$$

Here we have used the asymptotic estimate (compare with Belotserkovsky and Lifanov, 1993)

$$
\prod_{m=1}^{n}\left(1 + \frac{\beta}{m}\right) = \frac{n^{\beta}}{\Gamma(1+\beta)} + O\left(n^{\beta-1}\right), \qquad n \to \infty. \tag{9.60}
$$

By analogy, at $n \to \infty$

$$
\frac{\displaystyle\prod_{\substack{1\leq q\leq n\\q\neq j}} (x_j - t_q)}{\displaystyle\prod_{1\leq q\leq n-1} (x_j - x_q)} = \left(x_j - t_j\right)\frac{\displaystyle\prod_{q=1}^{j-1}\left(x_j - t_q\right)}{\displaystyle\prod_{q=1}^{j-1}\left(x_j - x_q\right)} \times \frac{\displaystyle\prod_{q=j+1}^{n}\left(x_j - t_q\right)}{\displaystyle\prod_{q=j+1}^{n-1}\left(x_j - x_q\right)}
$$

$$
= \frac{h}{2}\,\frac{\displaystyle\prod_{q=1}^{j-1}(q+1/2)h}{\displaystyle\prod_{q=1}^{j-1} qh} \times \frac{\displaystyle\prod_{q=1}^{n-j}(q-1/2)h}{\displaystyle\prod_{q=1}^{n-j-1} qh} \tag{9.61}
$$

$$
= \frac{h}{2}(n-j-1/2)\prod_{q=1}^{j-1}\left(1 + \frac{1}{2q}\right) \times \prod_{q=1}^{n-j-1}\left(1 - \frac{1}{2q}\right)
$$

$$
\sim \frac{h}{2}\,\frac{(b-x_j)\sqrt{x_j - a}}{\Gamma(1/2)\Gamma(3/2)\sqrt{b-x_j}} = \frac{h}{\pi}\sqrt{(x_j - a)(b-x_j)}.
$$

Hence expression (9.58) with $h \to 0$ tends to

$$
g(t_l) \sim \frac{1}{\pi\sqrt{(b-t_l)(t_l - a)}}\left[C - \frac{1}{\pi}\sum_{j=1}^{n-1} \frac{hf(x_j)\sqrt{(x_j - a)(b-x_j)}}{t_l - x_j}\right]
$$

$$
\sim \frac{1}{\pi\sqrt{(b-t_l)(t_l - a)}}\left[C - \frac{1}{\pi}\int_a^b \frac{f(x)\sqrt{(x-a)(b-x)}}{t_l - x}\,dx\right], \tag{9.62}
$$

which was to be proved. Note that in the last passage we have used quadrature formula (9.48).

Now let us consider the full equation

$$\int_a^b \left[\frac{1}{x-t} + K_0(x,t) \right] g(t)\, dt = f(x), \quad x \in (a,b). \tag{9.63}$$

Its general solution can be constructed by applying the inversion of the characteristic part (see Eq. (9.50)), which reduces Eq. (9.63) to a second-kind Fredholm integral equation:

$$g_1(x) + \int_a^b N_1(x,t)\, g_1(t)\, dt = f_1(x), \quad x \in (a,b), \tag{9.64}$$

with respect to the function $g_1(x) = \sqrt{(x-a)(b-x)}\, g(x)$. Here

$$N_1(x,t) = -\frac{1}{\pi^2\sqrt{(t-a)(b-t)}} \int_a^b \frac{K_0(\tau,t)\sqrt{(\tau-a)(b-\tau)}\, d\tau}{x-\tau},$$

$$f_1(x) = \frac{C}{\pi} - \frac{1}{\pi^2} \int_a^b \frac{f(\tau)\sqrt{(\tau-a)(b-\tau)}\, d\tau}{x-\tau}, \tag{9.65}$$

where C is again an arbitrary constant. It should be noted that the kernel of this Fredholm equation has a weak singularity.

Let us prove that a direct numerical solution to Eq. (9.63) can be constructed from the linear algebraic system

$$\begin{cases} h \sum_{j=1}^n \left[\dfrac{1}{x_i - t_j} + K_0(x_i, t_j) \right] g(t_j) = f(x_i), & i = 1, \ldots, n-1, \\ h \sum_{j=1}^n g(t_j) = C, \end{cases} \tag{9.66}$$

in the sense that its solution tends to the solution of equation (9.63) at any fixed point when $h \to 0$ (i.e., $n \to \infty$). Indeed, if we transfer the terms related to the regular kernel to the right-hand side and solve the resulting linear algebraic system with the characteristic matrix $1/(x_i - t_j)$, then we arrive at a finite-difference approximation of Eq. (9.64). The proof is completed by applying classical results on numerical solution of the second-kind Fredholm integral equation.

Helpful remarks

Approximation (9.48) shows that usual quadrature formulas, well known for regular integrals, can also be applied to Cauchy-type singular integrals with the chosen canonical sets of nodes.

9.5. Numerical Methods for Hyper-Singular Integral Equations

Consider a hyper-singular equation with a characteristic kernel

$$\int_a^b \frac{g(t)\, dt}{(x-t)^2} = f'(x), \quad x \in (a,b), \quad f(x) \in C_2(a,b). \tag{9.67}$$

Let us prove that a bounded solution of Eq. (9.67) is unique and is given as follows:

$$g(x) = \frac{\sqrt{(x-a)(b-x)}}{\pi^2} \int_a^b \frac{f(t)\,dt}{\sqrt{(t-a)(b-t)}\,(x-t)}. \tag{9.68}$$

Indeed, according to the definition of hyper-singular integrals (see Section 1.8), equation (9.67) is equivalent to

$$\frac{d}{dx}\int_a^b \frac{g(t)\,dt}{(x-t)} = -f'(x) \quad \sim \quad \int_a^b \frac{g(t)\,dt}{x-t} = -f(x) + C, \qquad x \in (a,b), \tag{9.69}$$

where C is an arbitrary constant. Now an inversion formula for the Cauchy characteristic integral operator (see Gakhov, 1966) determines the bounded solution as

$$g(x) = \frac{\sqrt{(x-a)(b-x)}}{\pi^2} \int_a^b \frac{f(t)}{\sqrt{(t-a)(b-t)}\,(x-t)}\,dt, \quad x \in (a,b), \tag{9.70}$$

and the constant C as

$$C = \frac{1}{\pi^2}\int_a^b \frac{f(t)}{\sqrt{(t-a)(b-t)}}\,dt. \tag{9.71}$$

Thus, as indicated in the Introduction, any bounded solution of Eq. (9.67) vanishes at $x \to a, b$.

To construct a direct collocation technique to solve Eq. (9.67) for arbitrary right-hand side, we introduce a similar but slightly different canonical partition compared with the case of singular integrals (see the previous section). Namely, let us divide the interval (a,b) into n small equal subintervals of the length $h = (b-a)/n$, by the nodes $a = t_0, t_1, t_2, \ldots, t_{n-1}, t_n = b, t_j = a + jh, j = 0, 1, \ldots, n$. The central points of each subinterval (t_{i-1}, t_i) are denoted by x_i, thus $x_i = a + (i - 1/2)h, i = 1, \ldots, n$.

If we try to arrange an approximation of integral in (9.67) by using finite sum, as in regular cases, then for $x = x_i$ we have

$$\int_a^b \frac{g(t)\,dt}{(x_i-t)^2} \approx \sum_{j=0}^n g(t_j)\int_{t_{j-1}}^{t_j}\frac{dt}{(x_i-t)^2} = g(t_i)\int_{-h/2}^{h/2}\frac{dt}{t^2}$$

$$+ \sum_{j\neq i} g(t_j)\left(\frac{1}{x_i-t_j} - \frac{1}{x_i-t_{j-1}}\right) = \sum_{j=1}^n g(t_j)\left(\frac{1}{x_i-t_j} - \frac{1}{x_i-t_{j-1}}\right), \tag{9.72}$$

where a value of the above hyper-singular integral of the function $1/t^2$ has been used. So, we try to approximate Eq. (9.67) by the linear algebraic system

$$\sum_{j=1}^n g(t_j)\left(\frac{1}{x_i-t_j} - \frac{1}{x_i-t_{j-1}}\right) = f'(x_i), \qquad i = 1, \ldots, n. \tag{9.73}$$

Further considerations are similar to those in Belotserkovsky and Lifanov (1993). Let us prove that, by assuming $x = x_l \in (a,b)$ is fixed, the difference between the solution $g(x_l)$ of system (9.73) and analytical solution (9.70) tends to zero, when $n \to \infty$. System (9.73) can be rewritten as

$$\sum_{j=1}^n g(t_j)\left(\frac{1}{x_j-t_i} - \frac{1}{x_j-t_{i-1}}\right) = f'(x_i), \qquad i = 1, \ldots, n, \tag{9.74}$$

and its principal determinant is

$$
\Delta =
\begin{vmatrix}
\left(\dfrac{1}{x_1 - t_1} - \dfrac{1}{x_1 - t_0} \right) & \cdots & \left(\dfrac{1}{x_n - t_1} - \dfrac{1}{x_n - t_0} \right) \\
& \cdots & \\
\left(\dfrac{1}{x_1 - t_n} - \dfrac{1}{x_1 - t_{n-1}} \right) & \cdots & \left(\dfrac{1}{x_n - t_n} - \dfrac{1}{x_n - t_{n-1}} \right)
\end{vmatrix}. \tag{9.75}
$$

Let us rewrite the ith row $(i = 2, \ldots, n)$ of Δ as the sum of all other rows, from $k = 1$ to $k = i$:

$$
\Delta =
\begin{vmatrix}
\dfrac{t_1 - t_0}{(x_1 - t_1)(x_1 - t_0)} & \cdots & \dfrac{t_1 - t_0}{(x_n - t_1)(x_n - t_0)} \\
& \cdots & \\
\dfrac{t_n - t_0}{(x_1 - t_n)(x_1 - t_0)} & \cdots & \dfrac{t_n - t_0}{(x_n - t_n)(x_n - t_0)}
\end{vmatrix}
= \dfrac{\displaystyle\prod_{j=1}^{n}(t_j - t_0)}{\displaystyle\prod_{i=1}^{n}(x_i - t_0)}
$$

$$
\tag{9.76}
$$

$$
\times
\begin{vmatrix}
\dfrac{1}{x_1 - t_1} & \cdots & \dfrac{1}{x_n - t_1} \\
\cdots & \cdots & \cdots \\
\dfrac{1}{x_1 - t_n} & \cdots & \dfrac{1}{x_n - t_n}
\end{vmatrix}
= \dfrac{\displaystyle\prod_j (t_j - t_0)}{\displaystyle\prod_i (x_i - t_0)}
\dfrac{\displaystyle\prod_{q<p}\prod (t_q - t_p) \prod_{q<p}\prod (x_p - x_q)}{\displaystyle\prod_{q,p}\prod (x_q - t_p)},
$$

where the lower limit in all products is 1 and the upper is n. A known value of the last determinant has been taken from Belotserkovsky and Lifanov (1993).

According to Cramer's rule, one needs to calculate the determinant Δ_l where the right-hand side column of (9.74) is substituted for the lth column of Δ in (9.75):

$$
\Delta_l =
\begin{vmatrix}
\left(\dfrac{1}{x_1 - t_1} - \dfrac{1}{x_1 - t_0} \right) & \cdots & f'(x_1) & \cdots & \left(\dfrac{1}{x_n - t_1} - \dfrac{1}{x_n - t_0} \right) \\
\left(\dfrac{1}{x_1 - t_2} - \dfrac{1}{x_1 - t_1} \right) & \cdots & f'(x_2) & \cdots & \left(\dfrac{1}{x_n - t_2} - \dfrac{1}{x_n - t_1} \right) \\
& \cdots & & \cdots & \\
\left(\dfrac{1}{x_1 - t_n} - \dfrac{1}{x_1 - t_{n-1}} \right) & \cdots & f'(x_n) & \cdots & \left(\dfrac{1}{x_n - t_n} - \dfrac{1}{x_n - t_{n-1}} \right)
\end{vmatrix}
$$

$$
\tag{9.77}
$$

$$
=
\begin{vmatrix}
\dfrac{t_1 - t_0}{(x_1 - t_1)(x_1 - t_0)} & \cdots & f'(x_1) & \cdots & \dfrac{t_1 - t_0}{(x_n - t_1)(x_n - t_0)} \\
\dfrac{t_2 - t_0}{(x_1 - t_2)(x_1 - t_0)} & \cdots & f'(x_1) + f'(x_2) & \cdots & \dfrac{t_2 - t_0}{(x_n - t_2)(x_n - t_0)} \\
\cdots & \cdots & \cdots & \cdots & \cdots \\
\dfrac{t_n - t_0}{(x_1 - t_n)(x_1 - t_0)} & \cdots & \displaystyle\sum_{k=1}^{n} f'(x_k) & \cdots & \dfrac{t_n - t_0}{(x_n - t_n)(x_n - t_0)}
\end{vmatrix},
$$

where we have used the same summation of rows as in the case of Δ.

The last determinant may be calculated arranging expansion by elements of the lth column as follows:

$$\Delta_l = \frac{\prod\limits_{j}(t_j - t_0)}{\prod\limits_{i \neq l}(x_i - t_0)} \sum_{m=1}^{n} \frac{\sum\limits_{k=1}^{m} f'(x_k)}{t_m - t_0} (-1)^{m+l} \frac{\prod\limits_{q<p;\,q,p\neq m}(t_q - t_p) \prod\limits_{q<p;\,q,p\neq l}(x_p - x_q)}{\prod\limits_{q\neq l}\prod\limits_{p\neq m}(x_q - t_p)}, \quad (9.78)$$

hence

$$g(x_l) = \frac{\Delta_l}{\Delta} = (x_l - t_0) \sum_{m=1}^{n} \frac{\sum\limits_{k=1}^{m} f'(x_k) \prod\limits_{p}(x_l - t_p) \prod\limits_{q}(x_q - t_m)}{(t_m - t_0)(x_l - t_m) \prod\limits_{p\neq m}(t_m - t_p) \prod\limits_{q\neq l}(x_q - x_l)}. \quad (9.79)$$

The last expression at $h \to 0 \sim n \to \infty$, under the condition $x_l \in (a,b)$ is fixed, can be estimated as follows:

$$\frac{\prod\limits_{p=1}^{n}(x_l - t_p)}{\prod\limits_{q\neq l}(x_q - x_l)} = (x_l - t_l) \frac{\prod\limits_{p=1}^{l-1}(x_l - t_p)}{\prod\limits_{q=1}^{l-1}(x_q - x_l)} \times \frac{\prod\limits_{p=l+1}^{n}(x_l - t_p)}{\prod\limits_{q=l+1}^{n}(x_q - x_l)} \quad (9.80)$$

$$= \frac{(-1)^n h}{2} \prod_{p=1}^{l-1}\left(1 - \frac{1}{2p}\right) \times \prod_{p=1}^{n-l}\left(1 + \frac{1}{2p}\right) \sim (-1)^n \frac{h\sqrt{b - x_l}}{\pi\sqrt{x_l - a}}, \quad n \to \infty,$$

if x belongs to the open interval (a, b) (i.e., $l \neq n$, $l \neq 1$). Here we have used the asymptotic estimate (compare with Belotserkovsky and Lifanov, 1993)

$$\prod_{m=1}^{n}\left(1 + \frac{\beta}{m}\right) = \frac{n^\beta}{\Gamma(1+\beta)} + O\left(n^{\beta-1}\right), \quad n \to \infty. \quad (9.81)$$

Further, by analogy

$$\frac{\prod\limits_{q}(x_q - t_m)}{\prod\limits_{p\neq m}(t_m - t_p)} = (-1)^n (t_m - x_m) \prod_{q=1}^{m-1}\left(1 + \frac{1}{2q}\right) \prod_{q=1}^{n-m}\left(1 - \frac{1}{2q}\right) \sim (-1)^n \frac{h\sqrt{t_m - a}}{\pi\sqrt{b - t_m}}. \quad (9.82)$$

Other terms in Eq. (9.79) can be simplified as follows ($h \to 0$):

$$(x_l - t_0) \to (x_l - a), \quad (t_m - t_0) \to (t_m - a), \quad h\sum_{k=1}^{m} f'(t_k) \to \int_a^{t_m} f'(t)\,dt = f(t_m), \quad (9.83)$$

hence expression (9.79) with $h \to 0$ tends to

$$g(x_l) \sim \frac{h}{\pi^2}\sqrt{(x_l - a)(b - x_l)} \sum_{m=1}^{n} \frac{f(t_m)}{\sqrt{(t_{m-a})(b - t_m)}\,(x_l - t_m)}$$

$$\sim \frac{\sqrt{(x_l - a)(b - x_l)}}{\pi^2} \int_a^b \frac{f(t)\,dt}{\sqrt{(t-a)(b-t)}\,(x_l - t)}. \quad (9.84)$$

Consider the full equation

$$\int_a^b \left[\frac{1}{(x-t)^2} + K_0(x,t) \right] g(t)\, dt = f'(x), \quad x \in (a,b), \tag{9.85}$$

where

$$K_0(x,t) = \frac{\partial K_1(x,t)}{\partial x}. \tag{9.86}$$

Its bounded solution can be constructed by applying inversion of the characteristic part, which reduces Eq. (9.85) to a second-kind Fredholm integral equation

$$g(x) + \int_a^b N_1(x,t)\, g(t)\, dt = f_1(x), \quad x \in (a,b), \tag{9.87}$$

where

$$N_1(x,t) = \frac{\sqrt{(x-a)(b-x)}}{\pi^2} \int_a^b \frac{K_1(\tau,t)\, d\tau}{\sqrt{(\tau-a)(b-\tau)}\,(x-\tau)}, \tag{9.88}$$

$$f_1(x) = \frac{\sqrt{(x-a)(b-x)}}{\pi^2} \int_a^b \frac{f(\tau)\, d\tau}{\sqrt{(\tau-a)(b-\tau)}\,(x-\tau)}. \tag{9.89}$$

Let us prove that if $f(x) \in C_1(a,b)$; $K_1(x,t) \in C_1[(a,b) \times (a,b)]$, then for any $x \in (a,b)$ the difference between solution $g(x)$ of the linear algebraic system

$$\sum_{j=1}^n \left[\frac{1}{x_i - t_j} - \frac{1}{x_i - t_{j-1}} + h K_0(x_i, t_j) \right] g(t_j) = f'(x_i), \quad i = 1, \ldots, n, \tag{9.90}$$

and the bounded solution of Eq. (9.85)–(9.86) tends to zero when $h \to 0$ (i.e., $n \to \infty$). Indeed, if one transfers the terms, related to the regular kernel, to the right-hand side and solves so written linear algebraic system with the characteristic matrix $1/(x_i - t_j) - 1/(x_i - t_{j-1})$, then one arrives at a finite-difference approximation of Eq. (9.87). The proof is finally completed if one applies classical results on numerical solution of the second-kind Fredholm integral equation.

An example for $K_0(x,t) = A(x-t)$, $f'(x) = -\pi \sim f(x) = -\pi x$ can be found in Iovane et al. (2003), where the numerical solution is compared with the exact one $g(x) = 8\sqrt{1-x^2}\,(4 + Ax)/(32 + A^2)$ in the case $A = 3$, $(a,b) = (-1,1)$.

Helpful remarks

In Eq. (9.84), when carrying out the last passage, we applied implicitly the quadrature formula (9.48) of the previous section in the case where the integrand possesses a square root singularity at the ends of the interval. It can be proved that the estimate (9.48) can be refined for this case as follows:

$$\int_a^b \frac{g(t)\, dt}{\sqrt{(t-a)(b-t)}\,(x_i - t)} = h \sum_{j=1}^n \frac{g(t_j)}{\sqrt{(t_j - a)(b - t_j)}\,(x_i - t_j)}$$

$$+ O\left(\frac{\ln n}{n^{1/2}\,(x_i - a)(b - x_i)} \right), \quad n \to \infty \sim h \to 0. \tag{9.91}$$

REFERENCES

Abramowitz M. and Stegun I. (1965), *Handbook of Mathematical Functions*, Dover: New York.

Achenbach J.D. (1973), *Wave Propagation in Elastic Solids*, North-Holland: Amsterdam.

Alexandrov A.D. and Zalgaller U.A. (1967), *Intrinsic Geometry of Surfaces*, Amer. Math. Soc.: Providence, RI.

Aleksandrov V.M. (1968), Asymptotic methods in contact problems of elasticity theory, *J. Appl. Math. Mech.*, **32**, 691–703.

Angell T.S. et al. (1989a), A constructive method for identification of an impedance scatterer, *Wave Motion*, **11**, 185–200.

Angell T.S. et al. (1989b), Target reconstruction from scattered far field data, *Annal. Telecomm.*, **44**, 456–463.

Apeltsin V.F. and Kyurkchan A.G. (1990), *Analytical Properties of Wave Fields*, Moscow State Univ.: Moscow (in Russian).

Atkinson F.V. (1949), On Sommerfeld's "Radiation condition", *Philos. Mag.*, **40**, 645–651.

Babeshko V.A. (1971), Convolution integral equations of the first kind on a system of segments, arising in elasticity theory and mathematical physics, *J. Appl. Math. Mech.*, **35**, No. 1.

Babich V.M. and Buldyrev V.S. (1989), *Asymptotic Methods in Short-Wavelength Diffraction Theory*, Springer-Verlag: Berlin, Heidelberg.

Banerjee P.K. and Butterfield R. (1981), *Boundary Element Methods in Engineering Science*, McGraw-Hill: London.

Bateman H. and Erdelyi A. (1953), *Higher Transcendental Functions, Vol. 1, 2*, McGraw-Hill: New York.

Bateman H. and Erdelyi A. (1954), *Tables of Integral Transforms, Vol. 1*, McGraw-Hill: New York.

Belotserkovsky S.M. and Lifanov I.K. (1993), *Method of Discrete Vortices*, CRC Press: Boca Raton.

Birman M.Sh. and Solomyak M.Z. (1979), Asymptotic behavior of the spectrum of differential equations, *J. Soviet Math.*, **12**, 247–283.

Blaschke W. (1930), *Differentialgeometrie und Geometrische Grundlagen Einsteins Relativitätstheorie, Vol. 1*, Springer-Verlag: Berlin.

Borovikov V.A. and Kinber B.Y. (1994), *Geometrical Theory of Diffraction*, IEEE Waves Ser. 37: London.

Boström, A. (1986), The null-field approach in series form—The direct and inverse problems, *J. Acoust. Soc. Amer.*, **79**, 1223–1229.

Bowman J.J., Senior T.B.A., and Uslenghi P.L.E. (1987), *Electromagnetic and Acoustic Scattering by Simple Shapes*, Hemisphere Publ.: New York.

Boyev N.V., Vatul'yan A.O., and Sumbatyan M.A. (1997), Reconstruction of the contour of obstacles from the scattered acoustic field in the high frequency range, *Acoustical Physics*, **43**, 391–394.

Boyev N.V. and Sumbatyan M.A. (1999), An inverse problem of high-frequency diffraction for nonconvex axially symmetric obstacles, *Russian Acoustical Physics*, **45**, 133–137.

Bratsun G.A. and Sumbatyan M.A. (1993), Investigation of stability of the method of auxiliary sources in practice, *Comput. Math. & Math. Phys.*, **33**, 125–127.

Brekhovskikh L.M. (1980), *Waves in Layered Media (2nd ed.)*, Academic Press: London.

Bremermann H. (1965), *Distributions, Complex Variables, and Fourier Transforms*, Addison-Wesley: Reading, MA.

Burov V.A. et al. (1986), Inverse scattering problems in acoustics (a review), *Russian Acoustical Physics*, **32**, No. 4.

Cappellini V., Constantinides A.G., and Emilani P. (1978), *Digital Filters and Their Applications*, Academic Press: London.

Carleman T. (1922), Sur la resolution de certaines equations integrales, *Arkiv. Mat. Astr. Phys.*, **16**, 1–19.

Collin R.E. (1960), *Field Theory of Guided Waves*, McGraw-Hill: New York.

Colton D. (1984), The inverse scattering problem for time-harmonic acoustic waves, *SIAM Review*, **36**, 323–350.

Colton D. and Kress R. (1983), *Integral Equation Methods in Scattering Theory*, John Wiley: New York.

Colton D. and Kress R. (1992), *Inverse Acoustic and Electromagnetic Scattering Theory*, Springer-Verlag: New York.

Colton D. and Monk P. (1985), A novel method for solving the inverse scattering problem for time-harmonic acoustic waves in the resonance region, *SIAM J. Appl. Math.*, **45**, 1039–1053.

Colton D. and Monk P. (1986), A novel method for solving the inverse scattering problem for time-harmonic acoustic waves in the resonance region II, *SIAM J. Appl. Math.*, **46**, 506–523.

Courant R. and Hilbert D. (1953), *Methods of Mathematical Physics, vol.1*, Interscience Publ.: New York.

De Boor C. (1978), *A Practical Guide to Splines*, Springer-Verlag: New York.

Druzhinina I.D. and Sumbatyan M.A. (1990), Numerical-analytical method in short-wave diffraction problems, *Soviet Physical Acoustics*, **36**, No. 2.

Druzhinina I.D. and Sumbatyan M.A. (1992), Two-dimensional shortwave diffraction on objects with arbitrary smooth boundary, *Soviet Physical Acoustics*, **38**, No. 3.

Erdelyi A. (1956), *Asymptotic Expansions*, Dover: New York.

Fedorjuk M.V. (1962), Stationary phase method for multiple integrals, *J. Comp. Math. Math. Phys.*, **2**, No. 1.

Fedorjuk M.V. (1977), *Steepest Descent Method*, Nauka: Moscow (in Russian).

Felsen L.B. and Marcuvitz N. (1973), *Radiation and Scattering of Waves*, Prentice Hall: Englewood Cliffs, NJ.

Fletcher C.A.J. (1984), *Computational Galerkin Methods*, Springer-Verlag: New York.

Forsythe G.E., Malcolm M.A., and Moler C.B. (1977), *Computer Methods for Mathematical Computations*, Prentice Hall: Englewood Cliffs, NJ.

Fridman V.M. (1962), On convergence of steepest descent type methods, *Usp. Mat. Nauk*, **17**, No. 3 (in Russian).

Gakhov F.D. (1966), *Boundary Value Problems*, Pergamon Press: Oxford.

Gel'fand I.M. and Shilov G.E. (1964), *Generalized Functions, Vol.1*, Academic Press: London.

Gill P.E., Murray W., and Wright M.H. (1981), *Practical Optimization*, Academic Press: London.

Gradshteyn I.S. and Ryzhik I.M. (1994), *Table of Integrals, Series, and Products (5th ed.)*, Academic Press: New York.

Groetsch C. (1984), *The Theory of Tikhonov Regularization for Fredholm Equations of the First Kind*, Pitman: London.

Grosjean C.C. (1965), On the series expansion of certain types of Fourier integrals in the neighbourhood of the origin, *Bull. Soc. Math. Belge*, **17**, 251–295.

Hackbusch W. (1995), *Integral Equations. Theory and Numerical Treatment*, Birkhauser Verlag: Basel.

Handelsman R.A. and Lew J.S. (1970), Asymptotic expansion of Laplace transforms near the origin, *SIAM J. Math. Anal.*, **1**, 118–130.

Hardy H. (1956), *Divergent Series*, Oxford University Press: London.

Hille E. and Tamarkin J.D. (1931), On the characteristic values of linear integral equations, *Acta Math.*, **57**, 1–79.

Hobson E.W. (1955), *The Theory of Spherical and Ellipsoidal Harmonics*, Chelsea Publ.: New York.

Hönl H., Maue A.W., and Westpfahl K. (1961), *Theorie der Beugung*, Springer-Verlag: Berlin.

Horn R.A. and Johnson Ch.R. (1986), *Matrix Analysis*, Cambridge Univ. Press: London.

Imbriale W.A. and Mittra R. (1970), The two-dimensional inverse scattering problem, *IEEE Trans. Antenn. Propag.*, **AP-18**, 633–642.

Iovane G., Lifanov I.K., and Sumbatyan M.A. (2003), On direct numerical treatment of hypersingular integral equations arising in mechanics and acoustics, *Acta Mechanica*, **162**, 99–110.

Jones D.S. (1952), Diffraction by a wave-guide of finite length. *Proc. Camb. Phil. Soc.*, **48**, 118–134.

Jones D.S. (1974), Integral equations for the exterior acoustic problems, *Quart. J. Mech. Appl. Math.*, **27**, 129–142.

Jones D.S. (1986), *Acoustic and Electromagnetic Waves*, Clarendon Press: Oxford.

Jorgenson R.E. and Mittra R. (1990), Efficient calculation of the free-space periodic Green's function, *IEEE Trans. Antenn. Propag.*, **38**, 633–642.

Kantorovich L.V. and Akilov G.P. (1982), *Functional Analysis*, Pergamon Press: Oxford.

Kantorovich L.V. and Krilov V.I. (1958), *Approximate Methods of Higher Analysis*, Noordhoff: Groningen.

Keller J.B. (1962), The geometrical theory of diffraction. *J. Optical Soc. Amer.*, **52**, 116–130.

Kirsch A. et al. (1988), Two methods for solving the inverse acoustic scattering problem, *Inverse Problems*, **4**, 749–770.

Kleinman R.E. and Roach G.F. (1982), On modified Green's functions in exterior problems for the Helmholtz equation, *Proc. Roy. Soc. London*, **A 383**, 313–332.

Koiter W.T. (1954), Approximate solution of Wiener–Hopf type integral equations with application, *Proc. Konikl. Nederl. Akad. Wet.*, **57**, 558–579.

Kristensson G. and Vogel C.R. (1986), Inverse problems for acoustic waves using the penalized likelihood method, *Inverse Problems*, **2**, 461–479.

Kupradze V.D. (1950), *Boundary Value Problems of the Theory of Oscillations and Integral Equations*, Gostekhizdat: Leningrad (in Russian).

Kurosh A.G. (1972), *A Course of Higher Algebra*, Mir Publ.: Moscow.

Kuznetsov N.V. (1966), Asymptotic distribution of eigenfrequencies of a plane membrane in the case when the variables can be separated, *Differential Equations*, **2**, 715–723.

Kuznetsov N.V. and Fedosov B.V. (1965), An asymptotic formula for the eigenvalues of the circular membrane, *Differential Equations*, **1**, No. 12.

Lavrentiev M.M. (1959), On integral equations of the first kind, *Russian Doklady*, **127**, No. 1.

Lavrentiev M.M. (1960), On integral equations of the first kind, *Russian Doklady*, **133**, No. 2.

Makai E. (1970), Complete orthogonal systems of eigenfunctions of three triangular membranes, *Studia Scientiarum Mathematicarum Hungarica*, **5**, 51–62.

Markushevich A.I. (1963), *The Theory of Analytical Functions*, Hindustan Publ.: Delhi-6.

Mathis A.W. and Peterson A.F. (1996), A comparison of acceleration procedures for the two-dimensional periodic Green's function, *IEEE Trans. Antenn. Propag.*, **44**, 567–571.

McNamara D.A.M., Pistorius C.W.I., and Malherbe J.A.G. (1990), *Introduction to the Uniform Geometrical Theory of Diffraction*, Artech House: Norwood.

Mikhlin S.G. (1964), *Variational Methods in Mathematical Physics*, Pergamon Press: Oxford.

Mikhlin S.G. (1965), *Problem of the Minimum of a Quadratic Functional*, Holden-Day: London.

Mindlin R.D. (1955), *An Introduction to the Mathematical Theory of Vibrations of Elastic Plates*, US Army Signal Corp. Eng. Lab.: Fort Monmouth, NJ.

Mittra R. and Lee S.W. (1971), *Analytical Techniques in the Theory of Guided Waves*, Macmillan: New York.

Morozov V.A. (1984), *Methods for Solving Incorrectly Posed Problems*, Springer-Verlag: Berlin, Heidelberg.

Morse P.M. and Feshbach H. (1953), *Methods of Theoretical Physics, (Parts 1,2)*, McGraw-Hill: New York.

Muskhelishvili N.I. (1965), *Singulare Integralgleichungen*, Akademie Verlag: Berlin.

Noble B. (1958), *Methods Based on Wiener–Hopf Technique*, Pergamon Press: New York.

Pissanetski S. (1988), *Sparse Matrix Technology*, Mir: Moscow.

Pogorelov A.V. (1961), *Differential Geometry*, Noordhoff: Groningen.

Pogorelov A.V. (1973), *Extrinsic Geometry of Convex Surfaces*, Amer. Math. Soc.: New York.

Pogorelov A.V. (1978), *The Minkowski Multidimensional Problem*, John Wiley: New York.

Polyanin A.D. and Manzhirov A.V. (1998), *Handbook of Integral Equations*, CRC Press: Boca Raton.

Preparata F.P. and Shamos M.I. (1985), *Computational Geometry: An Introduction*, Springer-Verlag: New York.

Ramm A.G. (1986), *Scattering by Obstacles*, Reidel Publ.: Dordrecht.

Rees C.S., Shah S.M., and Stanojevic C.V. (1981), *Theory and Applications of Fourier Analysis*, Marcel Dekker: New York.

Riekstinsh E.Ya. (1977), *Asymptotic Expansions of Integrals, vol.2*, Zinatne: Riga (in Russian).

Riekstinsh E.Ya. (1981), *Asymptotic Expansions of Integrals, vol.3*, Zinatne: Riga (in Russian).

Riesz F. and Sz.-Nagy B. (1972), *Lecons d'Analyse Fonctionelle*, Akademiai Kiado: Budapest.

Roger A. (1981), Newton-Kantorovich algorithm applied to an electromagnetic inverse problem, *IEEE Trans. Antenn. Propag.*, **AP-29**, 232–238.

Samarsky A.A. and Nikolaev E.S. (1978), *Methods of Solution of the Grid Equations*, Nauka: Moscow (in Russian).

Samko S.G. (2000), *Hyper-singular Integrals and Their Applications*, Gordon & Breach: London.

Scalia A. and Sumbatyan M.A. (1999), On high-frequency asymptotics in diffraction by finite-length waveguides: Open structures, *J. Eng. Math.*, **35**, 427–436.

Scalia A. and Sumbatyan M.A. (2001), Explicit short-wave asymptotics for diffraction by finite-length discontinuities in waveguides: Closed structures, *ZAMP*, **52**, 631–639.

Schoch A. (1941), Betrachtungen über das Schallfeld einer Kolbenmembran, *Akust. Z.*, **6**, 318–326.

Shenderov E.L. (1972), *Wave Problems of Hydroacoustics*, Sudostroyeniye: Leningrad (in Russian).

Sierpinski W. (1988), *Elementary Theory of Numbers*, North-Holland: Amsterdam.

Skudrzyk E. (1971), *The Foundations of Acoustics*, Springer-Verlag: Vienna/New York.

Smirnov V.I. (1964), *A Course of Higher Mathematics, Vol. 1,2*, Pergamon Press: Oxford.

Sumbatyan M.A. (1988), Development of the Schoch method for numerical study of the wave field of ultrasonic transducer, *Russian Acoustical Physics*, **34**, No. 1.

Sumbatyan M.A. (1992), A method of global random search in inverse problems with application to the problem of recognizing the shape of a defect, *J. Appl. Math. Mech.*, **56**, 779–781.

Sumbatyan M.A. (2000), Explicit solutions to acoustic problems in polyhedra, *Doklady Math.*, **62**, 125–128.

Sumbatyan M.A. (2001), A correct treatment of an ill-posed boundary equation in the acoustics of closed regions, *Comput. Math. & Math. Phys.*, **41**, 407–412.

Sumbatyan M.A. and Pompei A. (2001), Some explicit results for three-dimensional Helmholtz operator in polyhedra, *Rep. Math. Phys.*, **47**, 371–379.

Sumbatyan M.A., Pompei A., and Rigano A. (2000), New explicit solutions in acoustics of closed spaces on the basis of divergent series, *J. Acoust. Soc. Amer.*, **107**, 709–713.

Sumbatyan M.A., Solokhin N.V., and Trojan E.A. (1993), Reconstruction of convex flaws using backscattered ultrasound, *NDT& E Int.*, **26**, 227–230.

Sumbatyan M.A. and Troyan E.A. (1992), Reconstruction of the shape of a convex defect from a scattered wave field in the ray approximation, *J. Appl. Math. Mech.*, **56**, 464–468.

Thiele G.A. and Newhouse T.H. (1975), A hybrid technique for combining moment methods with the geometrical theory of diffraction, *IEEE Trans. Antenn. Propag.*, **AP-23**, 62–69.

Tikhonov A.N. and Arsenin V.Y. (1977), *Solutions of Ill-posed Problems*, Winston: Washington.

Tikhonov A.N. and Goncharsky A.V. (Eds.) (1987), *Ill-Posed Problems in the Natural Sciences*, Mir: Moscow.

Tikhonov A.N. and Samarsky A.A. (1977), *Equations of Mathematical Physics (5th ed.)*, Nauka: Moscow (in Russian).

Titchmarsh E.C. (1948), *Introduction to the Theory of Fourier Integrals (2nd ed.)*, Clarendon Press: Oxford.

Tobocman W. (1989), Inverse acoustic wave scattering in two dimensions from impenetrable targets, *Inverse Problems*, **5**, 1131–1144.

Tobocman W. (1986), Extension of the Helmholtz integral equation method to shorter wavelengths, *J. Acoust. Soc. Amer.*, **80**, 1828–1837.

Tobocman W. (1987), Extension of the Helmholtz integral equation method to shorter wavelengths. II, *J. Acoust. Soc. Amer.*, **82**, 704–706.

Vajnshtejn L.A. (1969), *Theory of Diffraction and Method of Factorization*, Golden Press: Boulder.

Voronin V.V. and Tsetsokho V.A. (1981), Numerical solution of an integral equation of the first kind with a logarithmic singularity by the interpolation-collocation method, *Comput. Math. & Math. Phys.*, **21**, No. 1.

Vorovich I.I. and Babeshko V.A. (1979), *Dynamic Mixed Boundary Value Problems of Elasticity Theory for Non-Classical Domains*, Nauka: Moscow (in Russian).

Vorovich I.I., Boyev N.V., and Sumbatyan M.A. (2001), Reconstruction of the obstacle shape in acoustic medium under ultrasonic scanning, *Inverse Probl. Eng.*, **9**, 315–337.

Vorovich I.I. and Sumbatyan M.A. (1990), Defect-image reconstruction from the scattered wave fields in an acoustic approximation, *Russian Mech. Solids*, **25**, 80–85.

Weyl H. (1912), Über die Abhäangigkeit der Eigenschwingungen einer Membran von der Bengrenzung, *J. Reine Angew. Math.*, **141**, 1–11.

Weyl H. (1952), Kapazitat von Strahlungsfelden, *Math. Ziet.*, **55**, 187–198.

Wiener N. (1934), *The Fourier Integral and Certain of its Applications*, Dover: New York.

Zemanian A.H. (1969), *Generalized Integral Transformations*, Interscience Publ.: New York.

Zhiglyavskii A.A. (1985), *Mathematical Theory of Global Random Search*, Izv. LGU: Leningrad (in Russian).

INDEX

V

velocity, 38–40, 75, 98, 133, 253
  group, 75, 81, 91
  phase, 39, 46
virtual image method, 119
  polygons, 119, 121
  polyhedra, 125, 128, 199

W

wave equation, 38–42
waveguide, 75, 99, 145–157
  closed, 151
  open, 145–151
wave front set, 39, 75

wave length, 40, 68, 142, 145, 149, 157, 170,
    180, 210–212, 217, 235, 262, 271
wave number, 39, 42–45, 57, 64, 74, 120, 126,
    131, 166, 209, 236, 239, 242, 253, 268
wave speed, 40–46, 75, 82, 87, 94–99, 221, 242
  longitudinal, 41, 82, 87, 166, 269
  transverse, 41, 82, 87, 166, 235, 269
well-posed problem, 219, 220, 235
  in the sense of Tikhonov, 219, 220
Weyl hypothesis, 125
Weyl–Carleman theory, 115–118, 131

Z

zeta function, 11, 87–90
  generalized, 11, 87–90